T0178640

Communications
in Computer and Information Science 1985

Rationale

The CCIS series is devoted to the publication of proceedings of computer science conferences. Its aim is to efficiently disseminate original research results in informatics in printed and electronic form. While the focus is on publication of peer-reviewed full papers presenting mature work, inclusion of reviewed short papers reporting on work in progress is welcome, too. Besides globally relevant meetings with internationally representative program committees guaranteeing a strict peer-reviewing and paper selection process, conferences run by societies or of high regional or national relevance are also considered for publication.

Topics

The topical scope of CCIS spans the entire spectrum of informatics ranging from foundational topics in the theory of computing to information and communications science and technology and a broad variety of interdisciplinary application fields.

Information for Volume Editors and Authors

Publication in CCIS is free of charge. No royalties are paid, however, we offer registered conference participants temporary free access to the online version of the conference proceedings on SpringerLink (http://link.springer.com) by means of an http referrer from the conference website and/or a number of complimentary printed copies, as specified in the official acceptance email of the event.

CCIS proceedings can be published in time for distribution at conferences or as postproceedings, and delivered in the form of printed books and/or electronically as USBs and/or e-content licenses for accessing proceedings at SpringerLink. Furthermore, CCIS proceedings are included in the CCIS electronic book series hosted in the SpringerLink digital library at http://link.springer.com/bookseries/7899. Conferences publishing in CCIS are allowed to use Online Conference Service (OCS) for managing the whole proceedings lifecycle (from submission and reviewing to preparing for publication) free of charge.

Publication process

The language of publication is exclusively English. Authors publishing in CCIS have to sign the Springer CCIS copyright transfer form, however, they are free to use their material published in CCIS for substantially changed, more elaborate subsequent publications elsewhere. For the preparation of the camera-ready papers/files, authors have to strictly adhere to the Springer CCIS Authors' Instructions and are strongly encouraged to use the CCIS LaTeX style files or templates.

Abstracting/Indexing

CCIS is abstracted/indexed in DBLP, Google Scholar, EI-Compendex, Mathematical Reviews, SCImago, Scopus. CCIS volumes are also submitted for the inclusion in ISI Proceedings.

How to start

To start the evaluation of your proposal for inclusion in the CCIS series, please send an e-mail to ccis@springer.com.

Federico Liberatore · Slawo Wesolkowski ·
Marc Demange · Greg H. Parlier
Editors

Operations Research and Enterprise Systems

11th International Conference, ICORES 2022, Virtual Event,
February 3–5, 2022, and 12th International Conference,
ICORES 2023, Lisbon, Portugal, February 19–21, 2023
Revised Selected Papers

 Springer

Editors
Federico Liberatore ⓘ
Cardiff University
Cardiff, UK

Slawo Wesolkowski
Department of National Defence
Ottawa, ON, Canada

Marc Demange
RMIT University
Melbourne, VIC, Australia

Greg H. Parlier
GH Parlier Consulting
Catonsville, MD, USA

ISSN 1865-0929 ISSN 1865-0937 (electronic)
Communications in Computer and Information Science
ISBN 978-3-031-49661-5 ISBN 978-3-031-49662-2 (eBook)
https://doi.org/10.1007/978-3-031-49662-2

This Springer imprint is published by the registered company Springer Nature Switzerland AG
The registered company address is: Gewerbestrasse 11, 6330 Cham, Switzerland

Paper in this product is recyclable.

Preface

This book includes extended and revised versions of selected papers from the 11th and 12th International Conferences on Operations Research and Enterprise Systems (ICORES 2022 and 2023). ICORES 2022 was held as an online event due to the Covid-19 pandemic, from 3–5 February 2022, and ICORES 2023 was held in Lisbon, Portugal, from 19–21 February 2023.

ICORES 2022 received 55 paper submissions of which 9% are included in this book, and ICORES 2023 received 55 paper submissions of which 15% are also included in this book. These papers were selected based on several criteria including reviews provided by program committee members, session chair assessments, and also program chair perspectives across all papers included in the technical program. The authors of these selected papers were then invited to submit revised and extended versions of their papers for publication in this book.

The purpose of the annual International Conference on Operations Research and Enterprise Systems (ICORES) is to bring together researchers, engineers and practitioners interested in both advances and applications in the field of operations research. Two simultaneous tracks are held, one covering domain independent methodologies and technologies and the other practical work developed in specific application areas.

The papers selected for this book contribute to the understanding of relevant trends of current research in Operations Research and Enterprise Systems, including: OR in Transportation, Predictive Analytics, Management Sciences, Maintenance, Logistics, Industrial Engineering and Energy and Environment, Optimization, Linear Programming, Decision Support Systems, Decision Analysis, Automation of Operations, Routing, Simulation, Stochastic Processes, Supply Chain Management and New Applications of OR.

We thank all authors for their contributions and our reviewers for ensuring a quality publication.

February 2023

Federico Liberatore
Slawo Wesolkowski
Marc Demange
Greg H. Parlier

Organization

Conference Chair

2022

Marc Demange   RMIT University, Australia

2023

Greg H. Parlier   GH Parlier Consulting, USA

Program Co-chairs

2022

Federico Liberatore   Cardiff University, UK
Greg H. Parlier   GH Parlier Consulting, USA

2023

Federico Liberatore   Cardiff University, UK
Slawo Wesolkowski   Department of National Defence, Canada

Program Committee

2022

Bernardetta Addis   Université de Lorraine, France
El-Houssaine Aghezzaf   Ghent University, Belgium
Javier Alcaraz   Universidad Miguel Hernández de Elche, Spain
Lionel Amodeo   University of Technology of Troyes, France
Gianpiero Bianchi   Italian National Institute of Statistics "Istat", Italy
Giancarlo Bigi   University of Pisa, Italy
Cyril Briand   LAAS-CNRS, France

2023

Roberto Cordone	University of Milan, Italy
Patrizia Daniele	University of Catania, Italy
Mirela Danubianu	"Stefan cel Mare" University of Suceava, Romania
Tatjana Davidovic	Mathematical Institute SANU, Serbian Academy of Sciences and Arts, Serbia
Clarisse Dhaenens	University of Lille, France
Abdessamad Didi	National Center for Energy and Nuclear Science and Technology, Morocco
Luis Miguel Ferreira	Universidade de Coimbra, Portugal
Michel Gendreau	Polytechnique Montréal, Canada
Giorgio Gnecco	IMT School for Advanced Studies Lucca, Italy
Francesca Guerriero	University of Calabria, Italy
Kenji Hatano	Doshisha University, Japan
Jillian Henderson	Defence Research and Development Canada, Canadian Department of National Defence, Canada
Hanno Hildmann	TNO, The Netherlands
Sin C. Ho	Chinese University of Hong Kong, China
Han Hoogeveen	Universiteit Utrecht, The Netherlands
Chenyi Hu	University of Central Arkansas, USA
Johann Hurink	University of Twente, The Netherlands
Josef Jablonsky	Prague University of Economics and Business, Czech Republic
Antonio Jiménez-Martín	Universidad Politécnica de Madrid, Spain
Yong-Hong Kuo	University of Hong Kong, China
Sotiria Lampoudi	Mast Reforestation, USA
Pierre L'Ecuyer	Université de Montréal, Canada
Pierre Lopez	LAAS-CNRS, Université de Toulouse, France
Renata Mansini	University of Brescia, Italy
Fabrizio Marinelli	Università Politecnica delle Marche, Italy
Concepción Maroto	Universidad Politécnica de Valencia, Spain
Sara Mattia	CNR-IASI, Italy
Carlo Meloni	Sapienza Università di Roma, Italy
Jairo Montoya-Torres	Universidad de La Sabana, Colombia
Muhammad Marwan Muhammad Fuad	Coventry University, UK
Tri-Dung Nguyen	University of Southampton, UK
José Oliveira	Universidade do Minho, Portugal
Mauro Passacantando	University of Milan-Bicocca, Italy
Shaligram Pokharel	Qatar University, Qatar
Celso Ribeiro	Universidade Federal Fluminense, Brazil
Michela Robba	University of Genoa, Italy

Andre Rossi	Université Paris-Dauphine, France
Fabrizio Rossi	University of L'Aquila, Italy
Laura Scrimali	University of Catania, Italy
Patrick Siarry	Université Paris-Est Créteil (LiSSi), France
Thomas Stützle	Université Libre de Bruxelles, Belgium
Wai Yuen Szeto	University of Hong Kong, China
Elena Tànfani	Università degli Studi di Genova, Italy
Chefi Triki	Hamad Bin Khalifa University, Qatar
J. M. van den Akker	Utrecht University, The Netherlands
Inneke Van Nieuwenhuyse	KU Leuven, Belgium
Maria Vlasiou	Eindhoven University of Technology, The Netherlands
Cameron Walker	University of Auckland, New Zealand
Santoso Wibowo	Central Queensland University, Australia
Yiqiang Zhao	Carleton University, Canada

Additional Reviewers

2022

Mohamed Bencheikh	Hassan II University Mohammedia, Morocco
Tatjana Jaksic-Kruger	Mathematical Institute SANU, Serbian Academy of Sciences and Arts, Serbia
Ornella Pisacane	Università Politecnica delle Marche, Italy

2023

Tomoya Kambara	Doshisha University, Japan
Ornella Pisacane	Università Politecnica delle Marche, Italy
Andrea Pizzuti	Università Politecnica delle Marche, Italy

Invited Speakers

2022

Nikolaos Matsatsinis	Technical University of Crete, Greece
David Ríos Insúa	ICMAT, Spain
Ivana Ljubic	ESSEC Business School of Paris, France
Yakov Ben-Haim	Technion - Israel Institute of Technology, Israel

2023

Han Hoogeveen   Universiteit Utrecht, The Netherlands
Rainer Schlosser   Hasso Plattner Institute, Germany
Joanna Józefowska   Poznan University of Technology, Poland

Contents

Applications

Methodologies and Technologies

Multiple Heuristics with Reinforcement Learning to Solve the Safe Shortest Path Problem in a Warehouse

Aurélien Mombelli[✉], Alain Quilliot, and Mourad Baiou

Université Clermont-Auvergne, CNRS, Mines de Saint-Étienne, Clermont-Auvergne-INP, LIMOS, 63000 Clermont-Ferrand, France
aurelien.mombelli@uca.fr

Abstract. Intelligent vehicles, provided with an ability to move with some level of autonomy, recently became a hot spot in the mobility field. Still, determining what can be exactly done with new generations of autonomous or semi-autonomous vehicles able to follow their way without being physically tied to any kind of track (cable, rail,. . .) remains an issue. We focus here on the top level of hierarchical decision level (distributes and schedules Pick up and Delivery tasks) and deal with the problem which consists in inserting an additional vehicle into an already working fleet and routing it while introducing a time-dependent estimation of the risk induced by the traversal of an arc at a given time. We propose a model and design a bi-level heuristic with dynamic programming and an A*-like heuristic along with three methods to generate speed functions which can all rely on a reinforcement learning scheme to route and schedule this vehicle.

Keywords: Shortest path · Risk aware · Time-dependant · A* · Reinforcement learning · Dynamic programming

1 Introduction

In an empty warehouse, an autonomous vehicle may travel at full speed toward its destination. However, if other autonomous vehicles are already working, travelling inside the warehouse implies avoiding congestion and costly accidents.

Monitoring a fleet involving autonomous vehicles usually relies on hierarchical supervision. The trend is to use three levels. At the low level, or *embedded level*, robotic-related problems are tackled for specific autonomous vehicles like controlling trajectories in real-time and adapting them to the possible presence of obstacles, see [23]. At the middle level, or *local level*, local supervisors manage priorities among autonomous vehicles and resolve conflicts in a restricted area [10,16] who worked on crossroad strategies. Then, at the top level, or *global level*, global supervisors assign tasks to the fleet and schedule paths. This level must take lower levels into account to compute its solution. See [4,6,13] for example or [22] who compute the shortest path thanks to the A* algorithm but assign each task to the fleet of autonomous vehicles using a multi-agent artificial intelligence to avoid conflict in arcs as much as possible.

© The Author(s), under exclusive license to Springer Nature Switzerland AG 2024
F. Liberatore et al. (Eds.): ICORES 2022/2023, CCIS 1985, pp. 3–25, 2024.
https://doi.org/10.1007/978-3-031-49662-2_1

Redirecting some vehicles to non-shortest paths may seem to increase the sum of the travel times but [20] showed that, in an airport, it actually decreased the sum of the travel times by decreasing the congestion time. With the same idea, several authors computed the shortest path thanks to the A* algorithm, first published by [19] in 1968. Then, if any conflict is detected, an avoidance strategy is applied [16].

This study puts the focus on the global level: routing and giving instructions to an autonomous vehicle in a fleet. An autonomous vehicle, idle until now, is chosen to carry out a new task. Since a vehicle is moving from some origin to some destination, performing some loading or unloading transaction and keeping on, one may think of related problems as a kind of constrained shortest path problems, see [12], if there is only one task or a PDP: Pick up and Delivery problem, see [17], if there are more tasks. But some specific features impose new challenges:

- The time horizon for autonomous or semi-autonomous vehicles is usually short and decisions have to be taken online, which means that decision processes must take into account the communication infrastructure, see [7], and the way the global supervisor can be provided, at any time, with a representation of the current state of the system and its short term evolution;
- As soon as autonomous vehicles are involved, safety is at stake (see [9]). The global supervisor must compute and schedule routes in such a way that not only tasks are going to be efficiently performed, but also that local and embedded supervisors will perform their job more easily.

To compute a safe shortest path, [8] used, a weighted sum of time and risks in Munster's roads in Ireland using an A* algorithm. In their case, the risk is a measure of dangerous steering or braking events on roads. But these techniques mostly cannot be applied here because the risk, in our case, is time-dependent. One can search for the optimal solution in a time-expanded network as did [14, 15]. A connection between two nodes in this network represents the crossing of an arc in the static network at a given time. Those kinds of networks are used, among other applications, for evacuation routing problems as did [11].

Here, the risk will not be a measure of dangerous events on the roads of a city but the expected repair costs in a destructive accident scenario, should it happens. To compute such expectation, planning of risks is to be computed from the already working fleet because the path they will follow and their speed functions are known. A risk planning procedure, we are provided with, is then used to transform previous information into risk functions over time that is necessary to compute the expected repair costs for the autonomous vehicle.

This article aims to answer the problem of finding a safe shortest path while considering a warehouse structure, paths followed by already working autonomous vehicles and a risk planning procedure. It is the continuation of a previous article presented at ICORES [1] with 3 new methods to generate decisions (i.e. speed functions) and a lot more experiences. First, a precise description of the problem is presented with structural results and complexity discussion. Then, the problem is tackled when the path is fixed and only the speed functions are unknown. In the fourth section, the whole problem is approached with two heuristics followed by, in the fifth section, learning processes. Lastly, numerical experiments are presented with a conclusion.

2   Detailed Problem

2.1   A Warehouse and the Risk Induced by Current Activity

A warehouse is represented as a planar connected graph $G = (N, A)$ where the set of nodes N represents crossroads and the set of arcs A represents aisles. For any arc $a \in A$, L_a represents the minimal travel time for an autonomous vehicle to go through aisle a. Moreover, two aisles may be the same length but one may stock fragile objects so that vehicles have to slow down.

Also, risk functions $R_a : t \mapsto R_a(t)$, generated from activities of aisle a, are computed using the risk planning procedure on experimentation in a real warehouse that we are provided with. It is important to note that the risk is not continuous. Indeed, there is, in an aisle, a finite number of possible configurations: empty, two vehicles in opposite directions, etc. (see Fig. 1). Each configuration is, then, associated with an expected cost of repairs in the event of accidents. Therefore, they are staircase functions evaluated in a currency (euro, dollars, etc.). Figure 2 shows an example of a risk function of an aisle.

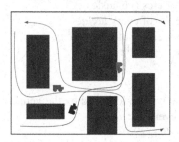

Fig. 1. At time t, 3 aisles have 1 vehicle each. At the next time, Blue and Purple join in the same aisle. One time after, all 3 vehicles join, generating high risks in this aisle (figure from [1]). (Color figure online)

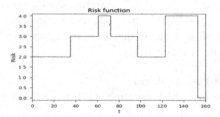

Fig. 2. Risk function of an aisle (figure from [1]).

From a risk function, we can have an estimation of the risk an autonomous vehicle takes in an aisle a between two times t_1 and t_2 with $v : t \mapsto v(t)$ as its speed function with Eq. 1.

$$risk(t_1, t_2, v) = H(v) \int_{t_1}^{t_2} R_a(t)\, dt \qquad (1)$$

We impose function H to be such that $H(v) \ll \frac{v}{v_{max}}$ in order to express the fact that a decrease of the speed implies a decrease of the risk. In further sections, H is set to $H : v \mapsto \left(\frac{v}{v_{max}}\right)^2$.

Remark 1. **Speed Normalization:** We only care here about traversal times of arcs $e \in A$, and not about their true length, in the geometric sense. So we suppose here that, for any arc e, $v_{max_e} = 1$. According to this we deal with reduced speed values $u \in [0, 1]$ and L_e means the minimal traversal time for arc e.

2.2 Our Problem: Searching for a Safe Shortest Path Inside the Warehouse

An idle autonomous vehicle must now carry out a new task inside the warehouse. Its task is to go from an origin node o to a destination node p. We must determine its path and its speed functions in each aisle of its path **while being provided with:**

- The warehouse structure: $G = (N, A)$ a planar connected graph with N the set of crossroads and A the set of arcs (i.e. aisles);
- The minimum travel time L_a of every arc a;
- The risk function $R_a : t \mapsto R_a(t)$ of every arc a;
- The origin node o and the destination node p;

Then, **we want to compute:**

- the path Γ from o to p that will be followed by the vehicle, along with entry time t_a and leaving time t_{a+1} of every arc a of Γ.
  If arc a is followed by arc b, $t_{a+1} = t_b$.
- the speed functions $v : t \in [t_a; t_{a+1}] \mapsto v_a(t)$ to apply when the vehicle is located inside every arc a of Γ.

As it is, we could have worked with a multi-objective problem. However, we want to compute a path and speed such that the autonomous vehicle is "safe". That means a maximum risk value constraint is added. The warehouse manager will impose a maximum value of risk R_{max} (quantified in currency, it can correspond to the cost of replacing a vehicle in the event of an accident) that an autonomous vehicle can take for a task. Then, the objective is to determine quickly:

SSPP: Safe Shortest Path Problem
Compute path Γ together with entry times t_a, leaving times t_{a+1} and speeds functions v_a
    such that:
    the arrival time in p is minimal;
    the global risk $\sum_{a \in \Gamma} risk\,(t_a, t_{a+1}, v_a) <= R_{max}$.

2.3 Some Structural Results

As it is stated, **SSPP** looks more like an optimal control problem than a combinatorial one. But, as we are going to show now, we may impose restrictions on speed function v, which are going to make the **SSPP** model get closer to a discrete decision model.

Proposition 1. *Optimal solution* (Γ, v) *of* **SSPP** *may be chosen in such a way that* v *is piecewise constant, with breakpoints related to the times* t_i *when vehicle* V *arrives at the end-nodes of arcs* $t_i, i = 1, \ldots, n$, *and to the breakpoints of function* $\Pi_i^e, i = 1, \ldots, n$.

Proof. Let us suppose that V is moving along some arc $e = e_i$, and that δ_1, δ_2 are 2 consecutive breakpoints in above sense. If $v(t)$ is not constant between δ_1 and δ_2 then we may replace $v(t)$ by the mean value v^* of function $t \mapsto v(t)$ between δ_1 and δ_2. Time value $Time(\Gamma, v)$ remains unchanged, while risk value $Risk(\Gamma, v)$ decreases because of the convexity of function H. So we conclude.

Proposition 2. *If optimal* **SSPP** *trajectory* (Γ, v) *is such that* $v(t) \neq 1$ *at some* t, *then* $Risk(\Gamma, v) = R_{max}$.

Proof. Let us suppose that path Γ is a sequence $e_1, \ldots, e_n$ of arcs of G. We proceed by induction on n.

- First case: $n = 1$.
  Let us suppose above assertion to be false. Breakpoints of $e = e_1$, may be written $t_0 = 0, t_1, \ldots, t_Q = Time(\Gamma, v)$, and we may set:
  - $q_0 = largest$ q such that $v < 1$ between t_q and t_{q+1};
  - $u_0 = related$ $speed$; $l_0 = distance$ $covered$ by V at time t_{q_0}.
  Let us increase u_0 by $\epsilon > 0$, such that $u_0 + \epsilon \leq 1$ and that induced additional risk taken between t_{q_0} and t_{q_0+1} does not exceed $R_{max} - Risk(\Gamma, v)$. Then, at time t_{q_0+1}, vehicle V covered a distance $l > l_0$. If $l < L_e$, then it keeps on at speed $v = 1$, and so arrives at the end of e before time t_Q, without having exceeded the risk threshold R_{max}. We conclude.

- Second case: $n > 1$.
  Let us suppose above assertion to be false and denote by R_1 the risk taken at the end of arc e, and by t_1 related time value. Induction applied to arcs $e_2, \ldots, e_n$, and risk threshold $R_{max} - R_1$ implies that the speed of V is equal to 1 all along the arcs $e_2, \ldots, e_n$. Let us denote by $\tau_0 = 0, \tau_1, \ldots, \tau_Q$ the breakpoints of e_1 which are between 0 and t_1 and let us set $\tau_{Q+1} = t_1$ and:
  - $q_0 = largest$ q such that $v < 1$ between τ_q and τ_{q+1};
  - $u_0 = related$ $speed$; $l_0 = distance$ $covered$ by V at time t_{q_0+1}.
  Then we increase u_0 by $\epsilon > 0$, such that $u_0 + \epsilon \leq 1$ and that induced additional risk taken between τ_{q_0} and τ_{q_0+1} does not exceed $(R_{max} - \frac{Risk(\Gamma, v)}{2}$. While moving at speed $u_0 + \epsilon$ along e_1, vehicle V faces 2 possibilities: either it arrives at the end of e_1 before time τ_{q_0+1} or it may keep on moving from time τ_{q_0+1} on along e_1 at speed $v = 1$. In any case, it reaches the end of e_1 at some time $t_1 - \beta, \beta < 0$, with an additional risk no larger than $(R_{max} - \frac{Risk(\Gamma, v)}{2}$. So, for any $i = 2, \ldots, n$ we compute speed value u_i such that moving along e_i at speed u_i between $t_{i-1} - \beta$ and t_{i-1} does not induce an additional risk more than $(R_{max} - \frac{Risk(\Gamma, v)}{2n}$. So we apply to V the following strategy: move as described above on arc e_1 and next, for any $i = 2, \ldots, n$, move along e_i at speed u_i between $t_{i-1} - \beta$ and t_{i-1} and next at speed 1 until the end of e_i. The additional risk induced by this strategy cannot exceed $(R_{max} - Risk(\Gamma, v))$. On another side, this strategy makes vehicle V achieve its trip strictly before time t_n. We conclude.

Proposition 3. *Given an optimal **SSPP** trajectory (Γ, v), with $\Gamma = \{e_1, \ldots, e_n\}$ and v satisfying Proposition 1. Let us denote by t_i the arrival time at the end of arc e_i. Then, for any $i = 1, \ldots, n$, and any t in $[t_{i-1}, t_i]$ such that $v = v(t) < 1$, the quantity $H'(v(t)).\Pi^{e_q}(t)$ is independent on t, where $H'(v)$ denotes the derivative of H in v.*

Proof. Once again, let us denote by t_i time when vehicle V arrives at the end of arc e_i. For a given i, we denote by $\delta_1, \ldots, \delta_{H(i)}$, the breakpoints of function Π^{e_i} which are inside interval $]t_{i-1}, t_i[$, by $\Pi_q^{i_q}$ related value of Π^{e_q} on the interval $]\delta_j, \delta_{j+1}[$, by $u_0, \ldots, u_q, \ldots, u_{H(q)}$, the speed values of V when it leaves those breakpoints, and by R_q the risk globally taken by V when it moves all along e_q. Because of Proposition 2, vector $(u_0, \ldots, u_{H(q)})$ is an optimal solution of the following convex optimization problem:

- Compute $(u_0, \ldots, u_{H(q)})$ such that $\sum_q u_q.(\delta_{q+1} - \delta_q)$ and which minimizes $\sum_q H(u_q)\Pi_q^{e_i}(\delta_{q+1} - \delta_q)$.

Then, Kuhn-Tucker conditions for the optimality of differentiable convex optimization program tell us that there must exists $\lambda \geq 0$ such that: for any q such that $u_q < 1, H'(u_q).\Pi_q^{e_i} = \lambda$. As a matter of fact, we see that λ cannot be equal to 0. We conclude.

Remark 2. In case $H(v) = u^2$, above equality $H'(u_q)\Pi_q^{e_i} = \lambda$ becomes $u_q R_q^{e_i} = \frac{\lambda}{2}$ where $u_q R_q^{e_i}$ means the instantaneous risk per distance $\frac{dR}{dL}$ value at the time when V moves along e_i between times δ_q and δ_{q+1}.

2.4   A Consequence: Risk Versus Distance Reformulation of the SSPP Model

Remark 1 leads us to define the Risk versus Time coefficient for arc e_i as the value $2H'(u_q)\Pi_q^{e_i}$ involved in Proposition 3. This proposition, combined with Proposition 1, allows us to significantly simplify **SSPP**: We define a *risk versus distance strategy* as a pair (Γ, λ^{RD}) where:

- Γ is a path, that means a sequence $\{e_1, \ldots, e_n\}$ of arcs, which connects origin node o do destination node d;
- λ_e^{RD} associates, with any arc e in Γ, *Risk versus Distance coefficient* $\lambda_e^{RD} = 2H'(v)R_e$. In case $H(v) = u_2$, we notice that this coefficient means the amount of risk per distance unit induced on arc e at any time t such that $v(t) < 1$, by any trajectory (Γ, v) which satisfies Proposition 3.

Let us suppose that we follow a trajectory (Γ, v) which meets Proposition 3, and that we know value λ_e^{RD} for any arc e of Γ. Since H is supposed to be convex and such that $H(v) \ll v$, we may state that H' admits a reciprocal function H'^{-1}. Then, at any time t when vehicle V is inside arc e, we are able to reconstruct value

$$v(t) : \begin{cases} H'^{-1}(\frac{\lambda_e^{RD}}{2R_e}), & \text{if } H'^{-1}(\frac{\lambda_e^{RD}}{2R_e}) < 1 \\ 1, & \text{otherwise} \end{cases} \tag{2}$$

According to this and Proposition 3, **SSPP** may be ewritten as follows with the notations $Risk(\Gamma, v)$ and $Time(\Gamma, v)$ as $Risk(\Gamma, \lambda^{RD})$ and $Time(\Gamma, \lambda^{RD})$:

Risk versus Distance SSPP Reformulation: Compute *risk versus distance strategy* (Γ, λ^{RD}) such that $Risk(\Gamma, \lambda^{RD}) \leq R_{max}$ and $Time(\Gamma, \lambda^{RD})$ is the smallest possible.

Then, a greedy algorithm can be derived from this reformulation on the shortest path:

RD Greedy Algorithm:

At any node u at time T and risk Π, the next arc of the path is traveled with $\lambda^{RD} = \frac{R_{max} - \Pi}{L^*_{u,p}}$ where $L^*_{u,p}$ is the distance between u and p in the shortest path.

2.5 Discussion About the Complexity

The time dependence of the transit network together with the proximity of the SSPP model with Shortest Path Constraint models suggests that SSPP is a complex problem. Sill, identifying the complexity of SSPP is not that simple, since we are dealing with continuous variables. Complexity also depends on function H, and so we suppose here that $H(v) = v^2$. We first may check that:

Proposition 4. *SSPP is in NP.*

Proof. It is clearly enough to deal with the case when $\Gamma = \{e_1, \ldots, e_n\}$ is fixed. Let us denote by $\Delta = \{0, d_1, \ldots, d_H\}$ the set of all breakpoints related to functions $\Pi^e, e \in \Gamma$, and by t_i the time when vehicle V will arrive at the end of e_i (we set $t_0 = 0$). If we know values $\{t_1, \ldots, t_n\}$, then we may retrieve values $\lambda^{RD}_e, e \in \Gamma$, through binary search. It comes that the core of our problem is about the computation of values $\{t_1, \ldots, t_n\}$. For any i, value t_i may be either equal to some value d_H or located inside some interval $]d_H, d_{H+1}[$. In order to make the distinction between those 2 configurations, we introduce the following function σ, which is going to characterize this logical positioning of values t_i with respect to Δ:

- If $t_i = d_H$ then we set $\sigma(i) = (h, 0)$;
- If $t_i \in]d_H, d_H + 1[$ then we set $\sigma(i) = (h, 1)$.

The number of possible functions σ is bounded by C_{H^n}. Now we notice that, once function σ is fixed, the problem becomes about computing values for those among variables $\{t_1, \ldots, t_n\}$ which are not non-instantiated, together with speed values for all consecutive intervals defined by elements of Δ and by those time values. This problem can be formulated as a cubic optimization problem, and one may check that, in case this problem has a feasible solution, then first order Kuhn-Tucker equations for local optimality determine exactly one local optimum. We conclude.

Conjecture 1. SSPP is in NP-Hard.

This conjecture is motivated by the fact that SSPP seems to be close to the constrained shortest path problems, which are most often NP-Hard [12]. Practical difficulty of SSPP may be captured through the following example (Fig. 3), which makes

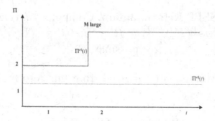

Fig. 3. Functions Π^{e_1} and Π^{e_2}.

appear that if (Γ, v) defines an optimal SSPP trajectory, the risk per distance value $\lambda_e^{RD} = 2.H'(v(t)).\Pi^e$ may be independent on t on arc e as told in Proposition 3, but cannot be considered as independent on arc e.

Path Γ contains 2 arcs, e_1 and e_2, both with length 1 and maximal speed 2. Function Π^{e_2} is constant and equal to 1. Function Π^{e_1} takes value 2 for $0 \leq t \leq 1$, and a very large value M (for instance 100) for $t > 1$ (see Fig. 3). $R_{max} = \frac{3}{4}$; Function H is: $v \mapsto H(v) = v^2$. Then we see that vehicle V must go fast all along the arc e_1, in order to get out of e_1 before this arc becomes very risky. That means that its speed is equal to 1 on e_1, and that its risk per distance value is equal to $\frac{1}{2}$. Next it puts the brake, in the sense that its speed remains equal to 1 but its risk per distance value decreases to $\frac{1}{4}$. It is easy to check that this routing strategy is the best one, with $Risk(\Gamma, v) = \frac{3}{4}$ and $Time(\Gamma, v) = 2$.

3   Solving the SSPP When the Path is Fixed

As the wanted output is made of two parts (the path and the speed functions on every arc of the path) and the second depends on the first, a sub-problem can be generated: let us denote by $t_i, i = 1 \ldots n$ the time when the vehicle finishes the traversal of the ith arc of Γ. Then, Γ being fixed, subproblem SSP(Γ) comes as follows:

SSP(Γ) Subproblem
Compute exit times $t_1, \ldots, t_n$ of Γ's arcs (and t_n is the arrival time in p).
such that:
the exit time t_n is minimal;
the global risk $\sum_{i=0}^{n-1} risk\,(t_i, t_{i+1}, v_i)$ is less than R_{max}.

As the exit time of the ith arc depends on the exit of $(i-1)^{th}$ arc, a Dynamic Programming forward scheme is well fitted because its time-space is the set of nodes that we are visiting one after another. The scheme we used is explained in Table 1 but the decision space and the filtering processes need more explanations and are detailed in the following sections.

3.1   Generate Decisions with Three Methods

As we saw in Sect. 2.4, there is a Risk versus Distance method to generate decisions but we propose two others: Risk versus Time and Distance versus Time.

Table 1. Dynamic Programming scheme used.

Time space I	$I = \{0, \ldots, n\}$ The ordered set of nodes of the path Do not mistake time in the sense of the vehicle with the dynamic programming time space The latter will be called "nodes" from now on
State space S	$s = (t, r) \in S$ t (resp. r) is a feasible date (resp. sum of the risks) at node i
Decision space	$\delta \in DEC(i, s)$
DEC(i,s)	At node i, v_i is a feasible speed function on the arc $(i, i+1)$ with t_{i+1} the exit date It means that $t_{i+1} - t \geq L_{(i,i+1)}$ and $r + risk\,(t, t_{i+1}, v_i) \leq R_{max}$
Transition space	$(i, s) \xrightarrow{t_{i+1}} (i+1, s')$ Transition from (t, r) to $(t_{i+1},\ r + risk\,(t, t_{i+1}, v_i))$
Bellman Principle	At node $i+1$ Only non-dominated feasible states are kept: $\forall (t_1, r_1), (t_2, r_2) \in S$, if $t_1 \leq t_2$ and $r_1 \leq r_2$, (t_2, r_2) is dominated
Search Strategy	Scanning time I forwardly and construct the feasible State space accordingly **Filtering processes** are discussed in Sect. 3.2

- **First Approach:** *The Risk versus Distance approach.*

Since $H(v) = v^2$, $H'(v(t))\Pi^e(t) = 2u(t)\Pi^e(t)$ for any t during the traversal of e. It comes that if we fix λ^{RD} the speed value $v(t)$ is given by: $v(t) = Inf(1, \lambda^{RD}/\Pi^e(t))$. Resulting state $(i + 1, r_2, t_2)$ will be obtained from λ^{RD} and (i, r_1, t_1) through the following iterative process:

Risk_Distance Transition Procedure:
Let us denote by L_e the length of arc e and by $T_1, \ldots, T_Q$ the breakpoints of Π^e which are larger than t_1 and by $\Pi^e_0, \ldots, \Pi^e_Q$ related Π^e values.
Initialization: $t_2 = t_1, r_2 = r_1, l = 0$ and $q = 0$;
While $L_e < l$, do:
$\quad v = \frac{\lambda^{RD}}{\Pi^e_q}$ if < 1 else 1
$\quad dt = \frac{L_e - l}{v}$ if $< T_{q+1} - T_q$ else $T_{q+1} - T_q$
$\quad t_2 = t_2 + dt$
$\quad l = l + v.dt$
$\quad r_2 = r_2 + v^2 \Pi^e_q dt$
$\quad q = q + 1$
If $r_2 < R_{max}$ then Success, else Fail
A special attention will be put for the first value of dt as it needs to be less than $T_{q+1} - t_1$ and not $T_{q+1} - T_q$.

- **Second Approach:** *The Risk versus Time approach.*

Since $H(v) = v^2$ we have that at any time t during the traversal of e, related risk speed $\frac{dR}{dT(t)}$ is equal to $v(t)^2 \Pi^e(t)$. It comes that if we fix λ^{RT} we get: $v(t) = Inf(1, \left(\frac{\lambda^{RT}}{\Pi^e(t)}\right)^{1/2})$.

Resulting state $(i+1, r_2, t_2)$ will be obtained from λ^{RT} and (i, r_1, t_1) through the same following iterative process a for the Risk versus Distance approach.

- **Third Approach:** *The Distance versus Time approach (mean speed approach).*

This method is a little different from the two other as a λ^{DT} will not give the speed function easily but only the value t_2: $t_2 = t_1 + \frac{L_e}{\lambda^{DT}}$. In order to determine the speed function $t \mapsto v(t)$ and the value r_2, the following quadratic program must be solved:

Distance_Time Transition Procedure:
Let us denote by L_e the length of arc e and by $T_1, \ldots, T_Q$ the breakpoints of Π^e which are larger than t_1 and by $\Pi_0^e, \ldots, \Pi_Q^e$ related Π^e values.
Then we must compute speed values $u_1, \ldots, u_Q \in [0, 1]$ such that:

$\sum_q u_q^2 \Pi_q^e (T_q - T_{q-1})$ is minimal.
$\sum_q u_q (T_q - T_{q-1}) = t_2 - t_1$

This quadratic convex program may be solved through direct application of Kuhn-Tucker 1st order formulas for local optimality. Then we get r_2 by setting: $r_2 = r_1 + \sum_q u_q^2 \Pi_q^e \cdot (T_q - T_{q-1})$. If $r_2 > R_{max}$ then the Mean Speed transition related to λ^{DT} yields a Fail result.

3.2 The Filtering Issue - Speeding up the Heuristic

There is a logical filtering that can be applied first: if the lower bound of the arrival time from a state (the time it took from o plus the time needed to p at v_{max}) is greater than an existing solution (from the RD Greedy algorithm for example). However, the logical filtering alone will not be very efficient. We propose two more filtering processes: one to limit the number of generated decisions and one to limit the number of states.

Limit the Number of Decisions. Starting from state $s = (t_i, r_i)$, a lot of decisions can be generated. But most of them are useless or not promising enough to be considered (too slow, too risky, slower and riskier than another state, etc.) and the optimal value λ_{opt} is unknown.

We propose to generate decisions by generating states for a few λ values between the low and high estimations of λ_{opt}: λ_{inf} and λ_{sup}.

Those values will be distributed between λ_{inf} and λ_{max} and led by λ_{midst} (half of them uniformly distributed between λ_{inf} and λ_{midst} and half of them between λ_{midst} and λ_{sup}).

Then, λ values need to be discussed as there are two possibilities: keeping them static for an instance or computing them for every state.

For static possibility, if our algorithms have already been used in the warehouse, their λ values will be kept as computed paths can overlap and will then be already learned for the next execution. For example, preprocessing can be applied to the warehouse by generating random paths solved by a greedy algorithm to generate one exit time and learn from all those decisions. The generation of random paths can end after a fixed number of generations or when the λ values seem to stabilise themselves.

The computed possibility relies on R_{max}, $L^*_{s,p}$ and the solution of the greedy algorithm and will compute λ values with the risk left, the distance left and the time left to the greedy solution from a current state.

Limit the Number of States. Now that, from every state of node $i - 1$, new states are generated, they are put together in an ordered set (λ_{midst} is used to order the set). Let us call this set S^i.

The Bellman principle and the logical filter are first applied: remove dominated states (states that are slower and riskier than another one) and those which finish after the computed upper bound. However, the number of generated states is still exponential. To keep a quick algorithm, the number of states in S^i will be bounded to S_{max}.

If $\#S^i > S_{max}$, which states must be removed? If the λ_{midst} value is very close to λ_{opt}, the S_{max} lowest value states are kept and all others are removed from S^i. However, if it is not, a high state can be better than the lowest state. Therefore, a method to determine whether λ_{midst} is a good approximation must be used.

We propose to compute the deviation of the state's risks of S^i from the travelled percentage of the path ($TravPer$) as in Eq. 3.

$$\Delta = \frac{\sum\limits_{(t,r)\in S^i} (\frac{r}{R_{max}} - TravPer)}{\#S^i} \tag{3}$$

If Δ's absolute value is high, generated states take, on average, too much risk or too little (meaning they can go faster).

States are, then, removed depending on Δ's value:

- If $|\Delta|$ is "high" (close to 1), λ_{midst} is supposed a bad approximation: $\left\lfloor \frac{\#S^i - S_{max}}{3} \right\rfloor$ states are removed from each third of S^i independently.
- If $|\Delta|$ is "medium": $\left\lfloor \frac{\#S^i - S_{max}}{2} \right\rfloor$ states are removed from the union of the 1st and 2nd third of S^i and the 3rd third of S^i independently.
- If $|\Delta|$ is "small" (close to 0), λ_{midst} is supposed a good approximation: the first S_{max} states are kept and all others are removed.

4 Solving the SSPP - Proposed Algorithms

Then, two algorithms can be created as follows:

- A decoupled method Algorithm 1: choose a path, choose speed values on it, modify the path locally, choose speed values again, keep the best, repeat.
- An A*-like method Algorithm 2: choose a node, expand it by one arc, choose the speed value on the new arc, push the new arrival node with the other unvisited nodes, repeat.

The A* algorithm is commonly used to search for the shortest path in a very large graph because a lower bound of expected value is associated with every unvisited node. In so doing, the most promising nodes will be visited first and all nodes having a greater lower bound than an existing solution will not be used at all. To solve the problem exactly, the A* algorithm applied in the time-expanded network of the warehouse is enough but is very slow and does not fit our requirements.

First, we introduce Algorithm 1, the decoupled heuristic to solve the SSPP.

Algorithm 1. SSPP - Decoupled method.

Require: o and p, the origin and destination node.
Require: S_{max} the maximum number of state to keep in the Dynamic Programming scheme.
Require: Γ an initial path between o and p.
$END = False$
While not END **Do**
    $END = True$
    Generate V: neighborhood of Γ
    For all $neighbor \in V$ **Do**
        Compute exit times of $neighbor$ with the Dynamic Programming scheme.
        If $neighbor$ finish earlier than Γ **Then**
            $\Gamma = neighbor$
            $END = False$
        End If
    End For
End While
Return Γ.

With no generation limit and no additional filter than those of the Dynamic Programming scheme, this heuristic can become an exact algorithm if the generated neighbourhood is large enough (modulo the time units precision).

The generation method used is: for every couple of nodes that are less or equal than two arcs away, pre-compute a path between them, other than the shortest one. The neighbour of a path is made by using the pre-computed non-shortest path of a portion of that path. The neighbourhood is then made of every possible neighbour of the path.

Finally, we introduce Algorithm 2, the heuristic based on A* algorithm. As the A* heuristic needs a function to estimate the value of the remaining path, we propose the function: $b_\lambda : x | (t, r, \Gamma) \mapsto sp(x) \left(\frac{f_\lambda(t,r)}{\text{length of } \Gamma} \right)$.

This function uses the value of the current path to estimate the value of the shortest path from the current node to the destination. A downside of this estimation is if the start of the path is very risky, the function will estimate the rest of the path to be very risky as well. That way, the A* like heuristic will abandon this state even if it is not true.

Algorithm 2. SSPP - A* like method.

Require: o and p, the origin and destination node.

Require: S_{max} the maximum number of states to keep in the Dynamic Programming scheme.

Let $sp : x \mapsto sp(x)$ be the length of the shortest path from x to p.

Let $b_\lambda : x|(t, r, \Gamma) \mapsto sp(x) \left(\frac{f_\lambda(t,r)}{\text{length of } \Gamma} \right)$ be an estimation of the remaining value to p.

Let $Dict$ be a dictionary indexed by nodes of priority queues which are ordered by $f_\lambda(x) + b_\lambda(x)$ in ascending order.

push $node = o|(t = 0, r = 0, \Gamma = [])$ in $Dict$.

At all times, $Best_{Dict}$ denotes the smallest value's state among heads of priority queues of $Dict$.

While node of $Best_{Dict}$ isn't p **Do**

    $current = Best_{Dict} = x_{i-1}|(t_{i-1}, r_{i-1}, \Gamma_{i-1})$.

    Pop $Best_{Dict}$ from corresponding priority queue.

    Generate elongations $(x_i|t_i, r_i, \Gamma_{i-1} + [x_{i-1}])_i$ from $current$.

    Push all elongations in their priority queues of D.

    For all priority queue $PQ \in Dict$

        If $\#PQ > S_{max}$ **Then**

            Filter PQ.

        End If

    End For

End While

Return Γ of $Best_{Dict}$.

With no generation limit and no priority queue filter, this heuristic becomes an exact algorithm (modulo the time units precision).

Because this heuristic was too slow, another filter was added: Each node is to be visited $2S_{max}$ times at most. However, short arcs are then privileged and R_{max} is reached before the end of the path.

Thankfully, a small change in the ordering of priority queues was enough to compensate. Priority queues are now separated in two halves: states $\{x|(t, r, \Gamma)\}$ that respect $r < R_{max} * \frac{t}{t+sp(x)}$ first. Each ordered by $f_\lambda(x) + b_\lambda(x)$ in ascending order.

5 Speeding Algorithms Through Statistical Learning Techniques

We consider here two ways of speeding our algorithms in order fit with a dynamic contexts. The first one impose to generate only three possible decisions and the second one is to learn better values of λ while solving an instance in a reinforcement learning way.

5.1 Bounding the Number of States Generated

Once acceleration parameter ρ has been tuned, controlling the size of decision set Λ means fixing the value of G_{max}. If we set $G_{max} = 1$ (greedy algorithm in the case of **DP_Evaluate** and shortest path algorithm in the case of **A*_SPR**), then the choice is about λ_{mean}^{RD}, which, in a first approach, should be equal to $\frac{R_{max}}{L_{o,d}^*}$. If $G_{max} \neq 1$, then we apply the following statistical learning process:

- We apply **DP_Evaluate** to instances which fit parameter ρ, while using some reference decision number G_{max}. For any instance I, we retrieve the optimal decision sequence $\{\lambda_1, \ldots, \lambda_n\}$.
- Then we compute, for every value λ_i in $\{\lambda_{min}, \ldots, \lambda_{max}\}$, the percentage $\tau(i)$ of occurrence of λ_i in those decision sequences.
- Finally, G_{max} being the target decision number, we split the decision range $[\lambda_{min}^{RD}, \lambda_{max}^{RD}]$ for possible decisions λ^{RD} into $G_{max} + 1$ intervals corresponding to same percentages of decisions λ_i. For instance, if $G_{max} = 3$, we split interval $[\lambda_{min}, \lambda_{max}]$ into 4 intervals $[\lambda_{min}, \lambda^{\frac{1}{4}}], [\lambda^{\frac{1}{4}}, \lambda^{\frac{1}{2}}], [\lambda^{\frac{1}{2}}, \lambda^{\frac{3}{4}}]$ and $[\lambda^{\frac{3}{4}}, \lambda_{max}]$ in such a way that:
  - 1/4 of decisions λ_i belong to interval $[\lambda_{min}, \lambda_1 4]$;
  - 1/4 of decisions λ_i belong to interval $[\lambda^{\frac{1}{4}}, \lambda^{\frac{1}{2}}]$;
  - 1/4 of decisions λ_i belong to interval $[\lambda^{\frac{1}{2}}, \lambda^{\frac{3}{4}}]$;
  - 1/4 of decisions λ_i belong to interval $[\lambda^{\frac{3}{4}}, \lambda_{max}]$.

Then restricted Λ to a 3-elements set: $\{\lambda^{\frac{1}{4}}, \lambda^{\frac{1}{2}}, \lambda^{\frac{3}{4}}\}$.

Bounding the Number of States Through Reinforcement Learning. Above method needs at least two resolutions of each instance to compute a good Λ set. Below, we propose to learn a good λ set during the resolution of the instance. For example, if the beginning of the path is very risky, the λ value will be slightly modified to authorized lower speeds for the rest of the path. Then, the learning process is as follow:

- Every time a states set is filtered, we compute, for every value λ_i in $\{\lambda_{min}, \ldots, \lambda_{max}\}$, a value Δ_i that is the sum of the percentage of risk in relation to the percentage of traveled path of every states generated by λ_i. This is the same deviation used to limit the number states in Sect. 3.2 but here, Δ_i is computed only with states generated by λ_i.
- Apply a modification on every λ_i depending on its respective above value. We choose to apply $\lambda_i = \lambda_i * (1 - 0.2 * \frac{\Delta_i}{\sum_j Dev_j})$.

6 Numerical Experiments

Goal: We perform numerical experiments with the purpose of studying the behavior of static **DP_Evaluate**, **Dec_SSPP** and **A*_SSPP** algorithms of Sect. 4. We pay special attention to the dependence of those algorithms to the choice of the decision mode (Risk versus Distance, Risk versus Time, Distance versus Time).

Technical Context: Algorithms were implemented in C++17 on an Intel i5-9500 CPU at 4.1 GHz. CPU times are in milliseconds.

Instances: We generated networks (N, A) as random planar graph create by a Delaunay triangulation. Those graphs are summarized through their number $|N|$ of nodes and their number $|A|$ of arcs. Length values $L_e, e \in A$, are uniformly distributed between 3

Table 2. Instance parameters table.

id	$|M|$	Freq	$\overline{R}$	α	L^*	Greedy
1	4	0.2	2	0.4	59	142.32
2	5	0.25	1.9	1	55	82.47
3	6	0.19	2	1.5	63	72.72
4	4	0.43	2	0.4	67	141.15
5	6	0.6	1.9	1	61	118.89
6	7	0.42	2	1.5	68	93.34
7	6	0.16	1.9	0.4	104	317.61
8	7	0.18	1.9	1	96	142.3
9	6	0.18	2	1.5	93	102.79
10	7	0.41	2	0.4	102	194.87
11	6	0.45	2	1	104	185.33
12	8	0.32	2	1.5	101	129.06
13	11	0.16	1.9	0.4	130	251.43
14	8	0.15	2	1	126	219.55
15	7	0.19	2	1.5	142	172.05
16	6	0.33	1.9	0.4	138	351.25
17	9	0.3	1.9	1	133	199.06
18	7	0.36	2	1.5	140	175.94

and 10. Function H is taken as function $u \mapsto H(u) = u^2$. Function Π^e are generated by fixing a time horizon T_{max}, a mean frequency $Freq$ of break points t_i^e, and an average value $\overline{R}$ for value $\Pi^e(t)$: More precisely, values Π^e are generated within a finite set $\{2\overline{R}, \frac{3\overline{R}}{2}, \overline{R}, \frac{\overline{R}}{2}, 0\}$. As for threshold R_{max}, we notice that if functions Π^e are constant with value $\overline{R}$ and if we follow a path Γ with length L_{diam}, the diameter of network G, at speed $\frac{1}{2} = \frac{v_{max}}{2}$, then the expected risk is $\frac{L_{diam}\overline{R}}{2}$. It comes that we generate R_{max} as a quantity $\alpha \frac{L_{diam}\overline{R}}{2}$, where α is a number between 0.2 and 2. Finally, since an instance is also determined by origin/pair (o, p), we denote by L^* the value $L^*_{o,p}$. Table 2 presents a package of 18 instances with their characteristics and the time value obtained by the greedy algorithm presented in Sect. 2.4.

Outputs related to the behavior of **DP_SSPP**.
We apply **DP_SSPP** while testing the role of parameters $\lambda = \lambda^{RD}, \lambda^{RT}, \lambda^{DT}$, as well as S_{max} and ρ. So, for every instance, we compute:

- in Table 3: The time value T_DP^{mode}, the percentage of risk R_DP^{mode} from R_{max}, and the CPU times (in milliseconds.) CPU^{mode}, induced by application of **DP_SSPP** on the shortest path between o and p with $\lambda^{mode} = \lambda^{RD}, \lambda^{RT}, \lambda^{DT}$, $S_{max} = +\infty$, $G_{max} = 21$ and $\rho = 8$;
- in Table 6: The time value T_DP^{mode}, the percentage of risk R_DP^{mode} from R_{max}, and the CPU times (in milliseconds.) CPU^{mode}, induced by application of **DP_SSPP** on the shortest path between o and p with $\lambda^{mode} = \lambda^{RD}, \lambda^{RT}, \lambda^{DT}$, $S_{max} = 21$, $G_{max} = 5$ and $\rho = 4$;
- in Table 7: For the specific mode λ^{RD}, related number $\#S$ of states per node i, together with time value T^{RD}, when $S_{max} = 3, 7, 11, 15, 21$, $G_{max} = 5$ and $\rho = 4$;

Table 3. DynProg - Impact of λ^{mode}, with $Smax = +\infty$, $Gmax = 21$ and $\rho = 8$.

id	T^{RD}	R^{RD}	T^{RT}	R^{RT}	T^{DT}	R^{DT}
1	102.3	98.3	118.5	99.9	126.2	78.9
2	66.1	96.1	68.1	99.1	62	87.8
3	63	84.1	63	84.1	72.3	49.5
4	129.4	79.8	136.6	89.7	131.7	96.3
5	72	97.2	70.2	98.2	75.3	92.7
6	78.7	99.8	77.6	98.1	76.2	99.6
7	279.1	97.3	278.1	99.5	284.6	99.2
8	116	96.3	116.2	94.3	118.5	99.9
9	94.7	98.9	93.2	97.9	93.2	86.2
10	176.7	92.6	167.4	99.3	174.7	99.8
11	123.5	92.8	123.4	93.5	123.2	92.6
12	110.3	99.8	109.6	99.7	111	99.9
13	208.1	97.4	208.6	97.9	226.5	99.1
14	149.4	98	148.1	99.7	148.9	97.6
15	157.3	96.9	161	89.7	152	89
16	254.8	99.9	259.9	90.1	251.5	95.4
17	161.1	97.4	155.7	99.5	155.1	99.9
18	146.1	99.8	147	98.5	146.5	99.5

Table 4. DynProg - Impact of ρ, with λ^{RD}, $Smax = +\infty$ and $Gmax = 21$.

ρ id	1.5 T^{RD}	#S	4 T^{RD}	#S
1	113.4	236.75	103.2	75.5
2	79.7	370.3	62	221.15
3	66.2	118.75	63	254.75
4	130.6	70.7	127.9	62.9
5	81.9	151.8	71.2	174.7
6	80.4	723.8	76.1	623.45
7	282.8	233.06	275.7	203.4
8	118.5	550.13	115.8	331.63
9	95.8	61.73	93.2	170.86
10	183.2	222.36	176.5	293.76
11	127.3	91.2	123.2	210.36
12	112.9	542	110.3	511.86
13	212.7	675.02	207.7	473
14	159.2	188.55	147.8	254.62
15	161.2	311.05	157.5	205.32
16	264.7	127.55	250.8	254.25
17	174.2	395.87	155.4	523.22
18	154.8	131.3	146	160.9

- in Table 4: For the specific mode λ^{RD}, mean number #S of states per node i, together with time value T^{RD}, when $S_{max} = +\infty$, $G_{max} = 21$ and $\rho = 1.5, 4$.

Comments:
There are two important points in those four tables:

Table 5. Instance parameters table.

id	$|N|$	$|M|$	Freq	$\overline{R}$	α	#SP	Greedy
1	20	87	0.2	2	0.4	59	142.3
2	20	87	0.25	1.9	1	55	82.4
3	20	87	0.19	2	1.5	63	72.7
4	20	87	0.43	2	0.4	67	141.1
5	20	90	0.6	1.9	1	61	118.8
6	20	90	0.42	2	1.5	68	93.3
7	30	147	0.16	1.9	0.4	104	317.6
8	30	138	0.18	1.9	1	96	142.3
9	30	144	0.18	2	1.5	93	102.7
10	30	153	0.41	2	0.4	102	194.8
11	30	147	0.45	2	1	104	185.3
12	30	150	0.32	2	1.5	101	129
13	40	201	0.16	1.9	0.4	130	251.4
14	40	204	0.15	2	1	126	219.5
15	40	204	0.19	2	1.5	142	172
16	40	201	0.33	1.9	0.4	138	351.2
17	40	201	0.3	1.9	1	133	199
18	40	204	0.36	2	1.5	140	175.9

Table 6. DynProg - Impact of λ^{mode}, with $Smax = 21$, $Gmax = 5$ and $\rho = 4$.

id	T^{RD}	R^{RD}	cpu^{RD}	T^{RT}	R^{RT}	cpu^{RT}	T^{DT}	R^{DT}	cpu^{DT}
1	140	93.9	0.28	132.8	99.6	0.27	142.3	87.8	0.53
2	72.3	99.5	1.57	78.7	96.1	0.46	79.6	88.5	1.19
3	63	84.1	1.22	63	84.1	1.14	72.6	55.1	1.04
4	140.3	96.6	0.34	137.1	91.7	0.19	229.1	91.3	0.18
5	90.2	90.4	0.98	76.2	98.2	1.1	85.6	96.6	1.34
6	81.4	99.9	1.36	78	99.6	1.72	78.2	97.8	1.99
7	280.5	98	1.21	283.3	98.4	1.01	287.5	82.4	1.12
8	120.2	99.2	1.5	121.2	96.5	1.44	125.8	99.6	1.03
9	95.7	97.2	0.99	93.2	97.4	0.88	94.6	83.5	1.06
10	187.2	94.7	1.08	188.9	97.6	1.07	194.8	95.8	1.11
11	124.9	96.2	1.08	124.5	93.8	1.14	125.1	95.9	1.74
12	126	95.9	1.5	111.2	99.2	2.84	115.6	98.4	3.43
13	214.4	73.4	3.2	212	90.6	1.74	255.3	89	1.5
14	156.9	93.5	1.18	149.2	99	1.31	153.5	77.7	1.4
15	158.2	96.3	1.04	161.2	90.6	0.87	172	96.1	0.83
16	276.2	92.1	0.87	292.4	80.7	10.35	351.2	79.4	1.01
17	167.9	93.5	1.61	161	99.8	1.61	178	95	2.65
18	148.9	98.5	1.11	148.6	98.4	1.16	153.4	99.2	1.8

• the length of the shortest path of the instance 3 is 63 time unit. Therefore solutions that go at full speed do not reach R_{max};

Table 7. DynProg - Impact of $Smax$, with λ^{RD}, $Gmax = 5$ and $\rho = 4$.

S_{max} id	3 T^{RD}	#S	7 T^{RD}	#S	11 T^{RD}	#S	15 T^{RD}	#S	21 T^{RD}	#S
1	142.3	1.1	141	2.45	140	3.4	140	3.75	140	3.95
2	82.3	2.5	72.3	4.09	78.3	5.5	78.3	6.45	72.3	7.55
3	63	4	63	7.75	63	10.05	63	11.65	63	12.95
4	187	1.45	150.1	3.55	140.8	4.34	140.3	4.84	140.3	5.4
5	109.1	3.6	109.1	6.4	90.2	8.8	90.2	10.15	90.2	12.3
6	111.1	4.15	111.1	8.3	110.8	11.3	82.8	14.05	81.4	18.75
7	445.6	2.1	316.7	3.73	326.1	4.66	286.9	6.6	280.5	7.93
8	171.5	2.36	146.6	4.8	120.2	6.93	120.2	8.63	120.2	10.63
9	101.2	1.9	95.9	3.86	95.8	5.73	95.7	7.66	95.7	9.26
10	190.1	2.2	187.6	3.93	187.2	5.06	187.2	6.86	187.2	8.53
11	200	2.33	139.6	4.2	144.1	6.33	124.9	7.2	124.9	8.5
12	144.4	2.86	140.1	5.76	137.2	7.73	111.7	9.63	126	12.4
13	281.9	3.55	313.6	7.45	215.2	8.6	221.7	11.62	214.4	13.25
14	207.3	2.4	156.9	4.17	156.9	6	156.9	7.47	156.9	9.52
15	160.8	1.57	159.9	3.6	158.2	5.77	158.2	6.17	158.2	8.92
16	353.6	1.5	332.4	2.85	304.6	4	304	5.09	276.2	6.32
17	205.9	2.85	176.1	5.32	167.9	7.57	167.9	9.17	167.9	11.1
18	190.4	1.92	148.9	3.7	148.9	4.92	148.9	6.72	148.9	8.12

Table 8. A* - λ^{RD}, $Smax = 21$, $Gmax = 5$ and $\rho = 4$.

id	T_A*	R_A*	#T	cpu	#S	dev
1	102.9	98.5	19	7.86	52.2	16.9
2	67.5	97.1	18	3.8	27.25	0
3	63	84.1	19	4.5	42.25	0
4	117.6	99	19	6.19	40.45	0
5	66.7	79.9	19	9.95	62.15	0
6	69.1	99.8	19	5.89	44.2	0
7	144.8	99.2	29	9.15	48.4	3.84
8	104.7	99	29	10.97	56.66	2.08
9	95	88.2	28	9.16	53	2.15
10	181.1	98.8	29	8.29	35.5	0
11	118.9	96.3	29	13.56	57.3	1.92
12	108.3	98	29	17.62	69.93	1.98
13	180.5	75.4	37	15.16	57.3	9.23
14	150.7	99.2	39	21.13	70.45	3.17
15	143.4	97.5	35	19.91	60.17	0
16	186.4	99.4	35	14.11	42.72	0
17	146.5	96.9	35	17.56	61.72	1.5
18	143.6	98.3	39	25.04	76	0

- the RD method always performs best or is near the best solution as it can be expected because of the RD reformulation, see Sect. 2.4;
- the DT method is designed to minimise its risks, then it may end without reaching R_{max} more often than the two other methods;
- on instance 11, the tree methods end without reaching R_{max}, so we can deduct that the last arc is not risky around the arrival date.

Table 9. Decoupled - λ^{RD}, $Smax = 21$, $Gmax = 5$ and $\rho = 4$.

id	T_Dc	R_Dc	$\#T$	cpu	$\#S$	dev
1	118	97.9	11	5.96	5.35	6.77
2	67.4	97.1	5	2.83	6.25	0
3	63	84.1	6	5.24	8.19	0
4	140.3	96.6	3	1.67	4.7	0
5	69.8	92	20	26.76	12.25	0
6	70.8	95.5	14	14.83	10.6	0
7	159.9	79.9	27	27.47	10.3	11.5
8	119.5	92.4	13	13.59	6.13	4.16
9	95	88.2	9	9.71	8.46	2.15
10	187.2	94.7	8	9.86	8.03	0
11	124.9	96.2	7	7.03	7.43	0
12	114.1	95.1	15	22.21	10.73	5.94
13	187.2	97.8	53	114.4	11.12	1.53
14	151.1	99.6	15	21.46	7.42	3.17
15	143.7	97.9	22	32.62	6.92	0
16	252.2	92.5	5	5.05	3.75	0
17	155.4	89.7	16	26.01	8.15	0
18	148.9	98.5	7	8.24	6.47	0

Table 10. Distribution of optimal decisions - λ^{RD}, $Smax = +\infty$, $Gmax = 21$ and $\rho = 8$.

id_λ	1	2	3	4	5	6	7
%	16.6	0.4	1.2	1.2	1.2	2.4	1.2
id_λ	8	9	10	11	12	13	14
%	1.4	0.9	1.6	7.4	15.2	11.3	7.9
id_λ	15	16	17	18	19	20	21
%	4.8	3.3	3.3	2.4	1.9	2.4	11

Then, we can see that the solutions in Table 6 are fairly close to the solutions in Table 3 but with just a few states kept at every node (hence the very short computing time).

Outputs related to the behavior of **A*_SSPP** and **Dec_SSPP**. We test the ability **A*_SSPP** and **Dec_SSPP**. to catch optimal solution, and observe the characteristics of resulting path. The instances used are shown with their parameters in Table 5. We rely on $\lambda = \lambda^{RD}$, $S_{max} = 21$, $G_{max} = 5$ and $\rho = 4$. For every instance, we compute:

- in Table 8: The time value T_A*, the percentage of risk R_A* from R_{max}, CPU time (in s.) cpu, the number $\#T$ of visited nodes, the number $\#S$ of generated states, and the deviation dev between the length of resulting path Γ and L^*, induced by **A*_SSPP** with $\lambda^{mode} = \lambda^{RD}$, $S_{max} = 10$ and $\rho = 4$;

Table 11. Decision Sets - λ^{RD}, $Smax = 21$, $Gmax = 5$ and $\rho = 4$.

id	mR	MR	mS	MS
1	3	9	19	35
2	2.8	6	21.8	36
3	0	0	21.66	40
4	3.75	9	18.75	34
5	2.83	7	22.5	33
6	5.85	10	28.28	58
7	1.33	3	20.83	38
8	6.28	12	17.57	25
9	2.16	3	28	49
10	4.85	23	30.28	74
11	3.66	9	27.16	48
12	2.37	9	24.5	51
13	6.09	21	26.36	47
14	4.62	19	18.62	34
15	0	0	25	45
16	6	18	24.16	45
17	6.44	16	25.44	39
18	3.42	8	26.42	39

- in Table 9: The time value T_{DC}, the percentage of risk R_Dc from R_{max}, CPU time (in s.) cpu, the number $\#T$ of trials, the number $\#S$ of generated states, and the deviation dev between the length of resulting path Γ and L^*, which derives from applying **Dec_SSPP** with $\lambda^{mode} = \lambda^{RD}$, $S_{max} = 10$ and $\rho = 4$.

Comments:

We see that **A*_SSPP** and **Dec_SSPP** are close to each other but the latter is behind most of the time. This is because **Dec_SSPP** explores more paths but generates fewer states and therefore obtains less good results when using the same path as **A*_SSPP**.

Outputs related to characteristics of the solutions. Once again, we rely on **DP_Evaluate**, and we look for information on the decision set Γ. We use $\lambda = \lambda^{RD}$, $S_{max} = 21$, $G_{max} = 3$ and $\rho = 4$. We consider the instances as a whole and compute:

- in Table 10: For every λ value, the percentage of optimal decisions related to λ in the optimal decision sequences computed by **DP_Evaluate**;
- in Table 11: The mean rank mR and the max rank MR of states (T, R) involved into those optimal trajectories, inside the state sets $STATE[i]$, $i = 1, \ldots, n$, according to the order induced by increasing T values, estimated as in Table 6. mS and MS respectively refer to the mean size of state set $State[i]$, $i = 1, \ldots, n$, and its maximal size.

Comments:

We can see that optimal decisions are often generated thanks to the middle λ values (from the 9th to the 15th value) and, in case of uncommon scenarios, the far left and far right values will come handy.

Outputs related to the learning process.

Lastly, we will use **DP_Evaluate** with the learning process of Sect. 5.1 while testing the

Table 12. DynProg with Learning - Impact of λ^{mode}, with $Smax = 9$, $Gmax = 3$ and $\rho = 4$.

id	T^{RD}	R^{RD}	cpu^{RD}	T^{RT}	R^{RT}	cpu^{RT}	T^{DT}	R^{DT}	cpu^{DT}
1	104.6	98.5	0.22	118.6	98.4	0.19	142.3	87.8	0.36
2	75	97.9	0.29	96.4	99.8	0.41	71.6	98.2	0.68
3	63	84.1	0.62	63	84.1	0.68	0	0	0
4	160.4	98.6	0.26	138.2	99.8	0.25	195.7	69.7	0.24
5	86.4	86.7	0.4	81.6	98.4	0.54	81.5	99.6	0.96
6	101.5	99.9	0.59	0	0	0	82.1	97.5	0.98
7	284.7	88.8	0.4	280	99.7	0.37	284.6	88.6	0.63
8	126.6	99.5	0.48	160.6	98.2	0.36	121.4	99.7	0.85
9	93.2	97.7	0.75	93.2	97.6	0.42	96	82.5	0.77
10	191.5	97	0.63	196.4	99.3	0.55	180.4	99.8	0.83
11	126.5	96.2	0.54	187.1	99.8	0.53	121.5	97.6	2.1
12	136.8	76.1	0.57	0	0	0	138.4	99.3	1.08
13	298.9	99.7	0.82	214	95.8	0.81	349.8	99.2	0.98
14	164.1	99.1	0.57	155.7	91.4	0.55	158.1	87.3	0.82
15	168.1	97.8	0.5	165.4	87.8	0.57	161.5	96.3	0.62
16	346.4	95.3	0.38	294.9	95.1	0.4	332.6	77.5	0.84
17	206.4	98.9	3.47	173.8	98.5	0.7	152	91.4	1.91
18	149.1	98.9	0.48	148.9	99.7	0.46	151.2	99.9	0.86

role of parameters $\lambda = \lambda^{RD}, \lambda^{RT}, \lambda^{DT}$, as well as S_{max} and ρ. So, for every instance, we compute in Table 12: The time value T_DP^{mode}, the risk value R_DP^{mode}, and the CPU times (in milliseconds.) CPU^{mode}, induced by application of **DP_SSPP** on the shortest path between o and p with $\lambda^{mode} = \lambda^{RD}, \lambda^{RT}, \lambda^{DT}$, $S_{max} = 9$, $G_{max} = 3$ and $\rho = 4$;

Comments:

Because of the short number of states kept (S_{max}) and generated (G_{max}), the method is the fastest of all and it performs as well as with $S_{max} = 11$ and $G_{max} = 5$ shown on Table 7. So it is the method that has performed best so far.

7 Conclusion

We dealt here with a shortest path problem with risk constraints, which we handled under the prospect of fast, reactive and interactive computational requirements. But in practice, a vehicle is scheduled in order to perform some kind of *pick up and delivery* trajectory while performing retrieval and storing tasks. It comes that a challenge becomes to adapt previously described models and algorithms to such a more general context. Also, there exist a demand from industrial players to use our models in order to estimate the best-fitted size of an AGV fleet, and the number of autonomous vehicles inside this fleet. We plan addressing those issues in the next months.

References

1. Mombelli, A., Quilliot, A., Baiou, M.: Proceedings of the 11th International Conference on Operations Research and Enterprise Systems. SCITEPRESS - Science and Technology Publications (2022). https://doi.org/10.5220/0010780700003117

2. Baïou, M., Quilliot, A., Adouane, L., Mombelli, A., Zhu, Z.: Algorithms for the safe management of autonomous vehicles, pp. 153–162 (2021). https://doi.org/10.15439/2021F18
3. Ren, Q., Man, K.L., Lim, E.G., Lee, J., Kim, K.K.: Cooperation of multi robots for disaster rescue. In: Proceedings of the International SoC Design Conference, ISOCC 2017 (2018). https://doi.org/10.1109/ISOCC.2017.8368834
4. Koes, M., Nourbakhsh, I., Sycara, K.: Heterogeneous multirobot coordination with spatial and temporal constraints, vol. WS-05-06, pp. 9–16 (2005)
5. Zhang, M., Batta, R., Nagi, R.: Modeling of workflow congestion and optimization of flow routing in a manufacturing/warehouse facility, vol. 55, pp. 267–280 (2009). https://doi.org/10.1287/mnsc.1080.0916
6. Vis, I.F.A.: Survey of research in the design and control of automated guided vehicle systems. Eur. J. Oper. Res. **170**, 677–709 (2006). https://doi.org/10.1016/j.ejor.2004.09.020
7. Vivaldini, K.C.T., Tamashiro, G., Martins Junior, J., Becker, M.: Communication infrastructure in the centralized management system for intelligent warehouses. In: Neto, P., Moreira, A.P. (eds.) WRSM 2013. CCIS, vol. 371, pp. 127–136. Springer, Heidelberg (2013). https://doi.org/10.1007/978-3-642-39223-8_12
8. Ryan, C., Murphy, F., Mullins, M.: Spatial risk modelling of behavioural hotspots: risk-aware path planning for autonomous vehicles. Transp. Res. Part A: Policy Pract. **134**, 152–163 (2020). https://doi.org/10.1016/j.tra.2020.01.024
9. Pimenta, V., Quilliot, A., Toussaint, H., Vigo, D.: Models and algorithms for reliability-oriented Dial-a-Ride with autonomous electric vehicles. Eur. J. Oper. Res. **257**, 601–613 (2017). https://doi.org/10.1016/j.ejor.2016.07.037
10. Philippe, C., Adouane, L., Tsourdos, A., Shin, H.S., Thuilot, B.: Probability collectives algorithm applied to decentralized intersection coordination for connected autonomous vehicles, vol. 2019-June, pp. 1928–1934 (2019). https://doi.org/10.1109/IVS.2019.8813827
11. Park, I., Jang, G.U., Park, S., Lee, J.: Time-dependent optimal routing in micro-scale emergency situation. In: Proceedings of the IEEE International Conference on Mobile Data Management, pp. 714–719 (2009). https://doi.org/10.1109/MDM.2009.122
12. Lozano, L., Medaglia, A.L.: On an exact method for the constrained shortest path problem. Comput. Oper. Res. **40**, 378–384 (2013). https://doi.org/10.1016/j.cor.2012.07.008
13. Le-Anh, T., De Koster, M.B.M.: A review of design and control of automated guided vehicle systems. Eur. J. Oper. Res. **171**, 1–23 (2006). https://doi.org/10.1016/j.ejor.2005.01.036
14. Krumke, S.O., Quilliot, A., Wagler, A.K., Wegener, J.-T.: Relocation in carsharing systems using flows in time-expanded networks. In: Gudmundsson, J., Katajainen, J. (eds.) SEA 2014. LNCS, vol. 8504, pp. 87–98. Springer, Cham (2014). https://doi.org/10.1007/978-3-319-07959-2_8
15. Franceschetti, A., Demir, E., Honhon, D., Van Woensel, T., Laporte, G., Stobbe, M.: A metaheuristic for the time-dependent pollution-routing problem. Eur. J. Oper. Res. **259**, 972–991 (2017). https://doi.org/10.1016/j.ejor.2016.11.026
16. Chen, L., Englund, C.: Cooperative intersection management: a survey. IEEE Trans. Intell. Transp. Syst. **17**, 570–586 (2016). https://doi.org/10.1109/TITS.2015.2471812
17. Berbeglia, G., Cordeau, J.-F., Gribkovskaia, I., Laporte, G.: Static pickup and delivery problems: a classification scheme and survey. TOP **15**, 1–31 (2007). https://doi.org/10.1007/s11750-007-0009-0
18. Amazon (2022). https://www.amazon.com/primeair
19. Hart, P.E., Nilsson, N.J., Raphael, B.: A formal basis for the heuristic determination of minimum cost paths. IEEE Trans. Syst. Sci. Cybern. **4**, 100–107 (1968). https://doi.org/10.1109/TSSC.1968.300136
20. Mo, D., Cheung, K., Song, S., Cheung, R.: Routing strategies in large-scale automatic storage and retrieval systems. In: IIE Annual Conference and Exposition 2005 (2005)

21. Chen, T., Sun, Y., Dai, W., Tao, W., Liu, S.: On the shortest and conflict-free path planning of multi-AGV systembased on Dijkstra algorithm and the dynamic time-window method. Adv. Mater. Res. **645**, 267–271 (2013). https://doi.org/10.4028/www.scientific.net/AMR.645.267

22. Wurman, P.R., D'Andrea, R., Mountz, M.: Coordinating hundreds of cooperative, autonomous vehicles in warehouses. AI Mag. **29**, 9–19 (2008). https://doi.org/10.1609/AIMAG.V29I1.2082

23. Martínez-Barberá, H., Herrero-Pérez, D.: Autonomous navigation of an automated guided vehicle in industrial environments. Robot. Comput.-Integr. Manuf. **26**, 296–311 (2010). https://doi.org/10.1016/j.rcim.2009.10.003

Minimizing the Non-value Task Times: A Pickup and Delivery Problem with Two-Dimensional Bin-Packing

Bárbara Romeira[1] [iD] and Ana Moura[2(✉)] [iD]

[1] Department of Economics, Management, Industrial Engineering and Tourism, University of Aveiro, Aveiro, Portugal
barbararomeira@ua.pt
[2] GOVCOPP – Systems for Decision Support Research Group, University of Aveiro, Aveiro, Portugal
ana.moura@ua.pt

Abstract. The current crisis that struck the automotive industry created an urgency for improvement initiatives throughout the value chain. And, although it is vital to improve the primary value chain, we should also look to improve the support services that provide for it. To this end, the current work focuses on the Manufacturing Tool Repair service, a support service in an automotive factory. This service is responsible for delivering repaired tools to the production lines and collecting the used ones. The time spent on this activity comprises 30% of the total shift time, which means that 142 min of a 480-min shift are used in this non-value-adding activity. To improve this service efficiency, we look for ways to increase the number of tools delivered in each trip, and consequently reduce the number of trips performed. For that, a mathematical model formulation for the Two-Dimensional Bin-Packing Problem was proposed. The approach was tested in 12 real-world instances and the results show that the model was able to optimally solve all within satisfying CPU time, thus allowing for real-time feedback. Also, we tested the efficiency of the model by increasing the pickup and delivery demand and the number of bins available to perform each trip.

Keywords: Two-Dimensional Bin-Packing · Vehicle routing problem · Simultaneous pickup and delivery

1 Introduction

The automotive industry has suffered greatly in the last couple of years. Decrease in sales, increase in raw material prices and COVID-19 led to an economic crisis within this industry. As a result, is highly imperative that these organizations increase the efficiency in all parts of the business, not just the production processes. Organizations must therefore look at their value chains and improve all services that support it. The current paper intents to do just that, by looking at one of those services – the Manufacturing tool repair (MTR) service of an automotive factory.

© The Author(s), under exclusive license to Springer Nature Switzerland AG 2024
F. Liberatore et al. (Eds.): ICORES 2022/2023, CCIS 1985, pp. 26–46, 2024.
https://doi.org/10.1007/978-3-031-49662-2_2

This work was initiated in [1], where a Vehicle Routing Problem with Simultaneous Pickup and Delivery, and Time windows (VRPSDPTW) was formulated to calculate the most time efficient route to Pickup and Delivery (P&D) the manufacturing tools from, and to the factory's production lines. The workers of the MTR service pickup used tools from the production lines, repair and calibrate them, and then deliver them to the respective production lines. The delivery and collection process is performed by two different vehicles and consumes, on average, 30% of the total working time of each shift. This means that 30% of the time is spent on a non-value-adding activity, causing delays in the repair and calibration, which are value-adding. To solve the problem a data processing framework and a Mixed-Integer Linear Programming (MILP) model for the VRPSDPTW were developed and implemented. The result was a 49% reduction of the travelling time for one vehicle and a 14% reduction for the other. Extending on the previous, the present work intends to further improve the P&D process by optimizing the vehicle volume utilization. Hence, the present work addresses the Two-Dimensional Bin-Packing Problem (2D-BPP) to obtain the most efficient allocation of the tools to the vehicles. As it will be presented ahead, two different vehicles are used for the transportation of manufacturing tools: a manual vehicle and an automatic vehicle. The paper will concentrate on the allocation of tools in the automatic vehicle, rather than in both vehicles. This is due to the fact that the manual vehicle's capacity is only influenced by the quantity of tools, not their positioning and volume. In summary, this paper adds to the current literature by:

- Proposing a formulation for the 2D-BPP that limits the number of levels in a bin;
- Extending the application of the 2D-BPP together with the VRPSDPTW in a real-world situation, considering the real-time demand at the production lines.

This paper is organized as follows: Sect. 2 offers a compilation of works related to the 2D-BPP. A detailed description of the MTR service current problems and of the solution proposed are presented in Sect. 3. In Sect. 4 the proposed model formulation is presented. In Sect. 5 the model is tested in different instances and the results are given and analyzed. Also in this section, the integration of the two models is presented with two real-world examples. Lastly, a summary of the main conclusions and future work is shown in Sect. 6.

2 Related Works

A combinatorial optimization problem is defined as the process of evaluating the solutions according to an objective function with the goal of finding the optimal solution [2, 3]. The Bin-Packing Problem (BPP) is one of the most widely studied combinatorial optimization problems in the literature. The BPP intends to pack a set of items into a finite number of fixed-capacity bins, with the goal of minimizing the number of bins used. According to the dimension of the items to be packed, the BPP can be defined as one-Dimensional Bin-Packing Problem (1D-BPP), two-Dimensional Bin-Packing Problem (2D-BPP), and three-Dimensional Bin-Packing Problem (3D-BPP) [2].

In this paper, we will be looking at the two-Dimensional Bin-Packing Problem (2D-BPP), where a set of m items, each with a positive width $w_1, \ldots, w_m$ and height

$h_1, \ldots, h_m$, are placed inside a number of finite bins with width W and height H, in such a way that the items are placed orthogonally and do not overlap. The objective of this problem is to minimize the number of bins used to pack all the items [4–6].

In this section a collection of works related to the 2D-BPP is presented. Table 1 offers a summary of the types of 2D-BPP published works and presented in this section and Table 2 presents a summary of the related objective functions. Table 3 categorizes each approach used in the works. The work presented in this paper also appears in all tables under the column Paper as "Our".

In [5] a 2D-BPP with due dates is solved using a sequential value correction heuristic, with the objective of minimizing the number of bins and the maximum lateness of the items inside the bins. The approach is compared to two benchmark algorithms, the multicrossover genetic heuristic (MXGA) presented in [7] and the Hybrid constraint and integer programming approach (CPMIP) proposed by [8]. The computational results clearly show that the approach in [5] outperforms the two benchmark approaches, obtaining better primal-dual gaps in smaller times. The authors also tested the approach in real world instances obtaining solutions in reasonable computational times.

Table 1. A summary of the 2D-BPP found in the literature.

Paper	2D-BPP	2D-VBPP	2D-VSBPP
[5]	x		
[6]	x		
[8]	x		
[9]	x		
[10]	x		
[11]		x	
[12]		x	
[13]		x	
[14]			x
[15]		x	
Our	x		

Polyakovskiy & M'Hallah [8], as stated above, proposed a CPMIP to solve the 2D-BPP, with the objective of minimizing the maximum lateness of the packed items. In their model they have several constraints regarding the overlapping of the items inside the bin and the rotation of those items is allowed. The authors tested the approach in the 1500 benchmark instances from [7], showing that their approach improves existing upper bounds on average 27.45% and obtains optimal solutions for 586 instances. Also, the CPMIP is compared to the MXGA from [7], obtaining better quality solutions in lower computational (CPU) times. The 2D-BPP with levels is addressed in [6] with the goal of minimizing the number of bins used. The authors used an Integer Linear Programming (ILP) model and added width and height constraints to each level of the bin. To test the ILP model 200 instances from [9] and 300 instances from [16]. Out of the 500 instances the ILP solved 316 instances optimally. Even when the optimal solution was not obtained the approach was able to get a solution within the imposed time limit for 470 instances. Martello and Vigo [9] implemented the first exact approach to solve the 2D-BPP using a Branch-and-Bound (BB) algorithm. They tested the algorithm with 36 benchmark instances from [17, 18] and [19], and 300 randomly generated instances proposed by [16]. The exact approach solved 34 out of the 36 benchmark instances optimally, and 211 out of the 300 instances. The authors in [10] simplified a 3D-BPP into a 2D-BPP by creating layers of even height. This simplification allowed them to apply a mixed-integer programming formulation to solve it. The formulation was then tested in a total of 20 instances where 250, 500, 750 and 1000 size boxes were randomly selected. According to the results 79% of the boxes are successfully grouped within, on average, 5.3 s and 157 s for 100 items.

Table 2. A summary of the objective functions found in the literature.

	[5]	[6]	[8]	[9]	[10]	[11]	[12]	[13]	[14]	[15]	Our
Maximize total area of boxes placed in the 1st section of the long layer					x						
Minimize total cost of all bins								x	x	x	
Minimize number of trucks						x					
Minimize number of refrigerated trucks						x					
Minimize number of refrigerated trucks with frozen products						x					
Minimize number of refrigerated trucks with standard products						x					
Minimize splitting						x					
Minimize number of bins	x	x		x			x				x
Minimize the maximum lateness of the items	x		x								

In contrast with the aforementioned authors in [11], a Two-dimensional Vector Bin-Packing Problem (2D-VBPP) is addressed using a heuristic algorithm and a Branch-and-Price (BP) algorithm. The proposed problem has several objectives as it can be seen by Table 2. Both algorithms were tested in 80 real world instances. The exact approach reaches the optimum for 68 out of the 80 instances within a 10-min CPU time, if splitting items is not allowed. With splitting items, the results show that the BP is still able to optimally solve 62 instances. The heuristic approach was able to find the optimal solution for 86% of the instances. Similarly, to the previous authors in [13] the

2D-VBPP was solved using an Integer Programming (IP) model for small instances and an Iterated local Search (ILS) for both smaller and bigger instances. The authors defined 40 instances with known optimal solution and randomly generated 120 instances. The ILS was compared to a Greedy-on-Weight-Range (GWR) heuristic also proposed by the authors of the paper. The computational results show that for small instances the ILS outperforms the GWR. For the randomly generated instances ILS gets better quality solutions with a small tradeoff in CPU time. Wei et al. [12] also used a BP approach to solve the 2D-VBPP and tested it in 400 instances from [20]. Then the computational results were compared to two BP from [20] and [21], and to two heuristics, a ILS proposed by [22] and a Consistent Neighborhood Search (CNS) from [23]. Compared to the BP algorithms the approach proposed by the authors was able to obtain the optimal solutions for 390 out of 400 instances, 6% more than in [20] and 35% more than in [21]. To compare the heuristics the authors only used 200 of the 400 benchmark instances. The results show that for both CNS and the proposed BP optimally solved 190 of the 200 selected instances. However, the latter used less CPU time in more than half of the test sets. In [15] the authors applied the BP algorithm to a 2D-VBPP with volumetric weight and general costs. The approach was tested in 90 test instances proposed in [24] and the computational results show that it solved optimally 86 of the 90 test instances, outperforming the approached tested in [24]. On top of this the total computational (CPU) time was reduced by 84.5%.

Table 3. A summary of the approaches used in the literature.

Paper	Exact	Heuristic	Hybrid
[5]		x	
[6]	x		
[8]			x
[9]	x		
[10]	x		
[11]	x	x	
[12]	x		
[13]	x	x	
[14]	x		
[15]	x		
Our	x		

Another type of the 2D-BPP was solved in [14], the Two-Dimensional Variable Sized Bin-Packing Problem (2D-VSBPP). As most of the aforementioned authors, a BP approach was used and tested with benchmark instances from [16] and [25]. The authors came to the conclusion that it was easier to solve the 2D-BPP.

3 Case Study

3.1 Manufacturing Tool Repair Service

The Manufacturing tool repair (MTR) service performs the repair and calibration of the manufacturing tools used to produce gearbox and motor metal components. On top of these activities, the MTR service is also responsible for:

- Picking the tools that need to be repaired from the production lines;
- Delivering, new or repaired and calibrated tools to the production lines;
- Helping production with possible problems related to manufacturing tools.

Note that, only the repair and calibration activities are considered value-added for this service. The picking and delivering activities are considered transport wastes and the resolution of problems are considered defects. Since each tool has a different lifespan, defined by the number of components produced, the repair and calibration frequency is, in normal conditions, the same for each tool. However, due to defects or misuse, it can happen that the manufacturing tool is not used until the end of its lifespan. In either case, the used tools have to be collected from the production lines and the new or repaired ones delivered. It's imperative that no production line stops due to lack of tools available for use.

The P&D points are defined in Fig. 1, as well as the routes currently followed by the MTR workers. All points in the routes are visited and this process takes an average of 18, 15 and 8 min for Routes A, B and C, respectively.

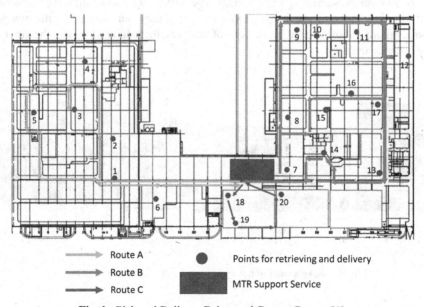

Fig. 1. Pick and Delivery Points and Current Routes [1].

Also, the pickup and delivery activity may be performed by one of two different vehicles: a manual vehicle and an automatic vehicle (Fig. 2).

The manual vehicle can carry a total of 90 tools, and it was adapted to transport tools which have special characteristics. Those tools will be defined henceforth as, extraordinary tools. In this case, the location of the tools in the vehicle is straightforward, because a tool corresponds to a space in the vehicle.

A   B

Fig. 2. Vehicles used by the MTR service: A – Manual vehicle; B – Automatic Vehicle [1].

For the automatic vehicle, the manufacturing tools are transported in baskets placed in a trailer attached to it. The trailer has 120x120 cm (Weight and Height) and it's divided in 3 levels, all with the same height and weight. The baskets have standardized dimensions and depending on the manufacturing tool transported they can carry a specific number of tools. Figure 3 shows the different types of brackets, their dimensions and the number of tools they can carry.

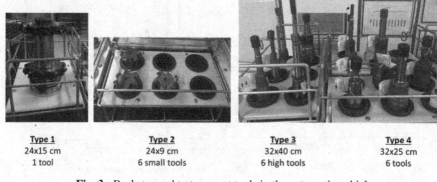

Type 1	Type 2	Type 3	Type 4
24x15 cm	24x9 cm	32x40 cm	32x25 cm
1 tool	6 small tools	6 high tools	6 tools

Fig. 3. Baskets used to transport tools in the automatic vehicle.

To deal with the current demand, the MTR service workers need to do each route at least 2 times, one within the first 20 min of the shift and another in the middle of the shift. However, depending on the production line, the MTR workers can go and check if any additional tools are necessary. Thus, sometimes they have an average of 6 additional trips per shift. This means that in each shift, 30% of the workers time is wasted in transport. On top this, this situation also causes delays in the repair and calibration processes, which hinders the service's efficiency and risks tool shortages in the production lines.

3.2 Solution Proposed

Since the pickup and deliver activities highly impact the service's response time the solution is to optimize them, which means:

- Reducing the number of pickup and delivery trips that the workers must do;
- Reducing the total time of the trip;
- Optimizing the automatic vehicle's volume utilization, so that it can collect and deliver more tools in the same trip.

To do so, we created a data collection and analysis framework, which gathers and processes data from the P&D points and MTR service. Then, the problem is modelled as a VRPSDPTW to get the optimized route and as a 2D-BPP to obtain the most efficient allocation of the baskets to the automatic vehicle. The first was addressed and presented in the work of Romeira and Moura [1], and the latter will be presented in the current work. To do the connection between the two models, some changes in the data pre-processing process presented in [1] had to be introduced in this paper, as it will be shown ahead.

The Data Pre-processing Process
The data pre-processing process presented in Fig. 4, is a matheuristic algorithm that uses the data retrieved from the e-Kanban system [26] to determine if a trip must occur. This can happen when a production line reaches the minimal stock of a specific tool or when the number of tools available to P&D is sufficient to perform a route. Note that, from here onwards the production lines will be referred as customers.

On one hand, when the minimal stock is reached the related customer is now considered a priority customer and, because of that, a trip alert is generated. After this alert, the system verifies if that tool is available for delivery in the MTR service. If the priority tool is not available an urgent alert is sent to the MTR Workers, so they can prepare the priority tools. In the event that the tool is available, the system calls the "Vehicle Selection" procedure to choose which vehicle will do the P&D route. Then, to take full advantage of the chosen vehicle's capacity the system runs the procedure "Availability of tools to P&D". This procedure determines if, for the same priority customers, there are any non-priority tools to be picked or delivered in that same route. Once the tools to P&D are sorted the MILP model will create the P&D route that minimizes the total travel time. After that, if the vehicle chosen for the route is the automatic vehicle, the system will use the 2D-BPP model to optimize the vehicle's space when it leaves the MTR service.

This model works by first performing the 2D-BPP for the depot node (MTR service), obtaining the most efficient way to allocate the baskets to the vehicle. Additionally, by using the value of the objective function (OF) it's possible to allocate additional baskets to the vehicle. Since we have 2 sides of the trailer each one with 3 levels, this means that if $OF < 6$ then there are still places available in the vehicle. Using this information, the system verifies if there are any additional tools to deliver the priority customers.

Once this is done, the final P&D route and Bin-Packing at the MTR service solution are available to the worker.

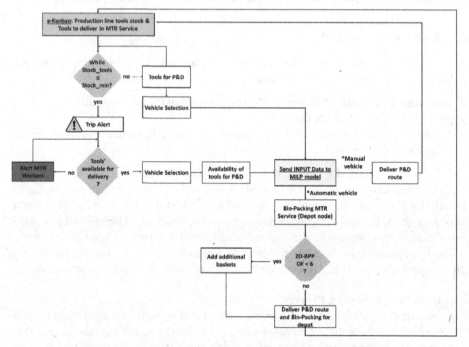

Fig. 4. The Data pre-processing process (Adapted from [1]).

On the other hand, when the number of tools is sufficient to do a trip the "Tools to P&D" procedure is performed. Then the appropriate vehicle is selected, using the procedure "Vehicle Selection". Once the vehicle is chosen the route is obtained with the MILP model and the Bin-packing model solved, if the vehicle selected is the automatic.

In summary, at the end of the pre-processing process we aim to obtain:

- The customers to visit in the route;
- The tools to pickup and delivery in each customer;
- The tools' location in the vehicle when using the automatic vehicle.

It is important to note that, to due the connection between the two proposed models, instead of using the number of tools as the input data for the VRPSDPTW, we now convert this number into number of baskets. The data processing does this conversation the following way:

- We have 1 tool that uses Type 2 baskets, this corresponds 1 basket in the VRPDSPTW model;
- We have 6 tools that use Type 2 baskets, this corresponds to 1 basket in the VRPDSPTW model;
- We have 8 tools that use Type 2 baskets, this corresponds to 2 baskets in the VRPDSPTW model;

- We have 2 tools that use Type 1 baskets, this corresponds to 2 baskets in the VRPDSPTW model;

Considering the number and dimensions of the baskets, the automatic vehicle's capacity limit is 22 baskets.

Vehicle Selection Procedure
The vehicle selection procedure selects the vehicle according to the type of tools and the vehicle's available at the time a route is triggered. First the procedure checks if the tools are extraordinary, if so, the manual vehicle is automatically selected. If not, the number of vehicles available is verified. When none of vehicles are available the procedure chooses the first vehicle that is confirmed to arrive at the MTR service. If only one vehicle is available, that vehicle is used for the trip. In the case that both vehicles are available the automatic vehicle is the one selected. In Fig. 5 the described procedure is summarized.

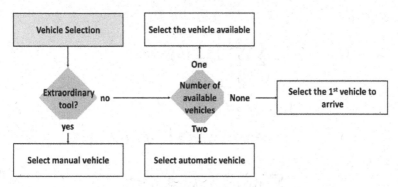

Fig. 5. The Vehicle Selection procedure.

4   2D-BPP Model

To allocate the baskets to the vehicles a linear programming model was developed based on the well-known 2D-BPP.

Considering $B = \{1, \ldots, k\}$ a set of k identical rectangular bins, where each Bin $k \in B$ has a width W and a height H, we have that for each bin k, there are $N = \{1, \ldots, i\}$ identical levels. Every item $j \in IT = \{1, \ldots, it\}$ has a width $w_j \leq W$, and a height $h_j \leq H$. The model is formulated in order to only place the items in the bin once and without overlap.

For the problem in the study, a bin corresponds a side of the automatic vehicle's trailer, which means $k \in B = \{1, 2\}$. Both bins have $H = 120$ cm and $W = 120$ cm. As said before, each side of the automatic vehicle has 3 levels all identical, thus the model assumes that $i \in N = \{1, 2, 3\}$. The items to be placed in the bins correspond to the baskets that contained the manufacturing tools. Depending on the type of basket we can have different widths and different heights for each one.

The first decision variable defines if basket j was placed in bin k and level i. Thus, the basket packing is:

$$x_{kij} = \begin{cases} 1 \text{ if basket } j \text{ is packed in bin } k \text{ level } i \\ 0 \text{ otherwise} \end{cases} \forall k \in B, \forall i \in N, \forall j \in IT \quad (1)$$

The second decision variable, z_{ki}, defines whether level i is allocated in bin k. So, the level packing is modeled by:

$$z_{ki} = \begin{cases} 1 \text{ if } i \text{ is allocated to bin } k \\ 0 \text{ otherwise} \end{cases} \forall k \in B, \forall i \in N \quad (2)$$

Thus, the proposed 2D-BPP model used is as follows:

$$\min \sum_{k=1}^{b} \sum_{i=1}^{n} z_{ki} \quad (3)$$

Subjected to

$$\sum_{k=1}^{b} \sum_{i=1}^{n} x_{kij} = 1 \, \forall j \in IT \quad (4)$$

$$\sum_{i=1}^{n} z_{ki} \leq 3 \, \forall k \in B \quad (5)$$

$$x_{kij} \leq z_{ki} \, \forall k \in B, \forall i \in N, \forall j \in IT \quad (6)$$

$$\sum_{j=1}^{it} w_j x_{kij} \leq W \, \forall k \in B, \forall i \in N \quad (7)$$

$$\sum_{i=1}^{n} h_j x_{kij} \leq H \, \forall k \in B, \forall j \in IT \quad (8)$$

Decision variables

$$z_{ki} \in \{0, 1\} \, \forall k \in B, \forall i \in N \quad (9)$$

$$x_{kij} \in \{0, 1\} \, \forall k \in B, \forall i \in N, \forall j \in IT \quad (10)$$

For the current model, the objective function (3) intends to minimize not only the number of bins used, but also the number of levels. This means that it is intended to minimize the number of vehicles used, taking into consideration that each vehicle is considered to have 2 bins (each side of the basket holder) and each of the bins has 3 layers. Constraint (4) is used to guarantee that each basket is only placed once in a bin and respective level. In other words, we guarantee that each basket is only packed once. To define the upper limit for the number of possible levels, which in this study is 3, constraint (5) is used. Constraints (7) and (8) guarantees that both width and height of a bin (the automatic vehicle's trailer) is not exceeded by the placed baskets. Equations (9) and (10) define the decision variable's domains.

5 Computational Results

The model was tested in 12 different instances from normal demand requests of the automatic vehicle. Each problem instance is defined by the number of brackets with repaired and calibrated tools that leave the MTR service to be delivered to the production lines. Table 4 presents the number of baskets of each type in the problem instances. The trailer in the automatic vehicle has height and weight equal to 120 cm. The baskets have the dimensions presented in Fig. 3.

Table 4. Number of baskets of each type in the problem instances.

Instance	N° baskets	Type 1	Type 2	Type 3	Type 4
#1	6	0	1	1	4
#2	10	1	4	3	2
#3	14	2	2	2	8
#4	15	0	6	2	7
#5	18	0	6	2	10
#6	21	0	8	2	11
#7	23	0	12	1	10
#8	23	1	9	4	9
#9	24	1	13	1	9
#10	25	2	14	0	9
#11	25	1	9	4	11
#12	26	0	17	0	9

5.1 Results

The model presented was implemented using the CPLEX Studio IDE 20.1.0, and the experiments were run on an Intel (R) CORE(TM) i7-10750H CPU 2,60 GHz with 16 Gb of memory.

Table 5 shows the results obtained for the test instances. The first column gives the test instances and the second, the OF. To better understand the OF results column S "Bins" and "Levels" were added. The first gives the number of trailer sides used and the latter the number of levels used. The "CPU" column gives the time needed to perform the Bin-Packing model. The last column has the GAP (in %) given by CPLEX, which gives the tolerance on the GAP between the best integer solution and the best node remaining.

The results show that the optimal solution was met for 12 out of the 12 test instances. All the solutions were obtained within a very good CPU time, on average 0.09 s, with the highest being 0.25 s. As we can see, even with a bigger number of baskets, the test instances are solved optimally and within the very small CPU times.

Table 5. Model results for the 2D-BPP.

Instances	OF	N° Trailers	N° Levels	CPU(s)	GAP (%)
#1	2	1	2	0.03	0
#2	3	1	3	0.25	0
#3	4	2	4	0.09	0
#4	4	2	4	0.05	0
#5	5	2	5	0.06	0
#6	6	2	6	0.05	0
#7	6	2	6	0.08	0
#8	6	2	6	0.05	0
#9	6	2	6	0.05	0
#10	6	2	6	0.13	0
#11	6	2	6	0.20	0
#12	6	2	6	0.05	0

Figure 6 and Fig. 7 present the representation of the solutions obtained for the first and last instances of the problem, respectively.

From the model results for the first instance, we know that only one side of the trailer is used, and the baskets are all placed in two levels as is shown in Fig. 6.

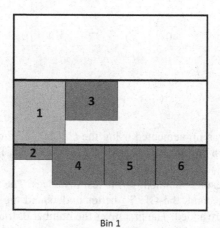

Bin 1

Fig. 6. Solution for the 1st instance.

For the 12th instance, we have a mixed of Type 2 and Type 4 baskets. Due to the number of baskets the solution uses the two sides of the trailer. Also, the baskets occupy all the available levels, as we can see by Fig. 7.

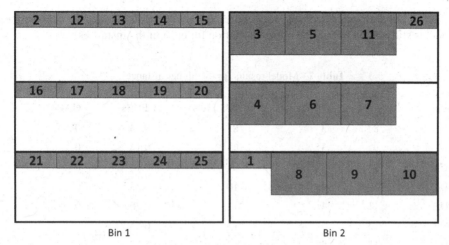

Fig. 7. Solution for the 12th instance.

5.2 Model Efficiency

Presently, the company in the study uses only one trailer with three levels. However, it's important to know if the proposed model would work in the case the company decides to increase the number of trailers used.

To that end, 10 randomly generated instances (presented in Table 6), infeasible on the original model and (since they have a number of baskets higher than the normal trailer capacity) were tested and the number of trailers increased to create a feasible solution. Also, it is important to highlight that the instances generated have noticeable different in the number of different types of baskets used.

Table 6. Number of baskets for each randomly generated instance, unfeasible by the original model.

Instance	N° baskets	Type 1	Type 2	Type 3	Type 4
#U1	32	1	16	2	13
#U2	35	2	17	4	12
#U3	35	8	4	13	10
#U4	40	12	10	18	0
#U5	45	5	20	2	18
#U6	50	5	14	9	22
#U7	60	17	3	12	28
#U8	100	0	8	45	47
#U9	120	20	0	90	10
#U10	155	43	98	0	14

The results obtained with the model are presented in Table 7. As we can see the model is still able to obtain the optimal solution for each of the instances.

Table 7. Model results for the higher instances.

Instance	OF	Nº Trailers	Nº Levels	CPU (s)	GAP (%)
#U1	8	3	8	0.23	0
#U2	9	3	9	0.13	0
#U3	9	3	9	0.13	0
#U4	10	4	10	0.33	0
#U5	11	4	11	0.25	0
#U6	13	5	13	0.55	0
#U7	15	5	15	0.27	0
#U8	31	11	31	5.66	0
#U9	34	12	34	7.92	0
#U10	32	11	32	11.86	0

The average CPU time for these instances is 2.73 s and, although it increases with the number of baskets to be transported, specifically in the last three instances, the results are still obtained within a very acceptable time, which permits the model to be used in a real time situation. Comparing instances U9 and U10 we can see that U9 uses more bins than U10 even though it transports less baskets. This happens because U9 has more baskets type 3 and those are bigger than the baskets type 2 used in instance U10.

Considering the results, the model proposed is still efficient to be used in case the company intends to increase the number of trailers to better fulfill the demand of the production lines and as an end objective, the reduction of the number of P&D trips.

5.3 Integrated Solution with VRPSDPTW and 2D-BPP

As stated before, this work is an extension of the work started in [1]. Therefore, to show how the two papers connect we will present in this sub-section an end-to-end route creation for pickup and delivery in two different real-world scenarios: when the route is triggered due to minimal stock levels or when the number of tools is enough to fill a vehicle.

Minimal Stock Trigger
The data collected from the e-Kanban system detected that in P&D points 3, 13 and 15 (see Fig. 1) tools A, B and C reached the minimal stock. Once the availability of the replacement tools is confirmed, the vehicle is selected. Having no extraordinary tools and both vehicles available, the automatic vehicle is selected. Table 8 gives us the number of tools to pickup and delivery in each point and the number and type of baskets used for each type of tool.

Table 8. Tools to pickup and delivery after the minimal stock trigger.

	P&D point	Tool Type	Tool quantity	Basket quantity	Basket Type
Pickup	3	A	4	1	Type 2
	13	B	4	1	Type 4
	15	C	6	1	Type 3
Deliver	3	A	6	1	Type 2
	13	B	8	2	Type 4
	15	C	3	1	Type 3

Following the diagram in Fig. 4, the system applies the procedure "Availability of tools to P&D", to guarantee that the vehicle's capacity is maximized. The procedure determines which additional tools can be delivered and collected from these points. Table 9 summarizes the tools that will be collected and delivered, as well as the number and type of baskets after the procedure "Availability of tools to P&D" is performed.

Table 9. Tools to pickup and delivery after "Availability of tools to P&D" procedure.

	P&D point	Tool Type	Tool quantity	Basket quantity	Basket Type
Pickup	3	A	4	1	Type 2
	13	B	4	1	Type 4
	15	C	6	1	Type 3
	3	E	1	1	Type 1
	13	F	4	1	Type 2
	15	G	6	1	Type 4
Deliver	3	A	6	1	Type 2
	13	B	8	2	Type 4
	15	C	3	1	Type 3
	3	D	4	1	Type 2

With this information, we apply the VRPSPDTW model to obtain the quickest route to pickup and delivery the selected tools, which is shown in Fig. 8. The blue circles represent the customers to be visited, and the blue square represents the depot. In green, we can see the number of baskets delivered and in grey, the pickup quantities.

The table in the figure contains the following information:

- Vehicle's arrival time at customer;
- Vehicle's departure time from customer;
- Number of baskets collected after leaving the customer (lp_i^k);
- Number of baskets to be delivered to the customer and the following customers (ld_i^k);

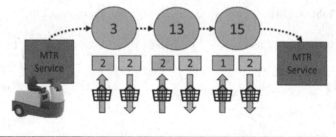

Arrival(min)	1,25	4,72	7,30	8,80
Departure(min)	2,47	5,90	7,91	-
Ip_i^k	2	4	1	-
Id_i^k	5	3	1	-
I_i^k	5	5	6	-

Fig. 8. Route obtained for the minimal stock trigger example.

- Total number of baskets after the vehicle leaves the customer (l_i^k).

It is important to note that, the model obtained the optimal solution, and it was calculated within 0.03 s. Having the final route, the 2D-BPP model is called to calculate the allocation of the baskets in the MTR node. The optimal solution is shown in Fig. 9 and it was calculated within 0.05 s.

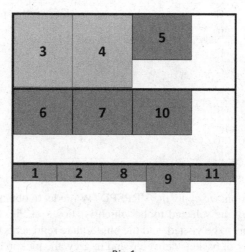

Bin 1

Fig. 9. 2-Dimensional Bin-Packing for the minimal stock trigger example.

The next step of the data pre-processing process includes the verification of the space still available in the vehicle to transport additional tools. This is done by analyzing the

OF obtained for the 2D-BPP. In the present example, the OF value is 3, which means that only one side of the trailer is used, and the other side is completely free. Thus, the system will look for more tools to deliver in the points with the minimal stock. Having none, the final P&D route and bin-packing solution are given to the MTR worker.

Tools to P&D Trigger
In contrast to the previous procedure, in this case the route creation is triggered when there are enough tools to fill a vehicle. The set of tools to be picked and delivered from Points 3, 4, 11, 13 and 15 triggered a new route. Table 10 gives the number of tools that triggered the route and the respective number and type of baskets where they are transported. Once this happens the vehicle is selected according to the procedure "Vehicle Selection".

Table 10. Tools to pickup and delivery after the "Tools to P&D" procedure.

	P&D point	Tool Type	Tool quantity	Basket quantity	Basket Type
Pickup	3	A	8	2	Type 2
	3	E	1	1	Type 1
	4	H	8	2	Type 4
	11	J	2	2	Type 1
	13	B	4	1	Type 4
	15	C	6	1	Type 3
	13	F	4	1	Type 2
	15	G	6	1	Type 4
Deliver	3	A	6	1	Type 2
	3	D	12	2	Type 2
	4	H	5	1	Type 4
	4	I	6	1	Type 3
	11	J	3	3	Type 1
	13	B	8	2	Type 4
	15	C	3	1	Type 3

Considering that the selected vehicle was the automatic the route is created using the VRPSPDTW model and the result presented in Fig. 10. The route has a total time of 10.89 min and was obtained 0.25 s with a GAP of 0%.

Having the P&D route the 2D-BPP model is applied to the depot node. Figure 11 shows the place of each basket in the trailer in order to optimize the trip. The model was able to obtain the solution within 0.06 s and with a GAP of 0%.

Note that the vehicle as still space for, at least, 4 more baskets. However, since the OF = 6 there will be no addition of tools to the vehicle for this trip.

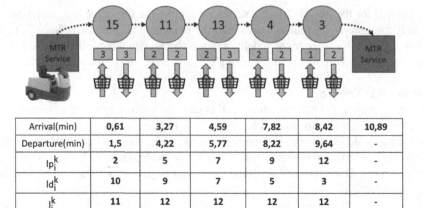

Arrival(min)	0,61	3,27	4,59	7,82	8,42	10,89
Departure(min)	1,5	4,22	5,77	8,22	9,64	-
lp_i^k	2	5	7	9	12	-
ld_i^k	10	9	7	5	3	-
l_i^k	11	12	12	12	12	-

Fig. 10. Route obtained for the Tools to P&D example.

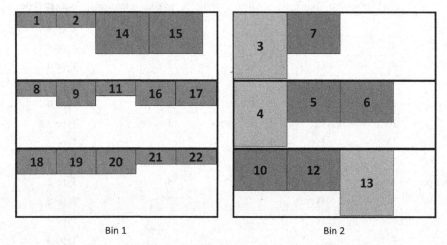

Fig. 11. 2-Dimensional Bin-Packing for the "Tools to P&D" trigger example.

6 Conclusions and Future Work

The inefficiencies associated to the Manufacturing Tool Repair service led to the development of a data collection and optimization framework for the pickup and delivery process of the manufacturing tools from and to the production lines. This framework was initiated in [1] with the definition of the pre-processing process and the implementation of a VRPSDPTW to obtain the quickest route to pick and delivery the manufacturing tools. The current work was the culmination of this work and aims to further improve this service's efficiency by implementing a 2D-BPP model. With this it's possible to increase the number of tools delivered, by the automatic vehicle. For this end, several changes were made in the pre-processing process, being them:

- Introduction of the "Vehicle Selection" procedure;

- Introduction of the Bin-Packing model in the MTR service node for the automatic vehicle.

The developed 2D-BPP model was then tested using 12 real-world test instances and 10 randomly generated instances, infeasible for the current vehicle's capacity. For the real test instances, the model is able to find the optimum for all instances, and it does so within a very satisfying CPU time (average 0.09 s. For the randomly generated instances, we assessed the efficiency of the model in case the company decides to invest in more trailers for the automatic vehicle. The results show that the model still performs very well. The optimal solution was still obtained for all test instances within a good CPU time.

Moreover, the integration of both the VRPSDPTW and the 2D-BPP was tested using the minimal stock trigger and the "Tools to P&D" trigger. The results support the use of the approach, since it is possible to obtain the solutions within seconds.

As future work we propose the improvement of the current 2D-BPP model by:

- improving the calculation of the free space in the vehicle, so we know exactly how many free spaces are available in the vehicle and their size;
- considering in the procedures "Availability of tools to P&D" and "Tools to P&D", the customers that triggered a new route and also other customers that are within a pre-defined distance limit from the triggering customers.

On top of this, we propose the implementation of the final matheuristic framework in the factory, in order to test its effectiveness and to calculate the actual efficiency improvements.

References

1. Romeira, B., Moura, A.: Optimizing route planning for minimising the non-added-value tasks times: a simultaneous pickup-and-delivery problem. In: 11th International Conference on Operations Research and Enterprise Systems, pp. 153–60 (2022). https://doi.org/10.5220/0010821000003117
2. Wang, N., Wang, J.-S., Zhang, Y.-X., Li, T.-Z.: Two-dimensional Bin-packing problem with rectangular and circular regions solved by genetic algorithm. Int. J. Appl. Math. **51**, 268–278 (2021)
3. Hoos, H.H., Stützle. T.: 1-Introduction. In: Stochastic Local Search, pp. 13–59. Morgan Kaufmann, Burlington (2005)
4. Zhang, Q., Liu, S., Zhang, R., Qin, S.: Column generation algorithms for mother plate design in steel plants. OR Spectr. **43**, 127–153 (2020). https://doi.org/10.1007/S00291-020-00610-Z
5. Arbib, C., Marinelli, F., Pizzuti, A.: Number of bins and maximum lateness minimization in two-dimensional bin packing. Eur. J. Oper. Res. **291**, 101–113 (2021). https://doi.org/10.1016/J.EJOR.2020.09.023
6. Lodi, A., Martello, S., Vigo, D.: Models and bounds for two-dimensional level packing problems. J. Comb. Optim. **8**, 363–379 (2004). https://doi.org/10.1023/B:JOCO.0000038915.62826.79
7. Bennell, J.A., Soon Lee, L., Potts, C.N.: A genetic algorithm for two-dimensional bin packing with due dates. Int. J. Prod. Econ. **145**, 547–560 (2013). https://doi.org/10.1016/J.IJPE.2013.04.040

8. Polyakovskiy, S., M'Hallah, R.: A hybrid feasibility constraints-guided search to the two-dimensional bin packing problem with due dates. Eur. J. Oper. Res. **266**, 819–839 (2018). https://doi.org/10.1016/J.EJOR.2017.10.046

9. Martello, S., Vigo, D.: Exact solution of the two-dimensional finite bin packing problem. Manage. Sci. **44**, 388–399 (1998). https://doi.org/10.1287/MNSC.44.3.388

10. de Carvalho, P.R.V., Elhedhli, S.: A data-driven approach for mixed-case palletization with support. Optim. Eng. (2021). https://doi.org/10.1007/S11081-021-09673-5

11. Heßler, K., Irnich, S., Kreiter, T., Pferschy, U.: Bin packing with lexicographic objectives for loading weight- and volume-constrained trucks in a direct-shipping system. OR Spectr. (2021). https://doi.org/10.1007/S00291-021-00628-X

12. Wei, L., Lai, M., Lim, A., Hu, Q.: A branch-and-price algorithm for the two-dimensional vector packing problem. Eur. J. Oper. Res. **281**, 25–35 (2020). https://doi.org/10.1016/J.EJOR. 2019.08.024

13. Hu, Q., Lim, A., Zhu, W.: The two-dimensional vector packing problem with piecewise linear cost function. Omega **50**, 43–53 (2015). https://doi.org/10.1016/J.OMEGA.2014.07.004

14. Pisinger, D., Sigurd, M.: The two-dimensional bin packing problem with variable bin sizes and costs. Discret. Optim. **2**, 154–167 (2005). https://doi.org/10.1016/J.DISOPT.2005.01.002

15. Wang, T., Hu, Q., Lim, A.: An exact algorithm for two-dimensional vector packing problem with volumetric weight and general costs. Eur. J. Oper. Res. **300**, 20–34 (2022). https://doi.org/10.1016/J.EJOR.2021.10.011

16. Berkey, J.O., Wang, P.Y.: Two-dimensional finite bin-packing algorithms. J. Oper. Res. Soc. **38**, 423–429 (1987). https://doi.org/10.1057/JORS.1987.70

17. Bengtsson, B.E.: Packing rectangular pieces—a heuristic approach. Comput. J. **25**, 353–357 (1982). https://doi.org/10.1093/COMJNL/25.3.353

18. Christofides, N., Whitlock, C.: An algorithm for two-dimensional cutting problems. Oper. Res. **25**, 30–44 (1977). https://doi.org/10.1287/OPRE.25.1.30

19. Beasley, J.E.: Algorithms for unconstrained two-dimensional guillotine cutting. J. Oper. Res. Soc. **36**, 297–306 (2017). https://doi.org/10.1057/JORS.1985.51

20. Caprara, A., Toth, P.: Lower bounds and algorithms for the 2-dimensional vector packing problem. Discret. Appl. Math. **111**, 231–262 (2001). https://doi.org/10.1016/S0166-218X(00)002 67-5

21. Alves, C., De Carvalho, J.V., Clautiaux, F., Rietz, J.: Multidimensional dual-feasible functions and fast lower bounds for the vector packing problem. Eur. J. Oper. Res. **233**, 43–63 (2014). https://doi.org/10.1016/J.EJOR.2013.08.011

22. Masson, R., Vidal, T., Michallet, J., Penna, P.H.V., Petrucci, V., Subramanian, A., et al.: An iterated local search heuristic for multi-capacity bin packing and machine reassignment problems: Expert Syst. Appl. **40**, 5266–5275 (2013). https://doi.org/10.1016/J.ESWA.2013. 03.037

23. Buljubašić, M., Vasquez, M.: Consistent neighborhood search for one-dimensional bin packing and two-dimensional vector packing. Comput. Oper. Res. **76**, 12–21 (2016). https://doi.org/10.1016/J.COR.2016.06.009

24. Hu, Q., Zhu, W., Qin, H., Lim, A.: A branch-and-price algorithm for the two-dimensional vector packing problem with piecewise linear cost function. Eur. J. Oper. Res. **260**, 70–80 (2017). https://doi.org/10.1016/J.EJOR.2016.12.021

25. Lodi, A., Martello, S., Vigo, D.: Heuristic and metaheuristic approaches for a class of two-dimensional bin packing problems. INFORMS J. Comput. **11**, 345–357 (1999). https://doi.org/10.1287/IJOC.11.4.345

26. Romeira, B., Cunha, F., Moura, A.: Development and application of an e-Kanban system in the automotive industry. In: IEOM Monterrey 2021 Conference, pp. 613–624 (2021)

Predictive Maintenance Optimization Under Stochastic Production in Complex Systems

Junkai He[1($\boxtimes$)] (ID), Selma Khebbache[1] (ID), Miguel F. Anjos[2] (ID),
and Makhlouf Hadji[1] (ID)

[1] IRT SystemX, 8 Avenue de la Vauve, 91120 Palaiseau, France
{junkai.he,selma.khebbache,makhlouf.hadji}@irt-systemx.fr
[2] School of Mathematics, University of Edinburgh, Peter Guthrie Tait Road,
Edinburgh, Scotland EH9 3FD, UK
anjos@stanfordalumni.org

Abstract. This paper focuses on predictive maintenance optimization under stochastic production in complex systems using prognostic Remaining Useful Life (RUL) information. At each stage of such a system, we consider redundant assets and use their RUL to guarantee system availability. However, the production capacity of our system is stochastic due to environmental and human factors. We aim at meeting client demands in a given optimization planning horizon while reducing the generated cost. We propose a deterministic mathematical model before providing a chance-constrained programming formulation to minimize the total cost. Two solution approaches for dealing with chance constraints are proposed to approximate the stochastic model in this maintenance optimization. Experimental results show the efficiency of the proposed model and chance-constrained approximation approaches.

Keywords: Predictive maintenance · Stochastic optimization ·
Chance constraints · Complex systems · Remaining useful life

1 Introduction

Lifetime of industrial assets in general decreases due to internal or external factors, which can impact the production of systems and cause economic problems. In order to reduce the deterioration rate of assets, maintenance is seen as an important solution by many enterprises because it can bring assets back to better conditions [1]. However, real-life maintenance faces two difficulties: on the one hand, traditional maintenance tools for single-asset systems are not suitable for contemporary complex systems because they require commonly more than one asset for production [2]. On the other hand, uncertainty in complex systems brings more challenges when taking maintenance and production decisions. Industrial managers and academic researchers have thus been focusing on

F. Liberatore et al. (Eds.): ICORES 2022/2023, CCIS 1985, pp. 47–68, 2024.
https://doi.org/10.1007/978-3-031-49662-2_3

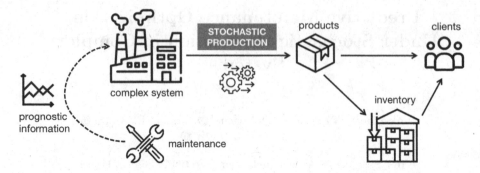

Fig. 1. Predictive maintenance is done for complex systems such that production and inventory can meet client demands. In this research, a stochastic production factor is considered when optimizing maintenance.

promising maintenance and production strategies for complex systems with the consideration of uncertainty.

Our recent research paper [3] has focused on overcoming the first challenge, which solves a predictive maintenance optimization problem for complex systems. More specifically, we consider a generic complex system with a series of processing stages and each stage contains multiple redundant assets. The productivity of this system depends on the conditions of assets and inventory is allowed as a buffer. To avoid a sudden failure, predictive maintenance is carried out on assets based on their prognostic information, i.e., *Remaining Useful Life* (RUL). The definition of an asset's RUL is the currently remaining time of operation before it fails [4]. In summary, the purpose of [3] is to use the obtained asset-level RUL information, coordinate the operations in different stages, and provide global maintenance and production decisions such that demands can be met in the planning horizon.

In this paper, we are dedicated to overcoming the second challenge and we thus extend our research [3] by considering stochastic production in maintenance optimization. Figure 1 illustrates the novelty part of this paper to reflect stochastic production that can occur due to a variety of factors, such as human operations, environmental issues, or electricity supply.

Hence, we extend the Deterministic Predictive Maintenance Optimization (DPMO) of [3] and propose a new *Stochastic Predictive Maintenance Optimization* (SPMO) problem in which production is stochastic. Our objective is to design a mathematical model to efficiently optimize the total cost (maintenance, system failure, inventory) while meeting client demands under stochastic production capacity. Our contributions to this paper are as follows:

- A new preventive maintenance optimization problem under stochastic production is studied to minimize the total cost in complex systems.
- A chance-constrained programming model is formulated for our stochastic optimization problem.

- Two linear approximations of the stochastic model are designed to attain approximate solutions for our problem.
- A comparison of the two provided approximations is made, and the impact of different probability choices in chance constraints are evaluated.

The remainder of this paper is organized as follows. We review related literature in Sect. 2 to position our research. Section 3 is dedicated to the DPMO problem in [3] and its *mixed-integer linear programming* (MILP) model, based on which the SPMO problem and a new chance-constrained programming model are proposed. In Sect. 4, we provide two approximation approaches to solve the chance-constrained model. In Sect. 5, we compare the proposed solution methods and evaluate the impact of probability in chance constraints. Conclusions and future research directions are provided in Sect. 6.

2 Related Work

In this section, we review previous works to situate our contributions to this paper. First, we review deterministic and non-deterministic maintenance literature related to our research. Then, we focus on maintenance optimization solution methods in terms of deterministic and stochastic circumstances.

2.1 Maintenance Optimization in Complex Systems

Deterministic Maintenance Problems. The RUL is an important indicator reflecting the current status of an asset [5]. It could be either obtained by defining the time length from the current time to the end-of-life of an asset. More frequently, it is defined as the time left before the health condition reaches a warning threshold. More than 270 papers have studied the prediction of asset's RUL [6]. However, in our paper, we focus on how to use the obtained RUL to decide maintenance planning. To our best knowledge, there are three distinct branches of RUL usage in the literature: (i) RUL-based inspection: Do et al. [7] utilized RUL prognostics for deciding the time point for the next coming inspection. (ii) RUL-based maintenance strategies: Chen et al. [8] decided different maintenance actions through the combinations of degradation levels and RUL values. (iii) RUL-based constraints: Camci et al. [4] used prognostic information to formulate failure rate constraints of components. Note that most existing works focus on the first two types while using RUL prognostics for formulating mathematical models is rare.

For describing the availability of a complex system through the conditions of its assets, Wu and Castro [9] proposed a linear combination of the degradation processes of several components. If this value exceeded a given threshold, preventive maintenance was performed. Lei and Sandborn [10] adopted a prognostic health analysis to predict the RUL of wind turbines. The authors assumed that turbines were dependent and system availability relied on the minimum RUL among them. Dong et al. [11] supposed that normal-distributed shocks occurred independently and described system reliability by conditioning on the numbers of arrived shocks.

Maintenance optimization in complex systems can also combine with some problems in other scopes so as to make global decisions. One mainstream branch is to simultaneously take maintenance and spare part ordering into account. Camci [12] studied corrective maintenance and spare part inventory strategies using prognostic information to minimize the failure risk. A genetic algorithm was proposed and computational results were compared with the ones obtained by the preventive maintenance strategy. Plenty of papers have considered spare part ordering, see the review article [1]. The integration of maintenance and production is also a main research direction because maintenance activities eventually impact the productivity of complex systems. Xiao et al. [13] assumed that a given set of jobs were processed on a series system. A genetic algorithm was used to determine the production schedule and the preventive maintenance interval to minimize the failures. Bahria et al. [14] devised an integrated approach to control the balance of maintenance, production, and quality in manufacturing. Different appropriate thresholds for conducting maintenance were discussed to guarantee the robustness of the production system.

Maintenance Problems with Uncertainty. Compared to deterministic maintenance optimization problems, the ones with uncertainty are rare. And even fewer are concerned with prognostic information, system availability, and maintenance integration. To the best of our knowledge, in the literature, uncertainty in maintenance optimization problems is usually considered as stochastic maintenance quality and stochastic maintenance duration. Khatabe and Aghezzaf [15] studied selective maintenance optimization when the quality of maintenance actions was stochastic. This quality was treated as a random variable with identified probability distributions. Shahraki et al. [16] also studied stochastic imperfect maintenance actions considering the dependency among components. Ghorbani et al. [17] assumed future operating conditions to be uncertain. The system was subject to several uncertain condition scenarios of exposure, conditional, usage, and stress. Each scenario was modeled with its associated occurrence probability. For stochastic maintenance duration, Khatabe et al. [18] studied a selective maintenance problem and maintenance duration was seen as a stochastic factor. Liu et al. [19] considered a multi-component maintenance and repair person assignment problem with stochastic maintenance duration.

From the literature, we observe that there is a lack of mathematical modeling based on RUL-based constraints. Moreover, the influence of individual RUL information on complex systems is seldom discussed. Hence, and to the best of our knowledge, the integrated optimization of maintenance and production in complex systems with backup assets has not been studied. Different from the above works, our recent research [3] used RUL information to plan predictive maintenance if the RUL of an asset reaches a given threshold. The system availability is related to the conditions (RUL values) of its assets. Besides, the integration of maintenance, production, and inventory is considered such that demands can be satisfied. However, the work in [3] is only valid in deterministic circumstances. In this paper, we extend it to a non-deterministic manner with the consideration of stochastic production capacity.

2.2 Maintenance Optimization Methods

Difference solution methods for maintenance problems in complex systems have been proposed, such as analytic, simulation, optimization [20,21]. In the following, we only discuss optimization-related references and clearly situate our contribution compared to these approaches.

Deterministic Optimization Methods. In the mentioned Camci [12] and Xiao et al. [13], problem-specific genetic algorithms were proposed to solve the problems. Rivera-Gómez et al. [22] considered a continuous production system with unexpected quality deterioration, with the purpose to reduce the occurred cost with a quality constraint. A non-linear programming model was formulated and solved for the problem. Zhou et al. [23] proposed an optimal preventive maintenance policy in order to get operational parameters for a production line. A non-linear model and a heuristic were designed to minimize the cost and guarantee the operating speed. Compared to the widely used non-linear formulation and (meta-) heuristics, linear formulation for maintenance optimization is very limited [4]. Our previous work [3] formulated a MILP model for the considered predictive maintenance optimization problem, with the purpose to provide optimal maintenance decisions. However, in order to solve the maintenance optimization problem under stochastic production in this work, we need to use corresponding stochastic optimization solution methods.

Stochastic Optimization Methods. For the mentioned stochastic maintenance optimization, Khatabe and Aghezzaf [15] proposed a stochastic non-linear programming model and used identified distributions to approximately solve the stochastic model. Khatabe et al. [18] further proposed a chance-constrained programming model, the expected value of stochastic duration was used for approximation. Then for Ghorbani et al. [17], Shahraki et al. [16], Liu et al. [19], the authors used the scenarios of stochastic factors for stochastic model approximation. From the existing stochastic solution methods for maintenance optimization, we only see chance-constrained programming in [18]. This kind of stochastic optimization method allows that some constraints in a model hold on reliability within the area, it is called stochastic probabilistic or chance-constrained programming (Birge [24], Charnes and Cooper [25] and Prékopa [26]). In our considered SPMO problem, we aim at guaranteeing with a probability that although production capacity is stochastic, the complex system can yield enough products to meet demands. As chance-constrained programming cannot be solved directly, we, therefore, concentrate on representing chance constraints by linear approximations with proper solution methods.

3 Problem Description and Formulation

In this section, we first recall the DPMO problem and its MILP model in our previous work [3]. Based on this, we then introduce stochastic production capacity into the model to obtain the SPMO problem, and formulate a chance-constrained programming approach for this problem.

3.1 DPMO and MILP Model

For sake of clarity, and to recall the DPMO problem and its MILP model, the used parameters and decision variables are provided in Table 1.

In the considered DPMO problem, the generic complex systems contains $|\mathcal{K}|$ processing stages and each stage k is configured with $|\mathcal{J}_k|$ assets. For example, we can see in Fig. 2 that $|\mathcal{K}| = 4$ and $|\mathcal{J}_1| = 2$ as there are two assets (in circle) belonging to stage 1. Each asset may have three states: working (in grey), standby (in white), and maintenance (in dashed fill), respectively. We assume that each asset is repairable, and that maintaining it does not affect the operation of a stage if there exists any available standby (redundant) asset. The planning horizon contains $|\mathcal{T}|$ periods (a period denotes one week for example). The purpose is to meet demands during this optimization planning horizon. In the following, we describe the main constraints to formulate the MILP model of the DPMO problem.

Table 1. Parameters and decision variables for DPMO.

Parameters	
$\mathcal{K}$	set of stages in complex systems
$\mathcal{J}_k$	set of assets in stage k
$\mathcal{T}$	set of periods, i.e., the planning horizon
$a_{k,j}$	decreased RUL per period if asset j in stage k is used
$b_{k,j}$	recovered RUL if asset j in stage k is maintained
$o_{k,j}$	original RUL of asset j in stage k
γ_k	RUL threshold of assets in stage k
m_k	maintenance duration of assets in stage k
q_{max}	production capacity of the system per period
d^t	demands in period t
c^m	unit maintenance cost
c^f	unit failure cost
c^i	unit inventory cost
Decision variables	
$X_{k,j}^t$	binary variable, $= 1$ if the RUL of asset j in stage k reach the threshold in period t, $= 0$ otherwise
$Y_{k,j}^t$	binary variable, $= 1$ if asset j in stage k operates in period t, $= 0$ otherwise
P_k^t	binary variable, $= 1$ if stage k is working in period t, $= 0$ otherwise
S^t	binary variable, $= 1$ if the system is working in period t, $= 0$ otherwise
$R_{k,j}^t$	RUL value of asset j in stage k in period t
Q^t	production amount of the system in period t
I^t	inventory in period t. Note that there is no inventory at the beginning

RUL Track Over Time. It is assumed that the prognostic RUL of asset j in stage k contains: the original RUL $o_{k,j}$ at the beginning of the horizon, the RUL usage rate $a_{k,j}$, and the recovered RUL $b_{k,j}$ if maintenance is done. Note that RUL can be alternatively depicted by values, quantiles, or probabilities. We choose the first one in our proposal, and consider the others for future work. As there are backup assets in each stage and only one asset is required to guarantee the performance of a stage, therefore, the model should keep track of the RUL of assets over the planning horizon. We recall the three conditions of assets in Fig. 2 to generate the following RUL-evolution formulas: (i) if an asset is used, its RUL value decreases based on a given usage rate; (ii) if an asset is in standby, its RUL will not change; (iii) if the RUL of an asset reaches the threshold, its RUL will stay at the threshold γ_k until maintenance is carried out. Note that any asset reaching the corresponding threshold can no longer operate and needs to be maintained. After maintenance, its RUL is restored to $b_{k,j}$. We assume that RUL threshold γ_k is provided by industrial experts.

$$
R_{k,j}^{t+1} = \begin{cases} b_{k,j} & \text{if asset } j \text{ in stage } k \text{ requires maintenance in period } t \\ R_{k,j}^{t} - a_{k,j} & \text{if asset } j \text{ in stage } k \text{ is used in period } t \\ R_{k,j}^{t} & \text{if asset } j \text{ in stage } k \text{ is in standby in period } t \end{cases}
$$

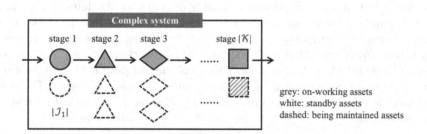

Fig. 2. The generic complex system has $|\mathcal{K}|$ processing stages and each stage have asset redundancy with a number of $|\mathcal{J}_k|$. For each period of the planning horizon, an asset has one of the three statuses: on-working, standby, or being maintained.

Complex System Availability and Production. The availability of the entire complex system is guaranteed if and only if all stages are processing, which further requires that one asset from each stage is working. The production amount of the complex system depends on asset conditions and is limited by a given upper bound. If the system cannot work in a period, there is no production and thus client demands may not be satisfied. To avoid this, we allow some possible inventory in preparation to make sure that the sum of production and inventory can meet demands.

The deterministic MILP model (denoted as **M1**) for the DPMO problem is formulated as follows. The objective function (1) is a sum of three parts: total maintenance cost $C^{maintenance}$, total system failure penalty $C^{failure}$, and total inventory expense $C^{inventory}$, specifically:

- $C^{maintenance}$ equals to unit maintenance cost c^m times the total number of maintenance actions in the planning horizon.
- $C^{failure}$ equals to unit system failure penalty c^f times the total number of occurrences of system failures.
- $C^{inventory}$ equals to unit inventory expense c^i times the inventory over time.

$$\textbf{Model M1:}\quad \min C^{maintenance} + C^{failure} + C^{inventory} \tag{1}$$

$$C^{maintenance} = c^m \cdot \sum_{k \in \mathcal{K}} \sum_{j \in \mathcal{J}_k} \sum_{t \in \mathcal{T}} X_{k,j}^t$$

$$C^{failure} = c^f \cdot \sum_{t \in \mathcal{T}} (1 - S^t)$$

$$C^{inventory} = c^i \cdot \sum_{t \in \mathcal{T}} I^t$$

The constraints of **M1** are presented as follows. Constraints (2) to (11) describe RUL evolution. Specifically, constraints (2) and (3) claim that maintenance is required for an asset if its RUL is no bigger than the threshold. Constraints (4) show the maintenance duration if an asset is under maintenance. Constraints (5) restrict that only one asset can be working in each stage if necessary. Constraints (6) to (11) present the evolution of asset's RUL under three conditions as we have defined before: constraints (6) and (7) show RUL track if an asset is maintained; constraints (8) and (9) show RUL track if an asset is used; constraints (10) and (11) show RUL track if an asset is in standby.

$$R_{k,j}^t > \gamma_k - \overline{M} \cdot X_{k,j}^t, \qquad \forall k \in \mathcal{K}, j \in \mathcal{J}_k, t \in \mathcal{T} \tag{2}$$

$$R_{k,j}^t \leq \gamma_k + \overline{M} \cdot (1 - X_{k,j}^t), \qquad \forall k \in \mathcal{K}, j \in \mathcal{J}_k, t \in \mathcal{T} \tag{3}$$

$$\sum_{t' \in [t, t+m_k-1]} Y_{k,j}^{t'} \leq 1 - X_{k,j}^t, \ \forall k \in \mathcal{K}, j \in \mathcal{J}_k, t \in [1, |T| - m_k + 1] \tag{4}$$

$$\sum_{j \in \mathcal{J}_k} Y_{k,j}^t \leq 1, \qquad \forall k \in \mathcal{K}, t \in \mathcal{T} \tag{5}$$

$$R_{k,j}^{t+1} \geq b_{k,j} - \overline{M} \cdot (1 - X_{k,j}^t), \qquad \forall k \in \mathcal{K}, j \in \mathcal{J}_k, t \in [1, |T| - 1] \tag{6}$$

$$R_{k,j}^{t+1} \leq b_{k,j} + \overline{M} \cdot (1 - X_{k,j}^t), \qquad \forall k \in \mathcal{K}, j \in \mathcal{J}_k, t \in [1, |T| - 1] \tag{7}$$

$$R_{k,j}^{t+1} \geq R_{k,j}^t - a_{k,j} - \overline{M} \cdot (1 - Y_{k,j}^t), \qquad \forall k \in \mathcal{K}, j \in \mathcal{J}_k, t \in [1, |T| - 1] \tag{8}$$

$$R_{k,j}^{t+1} \leq R_{k,j}^t - a_{k,j} + \overline{M} \cdot (1 - Y_{k,j}^t), \qquad \forall k \in \mathcal{K}, j \in \mathcal{J}_k, t \in [1, |T| - 1] \tag{9}$$

$$R_{k,j}^{t+1} \geq R_{k,j}^t - \overline{M} \cdot (X_{k,j}^t + Y_{k,j}^t), \qquad \forall k \in \mathcal{K}, j \in \mathcal{J}_k, t \in [1, |T| - 1] \tag{10}$$

$$R_{k,j}^{t+1} \leq R_{k,j}^t + \overline{M} \cdot (X_{k,j}^t + Y_{k,j}^t), \qquad \forall k \in \mathcal{K}, j \in \mathcal{J}_k, t \in [1, |T| - 1] \tag{11}$$

Constraints (12) to (16) describe stage availability and further system availability for each period via the RUL track of assets, based on which system production and inventory can be calculated. The premise that a stage operates normally is that one asset belonging to it is working. To this end, the availability of a stage can be described by constraints (12). System availability strictly requires that all the stages are available, which is expressed by constraints (13) and (14). The inventory in constraints (15) is calculated by summing up the production amount and the inventory in the last period after meeting the demands. Constraints (16) provide the upper bound of production amount in each period respecting system production capacity.

$$P_k^t \leq \sum_{j \in \mathcal{J}_k} Y_{k,j}^t, \forall k \in \mathcal{K}, t \in \mathcal{T} \tag{12}$$

$$\frac{1}{|K|} \sum_{k \in \mathcal{K}} P_k^t \leq S^t + \sum_{k \in \mathcal{K}} (1 - P_k^t), \quad \forall t \in \mathcal{T} \tag{13}$$

$$P_k^t \geq S^t, \forall k \in \mathcal{K}, t \in \mathcal{T} \tag{14}$$

$$I^t = Q^t + I^{t-1} - d^t, \quad \forall t \in \mathcal{T} \tag{15}$$

$$Q^t \leq q_{max} \cdot P_k^t, \forall k \in \mathcal{K}, t \in \mathcal{T} \tag{16}$$

The ranges of model **M1** are as follows, where $\mathbb{R}^*$ represents non-negative real numbers.

$$X_{k,j}^t, Y_{k,j}^t, P_k^t, S^t \in \{0, 1\}, \forall k \in \mathcal{K}, j \in \mathcal{J}_k, t \in \mathcal{T} \tag{17}$$

$$I^t, Q^t, R_{k,j}^t \in \mathbb{R}^*, \forall k \in \mathcal{K}, j \in \mathcal{J}_k, t \in \mathcal{T} \tag{18}$$

The above DPMO problem and its corresponding MILP model (studied in [3]) are only valid under deterministic environments. However, as indicated in Sect. 1, stochastic factors may bring more challenges when doing maintenance and production decisions for complex systems. In the following, we consider that production capacity is stochastic to extend the DPMO problem to the SPMO problem. Also, we formulate a chance-constrained programming model for it.

3.2 SPMO and Chance-constrained Programming Model

Stochastic production capacity of complex systems in the SPMO problem is denoted as $q_{max}(\tilde{\xi})$. The reason is that this capacity can be impacted by unpredictable elements such as human operations, environmental factors, etc. In the SPMO problem, we aim at guaranteeing that even under stochastic production capacity, client demands should be met in the planning horizon. To achieve this goal, we introduce chance constraints to formulate a chance-constrained programming model.

The original chance-constrained programming model was proposed by Charnes and Cooper [25]. The basic idea is that, in some models, constraints need not hold exactly but rather they may hold with some reliability or probability levels. For our considered SPMO problem, we propose chance constraints

(19) to guarantee that the event *"Given a stochastic production capacity, the production system can meet demands at every period"* holds a probability of α, where $0 \leq \alpha \leq 1$:

$$\mathbb{P}\left(q_{max}(\tilde{\xi}) \cdot P_k^t \geq Q^t\right) \geq \alpha, \ \forall t \in \mathcal{T} \tag{19}$$

Note that constraints (19) are the extended chance-constraints of deterministic constraints (16) in **M1**. With the above stochastic production capacity and chance constraints (19), a chance-constrained programming model (denoted by **M2**) for the SPMO problem can be formulated as follows.

$$\textbf{Model M2:} \quad \min C^{maintenance} + C^{failure} + C^{inventory}$$

$$\textbf{Subject to:} \quad (2) - (15), (17), (18), \text{and} (19)$$

Due to the complexity and non-linearity of chance constraints (19), model **M2** cannot be solved directly. In the next section, we present two methods to approximate **M2** so that approximate solutions can be reached.

4   Approximations of the Chance-Constrained Model

In this section, we propose two approaches to solve the chance-constrained programming model **M2** approximately. These approaches respectively assume that stochastic production capacity $q_{max}(\tilde{\xi})$ takes on discrete values or continuous values. We also explain the relationship between the proposed solution methods.

4.1   Approximation Using Discrete Information

In this subsection, we assume that $q_{max}(\tilde{\xi})$ can be expressed by a discrete set of scenarios. We consider a fixed number of scenarios $|\Omega|$ where scenario $\omega \in \Omega$ is associated with positive probability $\rho(\omega)$, and $q_{max}(\tilde{\xi})$ takes a value of $q_{max}(\omega)$ with a probability of $\rho(\omega)$. Note that, for each given scenario, the SPMO problem reduces to the corresponding DPMO problem.

Chance-constrained programming model **M2** can now be rewritten as the following expectation model (denoted as **EXP**):

$$\textbf{Model EXP:} \quad \min \left(C^{maintenance} + C^{failure} + C^{inventory}\right) / |\Omega|$$

$$C^{maintenance} = c^m \cdot \sum_{\omega \in \Omega} \sum_{k \in \mathcal{K}} \sum_{j \in \mathcal{J}_k} \sum_{t \in \mathcal{T}} X_{k,j}^t(\omega)$$

$$C^{failure} = c^f \cdot \sum_{\omega \in \Omega} \sum_{t \in \mathcal{T}} (1 - S^t(\omega))$$

$$C^{inventory} = c^i \cdot \sum_{\omega \in \Omega} \sum_{t \in \mathcal{T}} I^t(\omega)$$

Subject to: $(2)^\star - (15)^\star$

$$\sum_{\omega \in \Omega} \mathbb{P}_\omega \left(q_{max}(\omega) \cdot P_k^t(\omega) \geq Q^t(\omega) \right) \geq \alpha, \forall t \in \mathcal{T} \qquad (20)$$

where $\star$ denotes that decision variables in these constraints are related to a scenario-based dimension, and their ranges are as follows:

$$X_{k,j}^t(\omega), Y_{k,j}^t(\omega), P_k^t(\omega), S^t(\omega) \in \{0,1\}, \forall k \in \mathcal{K}, j \in \mathcal{J}_k, t \in \mathcal{T}, \omega \in \Omega, \quad (21)$$
$$I^t(\omega), Q^t(\omega), R_{k,j}^t(\omega) \in \mathbb{R}^*, \forall k \in \mathcal{K}, j \in \mathcal{J}_k, t \in \mathcal{T}, \omega \in \Omega. \quad (22)$$

The complete version of model **EXP** can be found in the Appendix for reference. This model is an expectation approximation of **M2** using scenarios, however, it still cannot be directly solved because of chance constraints (20).

For this reason, we introduce a new binary variable $W(\omega)$ defined as follows:

$$W(\omega) = \begin{cases} 1, & \text{if the event is satisfied under } \omega, \text{ i.e., } q_{max}(\omega) \cdot P_k^t(\omega) \geq Q^t(\omega) \\ 0, & \text{if the event is not satisfied under } \omega \end{cases}$$

Proposition 1. *Using binary variable* $W(\omega)$, *chance constraints (20) can be approximated by the following linear constraints (23) and (24), where* $\overline{M}$ *is a sufficient large number:*

$$Q^t(\omega) - q_{max}(\omega) \cdot P_k^t(\omega) \leq \overline{M} \cdot (1 - W(\omega)), \forall k \in \mathcal{K}, \forall t \in \mathcal{T}, \forall \omega \in \Omega \quad (23)$$
$$\sum_{\omega \in \Omega} \rho(\omega) \cdot W(\omega) \geq \alpha \qquad (24)$$

Proof. For scenario ω with a probability of $\rho(\omega)$, constraints (23) mean that if $W(\omega) = 1$, then the event in chance constraints (20) is satisfied under scenario ω. As chance constraints (20) make sure that for the scenarios satisfying the event, the sum of their probabilities is greater than or equal to the given probability. Therefore, we use constraint (24) to collect all scenarios with $W(\omega) = 1$ for ensuring this. ∎

Applying Proposition 1, our model **EXP** can be approximated by the following MILP model (denoted as **A1**).

Approximation A1: $\min \left(C^{maintenance} + C^{failure} + C^{inventory} \right) / |\Omega|$

$$C^{maintenance} = c^m \cdot \sum_{\omega \in \Omega} \sum_{k \in \mathcal{K}} \sum_{j \in \mathcal{J}_k} \sum_{t \in \mathcal{T}} X_{k,j}^t(\omega)$$

$$C^{failure} = c^f \cdot \sum_{\omega \in \Omega} \sum_{t \in \mathcal{T}} (1 - S^t(\omega))$$

$$C^{inventory} = c^i \cdot \sum_{\omega \in \Omega} \sum_{t \in \mathcal{T}} I^t(\omega)$$

Subject to: $(2)^\star - (15)^\star, (21) - (22), (23) - (24)$

where constraints (23) and (24) are the linear approximation of chance constraints.

We have thus obtained model **A1** which is a linear approximation of **M2** and can be solved directly by off-the-shelf solvers.

4.2   Approximation Using Continuous Information

In this subsection, we assume that $q_{max}(\tilde{\xi})$ is expressed using continuous values following normal distributions. Under this assumption, we obtain Proposition 2 below, and hence the second approximation of model **M2**. The choice of normal distributions is reasonable in this context because, in most cases, the stochastic production capacity will take values near its mean, so values further away from the mean should have lower probabilities.

Proposition 2. *If stochastic production capacity* $q_{max}(\tilde{\xi})$ *follows a normal distribution, i.e.,* $q_{max}(\tilde{\xi}) \sim \mathcal{N}(\mu, \sigma^2)$, *then chance constraints (19) can be approximated by linear constraints as follows:*

$$Q^t \leq \left(\sigma \cdot \Phi^{-1}(1 - \alpha) + \mu\right) \cdot P_k^t, \quad \forall t \in \mathcal{T}. \tag{25}$$

Proof. For $q_{max} \cdot P_k^t \geq Q^t$ in (19), it holds that $Q^t = 0$ if binary variable $P_k^t = 0$, and $q_{max} \geq Q^t$ if $P_k^t = 1$. We consider separately the two cases $P_k^t = 0$ and $P_k^t = 1$ to approximate (19):

– If $P_k^t = 0$, then it is obvious that (19) holds $Q^t = 0$ for any probability α.
– If $P_k^t = 1$, then $P_k^t \neq 0$ and (19) for $t \in \mathcal{T}$ can be rewritten as:

$$\mathbb{P}\left(q_{max}(\tilde{\xi}) \geq \frac{Q^t}{P_k^t}\right) \geq \alpha \tag{26}$$

As $q_{max}(\tilde{\xi}) \sim \mathcal{N}(\mu, \sigma^2)$, we have that $\frac{q_{max}(\tilde{\xi}) - \mu}{\sigma} \sim \mathcal{N}(0,1)$, and thus constraint (26) can be standardized as:

$$\mathbb{P}\left(\frac{q_{max}(\tilde{\xi}) - \mu}{\sigma} \geq \frac{\frac{Q^t}{P_k^t} - \mu}{\sigma}\right) \geq \alpha, \tag{27}$$

or equivalently as:

$$1 - \mathbb{P}\left(\frac{q_{max}(\tilde{\xi}) - \mu}{\sigma} \leq \frac{\frac{Q^t}{P_k^t} - \mu}{\sigma}\right) \geq \alpha. \tag{28}$$

Now, constraint (28) can be written as follows using the properties of the standard normal distribution:

$$\Phi\left(\frac{\frac{Q^t}{P_k^t} - \mu}{\sigma}\right) \leq 1 - \alpha, \tag{29}$$

where $\Phi(\cdot)$ denotes the probability distribution function of the standard normal distribution. Because the right-hand side of (29) is revealed as a given value, we have the following equivalent constraint:

$$\frac{\frac{Q^t}{P_k^t} - \mu}{\sigma} \leq \Phi^{-1}(1 - \alpha) \tag{30}$$

Using again the fact that $P_k^t \neq 0$ in this case, constraint (30) can be expressed as $Q^t \leq (\sigma \cdot \Phi^{-1}(1 - \alpha) + \mu) \cdot P_k^t$, which is (25) as desired. ∎

Using Proposition 2, chance-constrained programming model **M2** can be reformulated as another linear model (denoted as **A2**):

Approximation A2: $\min C^{maintenance} + C^{failure} + C^{inventory}$

Subject to: $(2) - (15), (17) - (18), (25)$

where constraints (25) are the approximated linear formulas that we have obtained in the proposition. Model **A2** is also a linear approximation of **M2**, which can be solved directly by off-the-shelf solvers.

4.3   Relationship Between the Approximation Approaches

The relationship between the two proposed approximation solutions is summarized in Fig. 3.

First, we have formulation **M1** that is a MILP model for the DPMO problem. Next, we consider stochastic production capacity $q_{max}(\tilde{\xi})$ and treat meeting demands as a probabilistic event, and this leads to a chance-constrained programming model **M2** for the SPMO problem. As **M2** cannot be solved directly, we proposed two approximation methods **A1** and **A2** that express $q_{max}(\tilde{\xi})$ in a discrete manner or a continuous manner, respectively.

The dashed line between the two approaches **A1** and **A2** means that they may be related in a specific way when the scenarios in **A1** are generated through normal distributions. For sake of clarity, we provide more details on their relationship in Sect. 5.

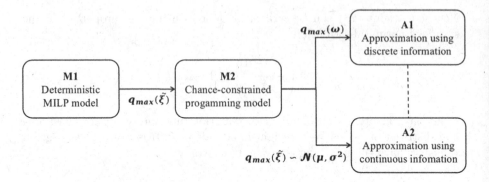

Fig. 3. Relationships among the proposed approaches.

5   Computational Experiments

In this section, we first introduce the settings of the parameters for the instances, stochastic production capacity, and result indicators for our computational experiments. We then compare the performance of the approximation methods **A1** and **A2**. Finally, we evaluate the influence of different probability levels (confidence levels) on the performance of the proposed methods.

The proposed simulations are carried out on a computer with Core I7 and an 8 GB RAM system. The deterministic and approximation models are solved using CPLEX 12.8 and Python V3.7.8.

5.1   Parameter Settings of Instances, Uncertainty, and Indicators

We carried out experiments using generated data with characteristics based on industrial applications. These parameters are generated according to the following rules.

- Number of stages $|\mathcal{K}|$ is up to 8.
- Number of assets $|\mathcal{J}_k|$ in each stage is generated as an integer within the interval $[1, 5]$. Note that $|\mathcal{J}_k| = 1$ means only one single asset without any backup (no redundancy) in stage k, while $|\mathcal{J}_k| \geq 2$ indicates that there is a redundancy of $|\mathcal{J}_k|$ of assets in process k.
- Number of periods $|\mathcal{T}|$ in the planning horizon is up to 20. We consider each period in our data to be one week.
- RUL decrease coefficient $a_{k,j}$ is set within the interval $[1, 3]$.
- RUL recovery factor $b_{k,j}$ is generated as a real number in the interval $[50, 100]$.
- Original RUL is generated uniformly from the continuous interval $[20, 100]$.
- RUL threshold is selected as a real number within the interval $[1, 10]$.
- Asset maintenance duration is set to 1 or 2.
- Deterministic production capacity of the complex system is set to 200.
- Demand is randomly generated uniformly from the interval $[120, 220]$.

– Unit maintenance cost, unit failure cost, and unit inventory cost are set to 50, 100, and 0.05, respectively.

For stochastic production capacity $q_{max}(\tilde{\xi})$, we use the following settings for the various models and approximations.

– For the deterministic model **M1**, production capacity q_{max} is set to 200 (as mentioned above).
– For the approximation **A1**, in order to be consistent with model **M1** and approximation **A2**, scenarios are generated following a normal distribution $q_{max}(\omega) \sim \mathcal{N}(200, 10^2)$, where 200 is the mean value as in **M1** and 10 is the standard deviation. The number of scenarios varies from 5 up to 30. We assume that the scenarios are equiprobable, i.e., $\rho(\omega) = \frac{1}{|\Omega|}$, and thus constraint (24) can be rewritten as $\sum_{\omega \in \Omega} \frac{W(\omega)}{|\Omega|} \geq \alpha$.
– For the approximation **A2**, we use the same normal distribution as for **A1** and set $q_{max}(\tilde{\xi}) \sim \mathcal{N}(200, 10^2)$.

To evaluate and compare the solution approaches **M1**, **A1**, and **A2**, we use the following computational settings and metrics.

– We set a CPU time limit of 2 h (7200 s) with a view to practical applicability.
– We use *obj* to refer to the optimal objective value (total cost) of each approach.
– We use *time(s)* to refer to the CPU time (in seconds) taken by each approach.
– We use *gap₁* to refer to the final gap between the upper and lower bounds computed by CPLEX, in the cases where this gap could not be reduced sufficiently within the given time limit, and the solver could only provide a near-optimal solution.
– We use *gap₂* to denote the percentage gap between the (near-)optimal objective values of **M1** and **A1**, or of **M1** and **A2**. Specifically,

$$gap_2(M1, A1) = \frac{obj_{A1} - obj_{M1}}{obj_{M1}} \times 100\%$$

and

$$gap_2(M1, A2) = \frac{obj_{A2} - obj_{M1}}{obj_{M1}} \times 100\%.$$

5.2 Comparison of Our Approaches

In this subsection, we set the planning horizon to 10 periods. The probabilities (confidence levels) are set to 80%, 85%, 90%, and 95% for both **A1** and **A2**. For the scenario-based approximation **A1**, we set the number of scenarios as 5, 10, and 30. Except for the values of probabilities and scenarios, the instances are exactly the same for **M1**, **A1**, and **A2**. The results are reported in Table 2 averaged over the 10 instances for each line in the table.

Table 2. Average performance of **M1**, **A1**, and **A2** over 10 instances.

Methods	α	$\Phi^{-1}(1-\alpha)$	$totalcost(gap_1)$	gap_2	$time(s)$
M1	–	–	150.95	–	0.9
A1 $\|\Omega\| = 5$	95%	–	151.50	0.36%	5.1
	90%	–	151.50	0.36%	5.1
	85%	–	151.50	0.36%	5.1
	80%	–	150.66 (33.19%)	–	7200
A1 $\|\Omega\| = 10$	95%	–	151.89	0.62%	10.5
	90%	–	151.37 (33.03%)	–	7200
	85%	–	151.37 (33.03%)	–	7200
	80%	–	150.95 (33.13%)	–	7200
A1 $\|\Omega\| = 30$	95%	–	151.17 (33.63%)	–	7200
	90%	–	150.94 (33.04%)	–	7200
	85%	–	150.84 (33.04%)	–	7200
	80%	–	150.68 (31.85%)	–	7200
A2	95%	−1.64	155.25	2.84%	0.3
	90%	−1.28	152.55	1.06%	0.3
	85%	−1.03	152.05	0.72%	0.3
	80%	−0.84	151.85	0.59%	0.3

From Table 2, we observe that the deterministic model **M1** provides the optimal solution with a total cost of 150.95 and computational time of 0.9 s. We take this as the benchmark for evaluating the performance of the approximations **A1** and **A2**:

– For the approximation **A1**, we observe that when $|\Omega| = 5$, the optimal costs for probabilities 95%, 90%, and 85% are the same. This is because all five scenarios must meet (no violation) constraints (23) in Proposition 1 when these probability levels are required. By contrast, for probability 80%, only four scenarios out of five have to respect constraints (23), and we see that the optimization problem is not solved within 7200 s. The best solution found has cost 150.66, with a gap of 33.19% between the upper bound and lower bound (provided by CPLEX). Similar observations hold for $|\Omega| = 10$ and $|\Omega| = 30$. Naturally, a greater number of scenarios can provide a more reliable approximation, but the computational time for **A1** is very high, even for small-size instances.

– For the approximation **A2**, we observe that the optimal costs change from 155.25 to 151.85 with the reduction of α from 95% to 80%. As our optimization minimizes the cost, it is to be expected that if we require the chance constraints to hold with higher probabilities, then this cost will increase. In other words, higher probabilities in the chance constraints mean that the event can be guaranteed even with a smaller stochastic production capac-

Table 3. Computational results with $|T| = 10$.

Methods	α	$\Phi^{-1}(1-\alpha)$	totalcost	gap_2	time(s)
M1	–	–	45.42	–	0.8
A2	99%	−2.33	51.53	13.5%	0.4
	98%	−2.05	49.45	8.9%	0.3
	97%	−1.88	48.61	7.0%	0.3
	96%	−1.75	48.27	6.3%	0.4
	95%	−1.64	47.98	5.6%	0.4
	94%	−1.55	47.71	5.0%	0.3
	93%	−1.48	47.47	4.5%	0.3
	92%	−1.41	47.47	4.5%	0.3
	91%	−1.34	47.26	4.1%	0.3
	90%	−1.28	47.06	3.6%	0.3
	89%	−1.22	47.06	3.6%	0.3
	88%	−1.17	46.88	3.2%	0.4
	87%	−1.13	46.88	3.2%	0.4
	86%	−1.08	46.72	2.9%	0.4
	85%	−1.03	46.72	2.9%	0.4
	100%	−∞	infeasible	–	–
	50%	0	45.42	0.0%	0.8
	0%	+∞	45.00	−0.9%	0.6

Table 4. Computational results with $|T| = 20$.

Methods	α	$\Phi^{-1}(1-\alpha)$	totalcost	gap_2	time(s)
M1	–	–	150.83	–	11.0
A2	99%	−2.33	163.07	8.1%	11.2
	98%	−2.05	159.10	5.5%	6.3
	97%	−1.88	157.78	4.6%	9.8
	96%	−1.75	156.54	3.8%	7.1
	95%	−1.64	155.95	3.4%	14.4
	94%	−1.55	155.41	3.0%	9.3
	93%	−1.48	154.94	2.7%	9.7
	92%	−1.41	154.94	2.7%	9.5
	91%	−1.34	154.51	2.4%	9.1
	90%	−1.28	154.12	2.2%	9.1
	89%	−1.22	154.12	2.2%	9.1
	88%	−1.17	153.76	1.9%	11.8
	87%	−1.13	153.76	1.9%	11.7
	86%	−1.08	153.43	1.7%	8.2
	85%	−1.03	153.43	1.7%	8.1
	100%	−∞	infeasible	–	–
	50%	0	150.83	0.0%	11.1
	0%	+∞	150	−0.6%	27.9

ity. This leads to lower production amounts, which may contribute increased inventory cost to the total cost. We note that while there are no significant differences between the maintenance decisions of **M1** and **A2**, what changes is the inventory cost, depending on varying production capacity. As regards computational time, **A2** can be solved for these instances within 1 s, just like **M1**.

Considering the limited capability for solving **A1**, we henceforth only consider **A2** in the next subsection, where we study the impact of probability α.

5.3   Impact of Probabilities in Chance Constraints

In this subsection, we study the impact of the choices of probability (confidence level) when using method **A2**. We test **M1** and **A2** using the same instances and the computational results (average of 10 instances) are reported in Tables 3 and 4 for $|T| = 10$ and $|T| = 20$, respectively. We do not test larger values of $|T|$ because solving either **M1** or **A2** is time-consuming for more that 25 periods in the planning horizon.

As shown in Tables 3 and 4, we have tested α values from 99% to 85% (because high probabilities are more realistic in practice) as well as three specific probabilities: 0%, 50%, and 100%. From Table 3, we observe that the benchmark model **M1** has an average optimal total cost of 45.42 and mean computational time of 0.8 s. As the probability in the chance constraints decreases, **A2** generally provides lower total costs, as discussed in Sect. 5.2. This trend is clearly illustrated by Fig. 4. The deviation between **M1** and **A2** also becomes smaller, as expected. In terms of computational time, **A2** under different probabilities remains very close to the benchmark. However, new observations can be made in Table 3 when α takes values of 0%, 50%, and 100%, specifically:

- If $\alpha = 100\%$, the value of item $\Phi^{-1}(1-\alpha)$ is negative infinity. Then constraints (25) in Proposition 2 become $Q^t \leq -\infty, \forall t \in T$, meaning that there is no production in any period. This is the theoretical significance of $\alpha = 100\%$, and it explains why **A2** with $\alpha = 100\%$ is always infeasible. In practical terms, it is obvious that holding a stochastic event with 100% probability is not possible, which is consistent with **A2** being infeasible for $\alpha = 100\%$.
- If $\alpha = 50\%$, the value of item $\Phi^{-1}(1 - \alpha)$ is 0, and therefore the stochastic factor takes the expected value. This is equivalent to the deterministic case, and indeed constraints (25) become equivalent to those in the deterministic model. Therefore, when $\alpha = 50\%$, **A2** gives the same results as **M1**.
- If $\alpha = 0\%$, the value of item $\Phi^{-1}(1 - \alpha)$ is $+\infty$, and constraints (25) become $Q^t \leq +\infty, \forall t \in T$, meaning that the system can produce without any limit. Therefore, there will be no inventory in any period, and the only contributions to the total cost are from the maintenance and failure costs. This is why the optimal cost of **A2** is smaller than **M1**.

All the conclusions obtained from Table 3 also hold for the results in Table 4. The purpose of the latter tests is to further confirm our observations. The only

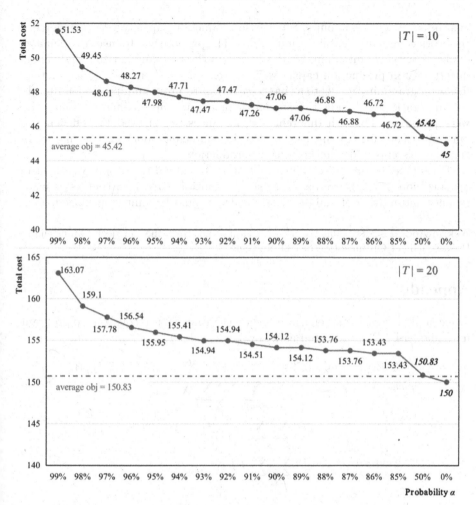

Fig. 4. Trends of the optimal costs of **A2** under different probabilities: the blue line above corresponds to the instances with $|T| = 10$, while the green line below corresponds to the instances with $|T| = 20$.

difference between the two tables is the length of the planning horizon, and a longer horizon implies greater optimal costs and computational times if all other size-related parameters remain the same. This observation was already made in our previous work [3].

6 Conclusions and Future Work

In this paper, we addressed predictive maintenance optimization under stochastic production in generic complex production systems. The RUL information of assets was used to manage the redundancy in each processing stage in order to guarantee system availability. Stochastic production capacity was considered in this work concerning real-life applications. The purpose was to meet the demands of clients with the minimum total cost within the planning horizon when facing uncertainty in production capacity. We extended a deterministic mixed-integer linear programming model to a chance-constrained programming model. In order to computationally handle this stochastic model, two approximation approaches were proposed. Through different test instances, we showed the efficiency of our approaches for solving the proposed maintenance optimization model for complex systems under uncertainty in production.

For future research directions, we propose to extend the proposed stochastic optimization problem from a single site to multiple sites such that asset maintenance and routing of maintenance resources could be jointly optimized.

Acknowledgements. This work is supported by the project Maintenance Prévisionelle et Optimisation of IRT SystemX.

Appendix

The complete version of scenario-based expectation chance-constrained programming model EXP is presented as follows.

$$Model\,EXP: \min \left(C^{maintenance} + C^{failure} + C^{inventory} \right) / |\Omega|$$

$$C^{maintenance} = c^m \cdot \sum_{\omega \in \Omega} \sum_{k \in \mathcal{K}} \sum_{j \in \mathcal{J}_k} \sum_{t \in \mathcal{T}} X_{k,j}^t(\omega)$$

$$C^{failure} = c^f \cdot \sum_{\omega \in \Omega} \sum_{t \in \mathcal{T}} (1 - S^t(\omega))$$

$$C^{inventory} = c^i \cdot \sum_{\omega \in \Omega} \sum_{t \in \mathcal{T}} I^t(\omega)$$

References

1. De Jonge, B., Scarf, P.A.: A review on maintenance optimization. Eur. J. Oper. Res. **285**(3), 805–824 (2020)
2. Zhu, Z., Xiang, Y., Zeng, B.: Multiasset maintenance optimization: a stochastic programming approach. INFORMS J. Comput. **33**(3), 898–914 (2021)
3. He, J., Anjos, F. M., Hadji, M., Khebbache, S.: Prognostic-based maintenance optimization in complex systems with resource limitation constraints. In: Proceedings of the 11th International Conference on Operations Research and Enterprise Systems, 3th–5th February 2022
4. Camci, F., Medjaher, K., Atamuradov, V., Berdinyazov, A.: Integrated maintenance and mission planning using remaining useful life information. Eng. Optim. **51**(10), 1794–1809 (2019)
5. Si, X.S., Wang, W., Hu, C.H., Zhou, D.H.: Remaining useful life estimation - a review on the statistical data driven approaches. Eur. J. Oper. Res. **213**, 1–14 (2011)

6. Lei, Y., Li, N., Guo, L., Li, N., Yan, T., Lin, J.: Machinery health prognostics: a systematic review from data acquisition to RUL prediction. Mech. Syst. Signal Process. **104**, 799–834 (2018)
7. Do, P., Voisin, A., Levrat, E., Lung, B.: A proactive condition-based maintenance strategy with both perfect and imperfect maintenance actions. Reliab. Eng. Syst. Saf. **133**, 22–32 (2015)
8. Chen, Z., Li, Y., Xia, T., Pan, E.: Hidden Markov model with auto-correlated observations for remaining useful life prediction and optimal maintenance policy. Reliab. Eng. Syst. Saf. **184**, 123–136 (2019)
9. Wu, S., Castro, I.T.: Maintenance policy for a system with a weighted linear combination of degradation processes. Eur. J. Oper. Res. **280**(1), 124–133 (2020)
10. Lei, X., Sandborn, P.A.: Maintenance scheduling based on remaining useful life predictions for wind farms managed using power purchase agreements. Renew. Energy **116**, 188–198 (2018)
11. Dong, W., Liu, S., Cao, Y., Javed, S.A., Du, Y.: Reliability modeling and optimal random preventive maintenance policy for parallel systems with damage self-healing. Comput. Ind. Eng. **142**, 106359 (2020)
12. Camci, F.: System maintenance scheduling with prognostics information using genetic algorithm. IEEE Trans. Reliab. **58**(3), 539–552 (2009)
13. Xiao, L., Song, S., Chen, X., David, W.C.: Joint optimization of production scheduling and machine group preventive maintenance. Reliab. Eng. Syst. Saf. **146**, 68–78 (2016)
14. Bahria, N., Chelbi, A., Bouchriha, H., Dridi, I.H.: Integrated production, statistical process control, and maintenance policy for unreliable manufacturing systems. Int. J. Prod. Res. **57**(8), 2548–2570 (2019)
15. Khatab, A., Aghezzaf, E.H.: Selective maintenance optimization when quality of imperfect maintenance actions are stochastic. Reliab. Eng. Syst. Saf. **150**, 182–189 (2016)
16. Shahraki, A.F., Yadav, O.P., Vogiatzis, C.: Selective maintenance optimization for multi-state systems considering stochastically dependent components and stochastic imperfect maintenance actions. Reliab. Eng. Syst. Saf. **196**, 106738 (2020)
17. Ghorbani, M., Nourelfath, M., Gendreau, M.: A two-stage stochastic programming model for selective maintenance optimization. Reliab. Eng. Syst. Saf. **223**, 108480 (2022)
18. Khatab, A., Aghezzaf, E.H., Djelloul, I., Sari, I.: Selective maintenance optimization for systems operating missions and scheduled breaks with stochastic durations. J. Manuf. Syst. **43**, 168–177 (2017)
19. Liu, L., Yang, J., Kong, X., Xiao, Y.: Multi-mission selective maintenance and repair persons assignment problem with stochastic durations. Reliab. Eng. Syst. Saf. **219**, 108209 (2022)
20. Xenos, D.P., Kopanos, G.M., Cicciotti, M., Thornhilla, N.F.: Operational optimization of networks of compressors considering condition-based maintenance. Comput. Chem. Eng. **84**, 117–131 (2016)
21. Ye, Y., Grossmann, I.E., Pinto, J.M., Ramaswamy, S.: Modeling for reliability optimization of system design and maintenance based on Markov chain theory. Comput. Chem. Eng. **124**, 381–404 (2019)
22. Rivera-Gómez, H., Gharbi, A., Kenné, J.P., Montaño-Arango, O., Corona-Armenta, J.R.: Joint optimization of production and maintenance strategies considering a dynamic sampling strategy for a deteriorating system. Comput. Ind. Eng. **140**, 106273 (2020)

23. Zhou, H., Wang, S., Qi, F., Gao, S.: Maintenance modeling and operation parameters optimization for complex production line under reliability constraints. Ann. Oper. Res. 1–17 (2019). https://doi.org/10.1007/s10479-019-03228-9
24. Birge, J. R., Louveaux, F.: Introduction to Stochastic Programming. Springer Science & Business Media, New York (2011). https://doi.org/10.1007/978-1-4614-0237-4
25. Charnes, A., Cooper, W.W.: Chance-constrained programming. Manag. Sci. **5**, 73–79 (1959)
26. Prékopa, A.: Contributions to the theory of stochastic programs. Math. Program. **4**, 202–221 (1973)

Automated City Segmentation for Pollution Threshold Attribution: The Example of New Cairo

Kareem Esam Eldin[1]([⊠])[ID], Youssef Khalil[1][ID], Mostafa ElHayani[2][ID], Mariam Zaky[1], and Hassan Soubra[1][ID]

[1] German University in Cairo, Cairo, Egypt
kareemelshafey98@gmail.com,
{mariam.aboelnour,hassan.soubra}@guc.edu.eg
[2] Technical University of Munich, Munich, Germany
mostafa.elhayani@tum.de

Abstract. Today the use of vehicles has greatly increased especially in cities where alternative transportation methods cannot really be relied on. Inevitably, this causes excessive increases in noise and air pollution. Our previous study [4] has exposed real-life measures of pollution levels in certain areas in Cairo, however, there is a lack in identifying certain maximum pollution thresholds to particular neighbourhoods or city segments. In this paper, we attempt to provide a solution to the pollution issue via threshold attribution and using New Cairo as an illustrative example. Our approach uses geographical coordinates and an input radius in meters to generate an interactive map. The map segments the city into different sections. Furthermore, we plot each section with corresponding pollution thresholds that appear when hovering over those areas. In Addition, we categorize each segment into one of three categories, where each category has a defined pollution threshold value.

Keywords: Pollution · IoT · Segmentation · ITS

1 Introduction

The WeForum Study in 2015 [9] estimated the number of vehicles to be around 1.1 Billion. The study also expected the number to increase to 1.5 Billion by 2025 and 2 Billion by 2040. This study indicates how overused vehicles are nowadays. The overuse shows that there will be a dramatic increase in pollution levels. Cars cause different types of pollution. Air Pollution affects the human heart and the respiratory system, which puts your life at risk. Noise Pollution can affect patients and doctors in hospitals or teachers and students in schools. In this paper, we aim to limit those two types of pollution, as they are the ones that affect human health with extreme affections.

The main goal of our research is to reduce noise and air pollution. This paper proposes an addition to a larger system. The approach aims to find a way to segment the city into different parts and define threshold values for each of those city segments. After calculating the thresholds, we compare them with data extracted by a mobile application and an IoT device that reads and calculate noise and air pollution in real-time [4]. This paper is organized as follows: Sect. 2 presents the literature review related to our work, Sect. 3 presents our methodology and describes the experimental set-up of our case study. Section 4 discusses the results. Finally, a conclusion follows in Sect. 5.

2   Literature Review

The main purpose of [7] was to prove that the sound energy will be reduced with the doubling of the distance from the sound source. The readings of the paper were not made with measurements, it was made out by calculations actually. The findings showed that doubling the distance reduces the sound energy by 6 DBs each doubling distance. The drawbacks of the paper are that the reduction in sound was made by calculations not by measurements. What we concluded from the paper is that "A doubling of the distance results in a level reduction of 6 dB as a typical characteristic of this type of source".

The main purpose of [2] was to find out the short term three dimensional variable at a vertical-horizontal axis of pollution related to vehicles emissions around highways. The experiment was made by using a drone (DJI M600 UAV) that had some portable sensors added to the drone that was added to monitor the air quality. They selected a highway in G3 (eastern Dezhou) which overlaps the G2 while passing through Dezhou. They have two lanes in each direction with a speed limit of 100–120 Km/h, the average vehicles passing daily is 717–1317 vehicles/hour which means 17208–31608 vehicles per day. The researcher have defined a route for the drone to pass on, the route is included in the paper to measure the air quality around the highway. Findings and results showed that most of the pollutants were higher at the sides of the road and then decreased when the distance was incremented. The field experiments showed that almost all the pollutants dropped to the background level after 100 m horizontally, which means when moving away from the source of pollution level drops. The drawbacks are no certain amount of air quality improved as mentioned in the paper. I could use the paper readings in increasing the distance to reduce the air pollution as the pollutants dropped in the areas away from the pollution source, values of pollution started dropping after 100 m horizontally.

The goal of [8] is to set a certain threshold for the allowed pollution levels for each of the following pollutants: PM2.5, PM_{10}, Ozone, NO2, SO2 and CO. The thresholds that was defined by the paper. The thresholds are as follows: for Particular Matter (PM2.5): $10 \mu\text{-g/m}^3$ (annual mean) and $25 \mu\text{-g/m}^3$ (24 h mean), Particular Matter (PM_{10}): $20 \mu\text{-g/m}^3$ (annual mean) and $50 \mu\text{-g/m}^3$ (24 h mean), Ozone: $100 \mu\text{-g/m}^3$ (8 h mean), Nitrogen Dioxide: $40 \mu\text{-g/m}^3$ (annual mean) and $200 \mu\text{-g/m}^3$ (1 h mean), Nitrogen Dioxide: $20 \mu\text{-g/m}^3$ (24 h mean) and $500 \mu\text{-g/m}^3$ (10 min mean) and Carbon Monoxide: 60mg/m^3 (30 min mean), 30mg/m^3 (1 h mean) and 10mg/m^3 (8 h mean). The drawbacks of this paper is that the defined thresholds are actually global thresholds, which means there is no specific threshold for each area. Actually there must be different thresholds for different areas (or segments). Hospitals for example will need the least air pollution than other areas (or segments). I could use the Threshold values defined in this paper to be used as our thresholds as it is a trusted source (World's

Health Organization). This paper defined thresholds for each of the pollutants: Particulate Matter (<2.5 Diameter), Particulate Matter (<10 Diameter), Ozone, Nitrogen Dioxide, Sulfur Dioxide and Carbon Monoxide.

The main purpose of [1] is to show that noise is one of the main factors that would negatively affect humans, causing multiple health problems like sleep annoyance and distribution, this study was made to determine noise pollution and reduce the negative health impact on humans. The thresholds was defined in different areas of the city, those areas and there thresholds are: 1) Outdoor living area LAeq[dB(A)] : 55 - greater than 55, 2) Inside bedrooms LAeq[dB(A)] : 30, 3) Outside bedrooms LAeq[dB(A)] : 45, 4) School classrooms preschoolers, indoors LAeq[dB(A)] : 35, 5) Preschool Bedrooms, indoor LAeq[dB(A)] : 30, 6) School, playground outdoor LAeq[dB(A)] : 55, 6) Hospital, wardrooms, indoors LAeq[dB(A)] : 30 and 7) Public addresses, indoors and outdoors LAeq[dB(A)] : 85. Drawbacks of the paper that the paper is missing some important values that are needed such as noise levels outside the hospitals. Other crucial values would be missing also, outside hospitals is just an example. The paper can help us in knowing the thresholds of noise in different places and the paper is made by the World's Health Organization (WHO), this is a trusted organization which defined a threshold table for the noise pollution covering different places. Each place has a threshold for the inside and the outside of the place. The thresholds defined in this paper could be used in our project for two things. First we could segment the areas according to the places that have a defined noise pollution threshold in this paper. Second we are going to use the outside threshold for the outside of each place in the segments.

The main aim of the paper [5] is to provide a new Intelligent Transportation System (ITS) that allows them to measure, record and monitor noise and air pollution caused by road vehicles. According to the pollution levels, the system should output the most efficient route. Using IOT Sensors which are placed in cars and the infrastructure, they were able to measure real-time data for both noise and air pollution. A defined city represents the city that is covered in their system and has real-time measurements. System was tested in different ways. One of the testing ways was testing getting the route from a not defined city to a defined city. The other way is the opposite, which is from inside the defined city to the outside of the city (undefined). Another way from a point in the defined city to another point in the defined city. Also they were testing the system on different levels of pollution. This paper is related somehow to our project. The system represented in this paper could be integrated with our machine to provide them with the thresholds defined for each segment of the city.

From the literature review, we can conclude that most proposed systems measure pollution levels in real-time or tackle pollution reduction on a general scale. We aim to define areas that need certain thresholds of maximum pollution levels.

3   Methodology

Our system segments cities according to buildings in each area. The segments defined by our system keep track of certain pollution levels that should not be exceeded. We use schools, universities, hospitals, and parks to define the different areas in the segmentation process. The thresholds are defined based on the WHO air pollution thresholds

(Particulate Matter 2.5, Particulate Matter 10, Ozone, Nitrogen Dioxide, Sulfur Dioxide, and Carbon Monoxide).

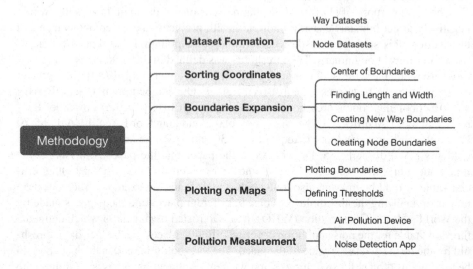

Fig. 1. Overview of the system proposed.

This section describes the steps to obtain segmented cities with different categories
and a defined threshold for each area. We further divide the section into the following
parts:

1. Dataset formation.
2. Sorting technique.
3. Increasing boundaries.
4. Plotting and visualization.
5. Pollution Measurement.

The overview of the system proposed is shown in Fig. 1.

First, the system uses a latitude and longitude input to define the area to segment.
Furthermore, a radius input decides how big the segmentation area is. The system then
generates a map for visualization. The map shows the different areas and their thresholds.

3.1   Dataset Formation

The data used is obtained using Overpass OpenStreetMap API. Datasets are obtained
from Overpass as two datasets for each of the categories. The first dataset contains
nodes data, while the second dataset includes ways data. The process is repeated for
each of the categories. In total, we collect six datasets. The categories for each city are
as mentioned in Fig. 2.

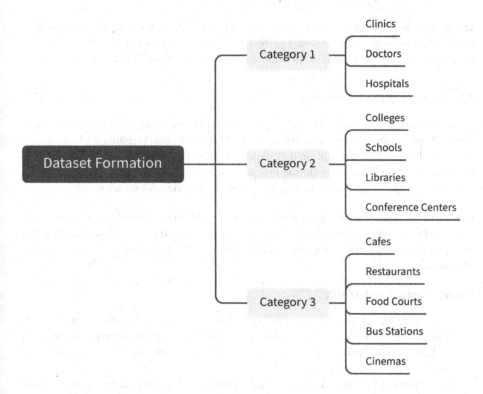

Fig. 2. Datasets are retrieved and divided into three categories.

3.2 Sorting Technique

Each way has a set of nodes. When connecting nodes we need to connect them with each other in the correct order. Connecting the set of nodes together without sorting would lead to unexpected intersecting lines in the middle of the shape. After sorting the set of nodes the boundaries are connected together using the correct sequence, giving the right shape.

Figure 3 shows the three main steps of Sorting Coordinates. The first step is finding the left and right-most points of a building. A line is drawn between both of them. The

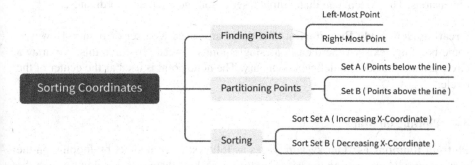

Fig. 3. The map for the sorting technique.

second step is dividing the points into two sets; Set A contains points below the line, while Set B contains those above. Lastly, Set A is sorted in increasing order based on the X-Coordinates, and Set B is sorted in decreasing order based on the X-Coordinates.

3.3   Increasing the Boundaries

As mentioned in [7], noise level is reduced by 6 dBs by doubling the distance away from the noise source. Also as mentioned in [2,6], air pollution is reduced when moving away from the source. The aim in this stage is to increase the boundaries of the areas to decrease both noise and air pollution inside the segments. As an input for this stage we take in the sorted points to begin the process of the boundaries extension.

Center of the Boundaries. The center point is needed for drawing the new boundaries, as it will be the center of the newly drawn rectangle. According to the outputs from the previous stage, we can obtain the center point by getting the mean of the X and Y coordinates.

Finding Length and Width

- To measure the length and width of any facility, we need to find the left most, right most, up most, and down most points. The width is calculated using the Right Most and Left Most points, while the height is calculated using the Up and Down most points.
- The Haversine Formula (Fig. 4) is used to calculate the distance between points to get the needed lengths and widths.

$$d = 2r \arcsin\left(\sqrt{\sin^2\left(\frac{\phi_2 - \phi_1}{2}\right) + \cos(\phi_1)\cos(\phi_2)\sin^2\left(\frac{\lambda_2 - \lambda_1}{2}\right)}\right)$$

Fig. 4. Formula used for distance calculation between geographic coordinates [3].

Creating New Way Boundaries. New boundaries are created to extend the shape into a rectangle. The center point, height, and width are given as inputs. The center is used to justify the rectangle center, while the width and height are used to get the rectangle dimensions. The system can then output a set of four geographic coordinates.

Creating New Node Boundaries. Buildings are saved as a set of points. However, some buildings exist in the dataset as single points instead. To solve this, we draw a circle around the node to define a boundary. The node point is used as the center of the circle.

3.4   Plotting on Maps

Plotting Boundaries. Our system works as follows; First, it starts by looping on the first category ways, getting the set of nodes for each of them, and passing through the sorting technique. For each category, each element is added to the array. Afterward, elements of the nodes dataset will be added to the array. For the map visualization, Scattermapbox is used. There are two visualization modes; The first is used to update the map based on categories by changing their colors. The second fully displays the map. The first mode uses the input mode (markers+lines), while the other uses (stamen-terrain).

Defining Thresholds

– Defining and Dividing Categories: According to both research papers done by WHO [1, 8]. One of those papers covers the topic of noise pollution thresholds for different areas, and the other covers the expected thresholds for components of air pollution. We grouped similar thresholds into three different categories. For each group, the threshold is defined by the minimum threshold of all elements belonging to that group.
– Adding Thresholds to Map: After defining different groups, they are plotted on the map with different colors to differentiate the criticality of the threshold values. The colors are Red, Orange, and Green. Red is the most critical, followed by orange and green being the least critical.

3.5 Pollution Measurement

The data-set generated by [4] is used to compare the actual pollution levels in real life and the thresholds produced by our system. IoT device and mobile application were used to measure air and noise pollution successively.

Fig. 5. Air Pollution Device.

Air Pollution Measurement. The computing unit of the IoT device shown in Fig. 5 consists of a sensor unit which consists of 3 sensors, MQ135 sensor which is used in measuring the air quality index, DHT11 to measure the temperature and the humidity and GP2Y1010AU0F to measure the concentration of PM_{10} in the air. The sensors are connected to Arduino Mega which opens a serial connection with a Node-MCU that has a built-in WiFi chip to transfer the data to their server. The device is stationed in two locations around New Cairo City, near by a very popular street in the area. 3307 samples were obtained from both locations, each sample contains 6 fields, AQI value, PM_{10} concentration, Temperature, humidity, and the location of the sample.

Noise Pollution Measurement. An android application was also implemented in [4] to accomplish a real-time noise pollution measurement. The application has been distributed among volunteers to measure noise levels across Cairo to acquire a large number of recordings and to be able to quantify noise pollution in a variety of locations.

4   Results

4.1   Area of Test

The area of test is New Cairo, Cairo, Egypt. The city's Longitude is 30.0288, Latitude is 30.0288 and the radius is 8000 m.

4.2   Datasets Formation

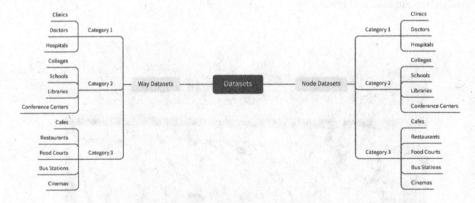

Fig. 6. Datasets are retrieved as Node Dataset and Way Dataset.

In this section, the datasets samples that were retrieved in the first stage are provided. Figure 6 shows the dataset retrieved.

Node Dataset. A sample of the Node Dataset is shown in Figure shown in Fig. 7. Columns give more information about each of the nodes. "Amenity" is a tag which could be, for example, a hospital. "Name" is the name of the node. "Latitude" and "Longitude" are the geographical coordinates of the node.

hospitalNode.csv hospitalWay.csv

1 to 5 of 5 entries Filter

	amenity	name	name:ar	latitude	longitude	id	name:en	emergency	contact:instagram
0	hospital	Al Tabib Center مركز الطبيب	Al Tabib Center مركز الطبيب	30.0039967	31.4224834	3395906136			
1	hospital			30.0593917	31.4918090	4615939994	المركز الطبي ٢		
2	clinic	family park		30.0769076	31.4994023	5359156224			
3	hospital	مستشفى	مستشفى	30.0147462	31.4181409	5817986010	Hospital	yes	
4	clinic	عيادة الدكتور عمرو سويدان	عيادة الدكتور عمرو سويدان	30.0283455	31.4861358	8501558934	Dr Amr Swidan Clinic		https://www.instagram.com/dr.amrswaidan/

Show 10 per page

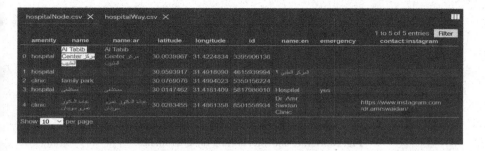

Fig. 7. A sample of 1st Category Nodes Dataset.

hospitalNode.csv hospitalWay.csv

1 to 4 of 4 entries Filter

	amenity	name	name:ar	name:en	id	addr:place	addr:place:ar	addr:place:en	construction
0	hospital	المنتجع الصحي للطب النفسي أ.د أحمد عكاشة	المنتجع الصحي للطب النفسي أ.د أحمد عكاشة	Psychological Health Resort of Prof Dr Ahmed Okasha	355428539				
1	hospital	مستشفى درة القاهرة	مستشفى درة القاهرة	Durrat Al Qahira Hospital	497855286	Durrat Al Qahira	درة القاهرة	Durrat Al Qahira	yes
2	hospital	مستشفى السلام الدولي	مستشفى السلام الدولي	Al Salam International Hospital	769942974				
3	hospital	مستشفى الدفاع الجوي التخصصي	المستشفى الجوي التخصصي	Air Force Specialized Hospital	773024488				

Show 10 per page

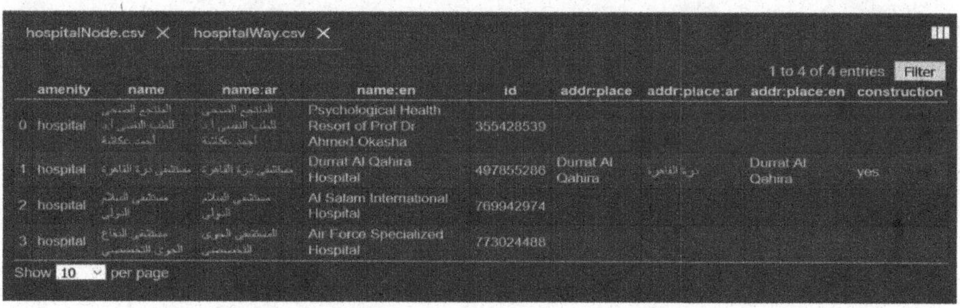

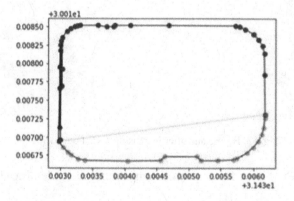

Fig. 8. A sample of 1st Category Ways Dataset.

Way Dataset. A sample of the Way dataset is shown in Fig. 8. The columns used are similar to those from the node dataset above.

4.3 Sorting Technique

Fig. 9. Sorting Technique Initial Output.

As shown in Fig. 9, the line is drawn between the left most point and the right most one, then set of points above and below the line are sorted according to our proposed approach. Figure 10 shows the difference between a non-sorted set of points and the results of sorting the points.

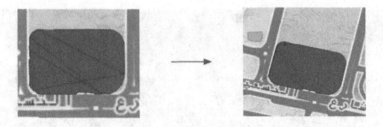

Fig. 10. Before and after sorting the coordinates.

4.4 Increasing the Boundaries

As explained, there are two types of boundaries, Way boundaries where all the points exist, and Node boundaries where building are only available as a single point.

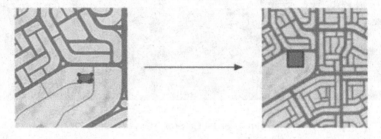

Fig. 11. Before and after increasing way boundaries.

Way Boundaries. Figure 11 shows the difference before and after increasing the boundaries.

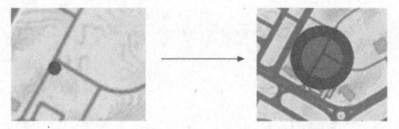

Fig. 12. Before and after increasing node boundaries.

Node Boundaries. Figure 12 shows the results of using a circle instead of a point for Node boundaries.

4.5 Plotting on Maps

Figure 13 shows the result of drawing all categories and boundaries on the map. As explained before, the areas were divided into three categories. Each category has areas that have similar threshold values. Red (Clinics, Hospitals); Orange (Colleges, Schools, Libraries). Green (Cafes, Restaurants, Food Courts).

Fig. 13. 3rd Category is plotted on the map. (Color figure online)

Fig. 14. Threshold values for 1st category for each air and noise pollution.

According to our study, air pollution thresholds are the same for all categories, but noise pollution should be different. For critical categories like the 1st and 2nd categories, we increase the boundaries more than usual to reduce the air pollution within the facility. Thresholds are shown when hovering above the area, as shown in Fig. 14.

4.6 Air Pollution

Overview. Collecting the data regarding the air pollution measurement has been done in 2 locations inside Cairo [4], nearby a famous street in New Cairo city with is the 90th street, the first location has been chosen based on its surrounded area. The other is located in front of a construction area that would increase air pollution. A total of 1784 samples were collected at the first location and 1523 samples at the second location.

Data-Set. Figure 15 shows a sample of the dataset.

Index	AQIValue	Pm10	Temperature	Humidity	Location	vDate
0	55	43.06	29	57	60b56e6383b6a02a48c49b5c	2021-07-09T01:45:14.000Z
1	48	47.21	29	57	60b56e6383b6a02a48c49b5c	2021-07-09T01:46:14.000Z
2	49	49.7	29	57	60b56e6383b6a02a48c49b5c	2021-07-09T01:49:14.000Z
3	98	47.21	29	57	60b56e6383b6a02a48c49b5c	2021-07-09T01:50:15.000Z
4	30	39.74	29	57	60b56e6383b6a02a48c49b5c	2021-07-09T01:51:15.000Z

Fig. 15. Sample of the Air pollution measurements data.

4.7 Noise Pollution

Overview. Noise pollution measurements were done using a mobile application, distributed to volunteers, and reading noise pollution levels from streets known to have high traffic congestion.

Dataset. Figure 16 shows a sample of the collected data.

59	30.0159816 , 31.408593	مدينة.. الواحة، الحي العاشر، مدينة نصر, Al H…	89.0608	2021-06-05 17:56:02.669163	SM-N975F
60	30.0165899 , 31.4247883	29 Street 3, First New Cairo, Cairo Governorate, Egypt	83.0062	2021-06-05 18:13:45.523859	SM-N975F
61	30.0161484 , 31.4402448	78 South Teseen Road 5"District. First New …	82.9785	2021-06-05 18:20:58.892792	SM-N975F
62	30.0159042 , 31.4230899	7 Al Shouyfat, First New Cairo, Cairo Governorate, Egypt	76.4717	2021-06-05 12:56:56.679037	SM-N975F
63	30.0161484 , 31.4402448	78 South Teseen Road 5"District. First New …	88.5701	2021-06-05 18:24:58.888384	SM-N975F
64	30.0158647 , 31.4348348	7 Al Shouyfat, First New Cairo, Cairo Governorate, Egypt	86.2075	2021-06-05 13:16:29.792172	SM-N975F
65	30.0169108 , 31.4319132	31 Street 2, First New Cairo, Cairo Governorate, Egypt	78.1694	2021-06-05 18:34:28.557908	SM-N975F
66	30.0169108 , 31.4319132	31 Street 2, First New Cairo, Cairo Governorate, Egypt	79.2982	2021-06-05 18:33:28.553857	SM-N975F
67	30.0159042 , 31.4230899	7 Al Shouyfat, First New Cairo, Cairo Governorate, Egypt	83.0957	2021-06-05 13:00:56.663909	SM-N975F

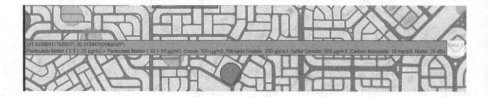

Fig. 16. Noise Pollution records in the Database.

4.8 Comparison of Measurements and Threshold

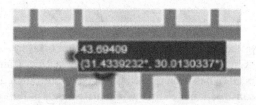

Fig. 17. Noise & Air Pollution Thresholds Defined.

Fig. 18. Noise Pollution Measurements.

Noise Pollution. According to Fig. 18, the noise reading is 43 DBs. However, Fig. 17 shows that the threshold is 35 DBs. This indicates that the noise pollution is above the threshold value.

Air Pollution. As mentioned in the results section in [4], at the same location the peak hours were in the evening. Three scenarios were introduced, two of them were within a peak hour in the weekend with average PM_{10} concentrations 70.07 $\mu g/m^3$ and 44.91 $\mu g/m^3$. The third scenario was an exceptional case in which it was during the EURO 2020 final game divided into 3 parts; during the game, during the ceremony, after the ceremony, regarding the first part, the average concentration of PM_{10} was 85.888 $\mu g/m^3$, the second part (during the ceremony) the average concentration of PM_{10} was 83.1559 $\mu g/m^3$, the last part (after the ceremony) the average concentration of PM_{10} was 82.143 $\mu g/m^3$.

5   Conclusion

In this paper, we proposed a Noise and Air pollution management system to segment a city into different sectors and define thresholds for each of those parts. The proposed system provides different segments of a city, categorizing them into one of three categories and defining a threshold for each type. The thresholds are defined based on the WHO air pollution thresholds (Particulate Matter 2.5, Particulate Matter 10, Ozone, Nitrogen Dioxide, Sulfur Dioxide, and Carbon Monoxide). We tested our system in New Cairo, Cairo, Egypt. Furthermore, we compared our results to actual measurements taken in a previous study [4]. We observe that the pollution measurements exceeded the required thresholds. The observations show that actions to reduce pollution levels are needed. Our system could be used as a part of a larger ITS that should help in the reduction of pollution.

References

1. Byrne, L., Angus, D., Wiles, J.: Guidelines for community noise
2. Cao, R., Li, B., Wang, H.W., Tao, S., Peng, Z.R., He, H.: Vertical and horizontal profiles of particulate matter and black carbon near elevated highways based on unmanned aerial vehicle monitoring (2020)
3. Haversine formula. https://github.com/DaniilSydorenko/haversine-geolocation
4. Khalil, Y., Zaky, M., ElHayani, M., Soubra, H.: Real life pollution measurement of Cairo. In: Proceedings of the 11th International Conference on Operations Research and Enterprise Systems - ICORES, pp. 222–230. INSTICC, SciTePress (2022). https://doi.org/10.5220/0010896000003117
5. Zaky, M.O., Soubra, H.: An intelligent transportation system for air and noise pollution management in cities (2021)
6. Pan, S., Choi, Y., Roy, A., Jeon, W.: Allocating emissions to 4 km and 1 km horizontal spatial resolution and its impact on simulated NOx and O3 in Houston, TX (2017)
7. Rathe, E.J.: Note on two common problems of sound propagation (1969)
8. Exposure to air pollution: A major public health concern (2010)
9. Smith, M.N.: The number of cars worldwide is set to double by 2040 (2016)

Customer Satisfaction and Company Revenue: A Solution Approach for the Passenger Management on Buses

Francesca Guerriero[✉][iD], Martina Luzzi[✉][iD], and Giusy Macrina[✉][iD]

Department of Mechanical, Energy and Management Engineering, University of Calabria, Arcavacata, Italy
{francesca.guerriero,martina.luzzi,giusy.macrina}@unical.it

Abstract. In this work we investigate the bus group assignment problem considering the maximization of both the revenue derived from the tickets sale and the service level offered to the customers. The service level is defined as the capacity of the company to satisfy customers' demands by taking into account the seating configuration, i.e. guaranteeing that all the members of the same group are seated close to each other. We propose a dynamic formulation of the problem, then, a linear programming approximation. Furthermore, we implement a booking limit revenue management policy to solve the problem and manage seat inventory. Empirical tests are made in order to compare the performance of the booking limit policy with a first came first served strategy, and a model solution in a context of perfect knowledge of the occurred demand. The computational results show that the booking limit policy performs better than the first came first served strategy and in many cases equals the performance of the perfect knowledge strategy.

Keywords: Seat allocation problem · Passenger transportation · Service level · Group assignment

1 Introduction

Passenger transportation sector is constantly growing. According to the report published by the Business Research Company, the global transit and ground passenger transport market size is expected to grow from 578,44 billion in 2021 to 638,28 billion in 2022 and will reach 911,78 billion in 2026 (www.thebusinessresearchcompany.com).

Since transport companies offer a set of perishable products over a finite time horizon, the management of vehicle capacity is crucial. The seat allocation is one of the main problem that every provider of passenger transport service has to deal with before selling the tickets, it consists in finding the optimal assignment of passengers to the available seats.

It is a well known problem, in fact, several researchers have studied and proposed numerous variants as well as several approaches to solve them.

Yazdani et al. [1] propose an algorithm to solve a real-time seat allocation problem for minimizing boarding/alighting time on trains, improving the quality of the service and guarantee the passengers' safety.

F. Liberatore et al. (Eds.): ICORES 2022/2023, CCIS 1985, pp. 82–95, 2024.
https://doi.org/10.1007/978-3-031-49662-2_5

Sumalee et al. [2] propose a stochastic dynamic transit assignment model with an explicit seat allocation process applicable to a general transit network, to estimate the probability of a passenger waiting at a station or onboard to get a seat. The authors develop an heuristic algorithm and the numerical tests show significant influences of the seat allocation model on the equilibrium departure time and the route choices of passengers.

Yan et al. [3] develop a probabilistic nonlinear programming seat allocation model for high-speed railway passenger service network based on flexible train composition that is able to provide a decision-making basis for discount sales and ticket allocation.

Milne and Salari [4] investigate the optimization of assigning passengers on airplanes based on their carry-on luggage by proposing a mixed integer programming model that determines the number of luggage to be carried by passengers assigned to each seat. Numerical results indicate that the proposed approach allows to reduce the time to complete the boarding of the plane, and the improvement is greatest when a high number of luggage is carried into the plane. The optimal distribution of luggage assigns passengers with few carry-on bags to the rows of the plane closest to the entrance.

Notomista et al. [5] propose a novel seat allocation algorithm to minimize the boarding time using online seat assignment based on passenger classification. Extensive simulations show that a mean reduction of the boarding time, about the 15%, is achieved compared to existing boarding strategies.

A variant of this problem, widely investigated by the scholars, is the seat reallocation problem, that arises when the vehicle (typically an aircraft) is changed just before departure. For example, Dae Ko et al. [6] propose an efficient airline seat reallocation algorithm to minimize customers' dissatisfaction. Another variant of the seat allocation problem is the group seat allocation problem whose objective is finding the optimal allocation of groups of passengers.

Tajima and Misono [7] use a set packing formulation to solve airline seat allocation and reallocation problems considering groups of passengers. Both problems aim at maximizing customers' satisfaction, i.e., guaranteeing that the members of the group are seated close one to each other.

Song et al. [8] develop a seat allocation model for group demand in air travel markets aimed at maximizing the total revenue. The same authors also provide an extension of this work that consists in an integrated stochastic programming seat allocation model for individual and group demand in airlines [9].

Although the seat allocation problem is widely used in the context of passenger transportation, it is worth to notice that it can be applied in many other fields.

Kwag et al. [10] investigate the seat allocation problem for an e-sports gaming center where generally groups of people play games while eating. Lee [11] proposes a min-cut algorithm for the arrangement problem of the seats in wedding hall aimed at minimizing the loss of guest relations for a complex guest relations network.

From this analysis, we can notice that even if the seat allocation problem is widely addressed for airline and railway, scarce attention is dedicated to the bus sector despite its market growth. Thus, in our work we investigate the seat allocation problem applied in the bus sector, considering groups of passengers and level of service.

In particular, we propose an extension of the Bus Group Assignment Problem under Covid-19 social distancing introduced by [12]. Since the majority of rules imposed during the pandemic time to prevent the contagion are relaxed or removed, some features of the previous work are been readjusted and changed. In particular, we do not consider social distancing anymore but we concentrate the attention on the management of groups of passengers on buses with the purpose to maximize total revenue, optimize the capacity of the buses and maximize the customer satisfaction respecting the service level.

The novelty of this work is not only the application field, but also the introduction of the service level as the capacity of the company to respect a closeness criterion to assign the seats of the members of the same group. To the best of our knowledge, only Tajima and Misono [7] consider a similar definition of the service level; however, our work differs from this paper for several aspects. In fact, Tajima and Misono [7] consider the airline sector, that is very different from the bus one. In fact, the aircraft has one origin and one destination, while a bus may have several origins and destinations. Moreover, their fleet is composed of only one vehicle, while in our work we consider more than one bus. Finally, they adopt a different approach, i.e., a set packing formulation, in which the objective function is the minimization of a cost function related to the passengers.

The rest of the paper is organized as follow: Sect. 2 describes the features of the problem, Sect. 3 contains the dynamic formulation of the problem and Sect. 4 its linear approximation. Section 5 is devoted to the explanation of the booking limit policy that is used to solve the proposed problem. In Sect. 6 computational experiments and numerical results are discussed and Sect. 7 is dedicated to sum up conclusions.

2   Problem Description

In this section, we present a mathematical model for managing groups of passengers, characterized by a different size, on a bus, to maximize the total revenue deriving from the sale of travel tickets and ensuring, if possible, a certain *service level*. The *service level* can be defined as the ability of the company to satisfy the requests and the expectations of the customers, i.e., being seated close to each other, at the right time and with the right products or services. The model can be seen as a variant of the bus group assignment problem under COVID-19 social distancing, proposed in [12].

In particular, we consider a bus company that sells, on a given time horizon, a set of products to several groups of customers. The products offered by the company are transportation services from a given set of origins to a given set of destinations, performed by its buses. At each time of the planning horizon, the company has to decide how to manage the overall capacity in the most profitable way.

Let:

- $E = \{e_1, \ldots, e_n\}$ be the set of n origins
- $F = \{e_1, \ldots, e_n\}$ be the set of q destinations

We may indicate as (e, f) with $\{(e, f) : f > e, e \in E, f \in F\}$ the generic product, i.e., the OD transportation service from the bus station e to the bus station f. In addition, let:

– $(EF) = (e_1, f_1), \ldots, (e_1, f_q), (e_2, f_1), \ldots, (e_n, f_q)$ be the set of all the product sorted in an increasing order of the origin station

We introduce an index p with $p = (1, \ldots, |(EF)|)$ used to identify the products offered by the company. The company has a set of buses, each of them is characterized by:

– m rows, i with $i = 1, \ldots m$
– l seats on each row, each row has the same number of seats

The company offer several lines, each line is performed by a bus.
Let W with $w = 1, \ldots W$ be the number of lines of the bus company. Each line is characterized by:

– a given number of stops denoted as $S_w + 1$ including the starting and the terminal bus stations
– legs between each two bus stations denoted as S_w

We also indicate as q_{wi} the capacity of the row i on the line w, that is the seating capacity on each row of the bus, which performs the line w.

All the products (i.e., Origin-Destination (OD) transportation service) produced by each line w, $w = 1, \ldots, W$ are stored in a sequence according to the incremental order of f and e, that is $(EF)^w = (1, 2)^w, \ldots, (1, S_w + 1)^w, (2, 3), \ldots, (S_w, S_w + 1)$. After numbering all the products in $(EF)^w$ from left to right, we can get the product sequence indexed by the serial number $p^w = (1, \ldots, |(EF)^w|)$.

In order to model the information related to legs, OD transport service and lines we introduce the binary matrix $H^w = h^w_{s p^w}, s = 1, \ldots, S^w, p^t = w \ldots, |(EF)^w|$. Each element of the matrix is equal to 1 if product p^w generated by the line w uses leg s and zero otherwise. Thus, each column of matrix H^w contains all the information related to the legs involved in the OD transportation services provided by the line w. Table 1 represents an example of the matrix H^w for a line with 28 products and $S^w = 7$ legs. We indicate with $k = 1, \ldots, K$ the group size and we assume that alternative products are available, that is different lines can deliver the same OD transportation service from e to f. In order to handle this specific situation, thus to keep track of which lines can deliver a specific product we introduce a binary parameter γ^w_p that denotes the relationship between product p and line w. In particular, $\gamma^w_p = 1$, if the transportation service p can be delivered by the line w, that is $(e, f) \in (EF)^w$, and zero otherwise. At each time period of the booking horizon, the company has to decide on accepting/denying the request of a group of customers asking for a product p, that is an OD transportation service. In addition, we assume that booking requests made for groups with k greater than one cannot be partially accepted.

The service level depends on how the members of the groups of passengers are allocated in the bus. Thus, it is defined on the basis of a closeness criterion, that could be fully, partially or not respected. In particular, it is fully respected if all the passengers of the group are seated closer one to each other; it is partially respected if only part of the group is seated close one to each other, otherwise, if all the customers' seats are scattered in the bus, the criterion is not respected. If the closeness criterion is either partially or not respected, the service level is low, and thus the revenue deriving from

Table 1. Representation of the matrix H, for a line with 28 products and 7 legs.

Products	Legs (1,2)	(2,3)	(3,4)	(4,5)	(5,6)	(6,7)	(7,8)
1,2	1	0	0	0	0	0	0
1,3	1	1	0	0	0	0	0
1,4	1	1	1	0	0	0	0
1,5	1	1	1	1	0	0	0
1,6	1	1	1	1	1	0	0
1,7	1	1	1	1	1	1	0
1,8	1	1	1	1	1	1	1
2,3	0	1	0	0	0	0	0
2,4	0	1	1	0	0	0	0
2,5	0	1	1	1	0	0	0
2,6	0	1	1	1	1	0	0
2,7	0	1	1	1	1	1	0
2,8	0	1	1	1	1	1	1
3,4	0	0	1	0	0	0	0
3,5	0	0	1	1	0	0	0
3,6	0	0	1	1	1	0	0
3,7	0	0	1	1	1	1	0
3,8	0	0	1	1	1	1	1
4,5	0	0	0	1	0	0	0
4,6	0	0	0	0	1	0	0
4,7	0	0	0	0	1	1	0
4,8	0	0	0	0	1	1	1
5,6	0	0	0	0	1	0	0
5,7	0	0	0	0	1	1	0
5,8	0	0	0	0	1	1	1
6,7	0	0	0	0	1	1	0
6,8	0	0	0	0	0	1	1
7,8	0	0	0	0	0	0	1

the passengers request decreases. In other words, it is assumed that the company incurs a penalty, when the service level is not fully satisfied.

It is worth observing that several configurations can be used to allocate different sized groups in the bus, in such a way to provide a satisfactory service level for the passengers, without reducing the revenues.

To make an example, let us assume that each row of the bus contains 4 individual seats. In this case, the feasible assignment configurations that fully respect the service level, depending on the group size (i.e., $k = 2, 3, 4$), are shown in Figs. 1–3. It is worth observing that for a single customer the service level is always fully satisfied.

In Figs. 1, 2 and 3 the orange squares represent a layout configuration, where the allocation of groups ensures a fully service level and satisfies the capacity constraint (seats availability).

To model all the possible configurations, including those in which the service level in not fully satisfied, we introduce, for each line w, a matrix B^w, whose single element, denoted by b_{ij}^w, represents the number of seats of the row i used in the configuration j. The dimension of the matrix B^w depends on both the possible group size and the number of rows composing the bus. In particular, B^w can be viewed as partitioned in K sub-matrices B_k^w, $k = 1, \ldots, K$, one for each possible group size: B_k^w contains m rows and n_k columns, where n_k represents the number of possible configurations to allocate a group of k people in a bus with m rows.

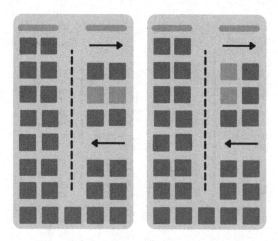

Fig. 1. Example of a possible bus layout configuration considering the allocation of a group of 2 people ensuring a fully service level. The orange squares represent the occupied seats. (Color figure online)

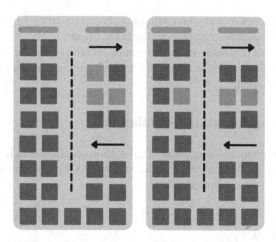

Fig. 2. Example of possible bus layout configuration considering the allocation of a group of 3 people ensuring a fully service level. The orange squares represent the occupied seats. (Color figure online)

Table 2. Representation of the matrix B^w with $k = 1, \ldots, 4$ and $m = 4$.

k = 1	k = 2	k = 3	k = 4
1 0 0 0	2 0 0 0 1 0 0 1 0 1	3 0 0 0 2 0 0 1 0 0 1 0 2 0 2 1 0 1 1 1	4 0 0 0 2 0 0 3 0 0 1 0 0 2 0 1 0 1 0 1 2 0 2 3 0 3 1 0 1 2 2 1 1 1 1
0 1 0 0	2 0 0 1 1 0 0 1 0	0 3 0 0 1 2 0 2 1 0 1 1 0 2 0 0 1 0 0 1	0 4 0 0 2 2 0 1 3 0 3 1 0 1 2 2 1 1 1 1 0 2 0 0 3 0 0 1 0 0 1 0 0 2 1
0 0 1 0	0 0 2 0 0 1 1 1 0 0	0 0 3 0 0 1 2 0 2 1 1 1 1 0 0 2 0 0 1 0	0 0 4 0 0 2 2 0 1 3 0 3 1 1 1 1 2 2 1 1 2 0 0 1 0 0 3 0 0 1 0 2 1 0 0
0 0 0 1	0 0 0 2 0 0 1 0 1 1	0 0 0 3 0 0 1 0 0 2 0 1 0 1 1 0 2 2 1 1	0 0 0 4 0 0 2 0 0 1 0 0 3 0 1 0 1 0 2 1 0 2 2 0 1 1 0 3 3 1 1 1 2 1 2

Table 2 gives a representation of the matrix B^w, when $k = 1, \ldots, 4$ and the buses contain $m = 4$ rows. In this specific scenario, B^w is partitioned in $k = 4$ sub-matrices, one for each value of the group size: $B_1^w \in N^{4x4}$, $B_2^w \in N^{4x10}$, $B_3^w \in N^{4x20}$, $B_4^w \in$

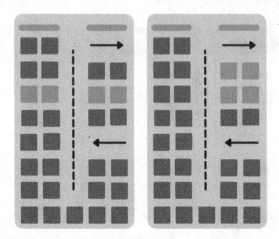

Fig. 3. Example of possible bus layout configuration considering the allocation of a group of 4 people ensuring a fully service level. The orange squares represent the occupied seats. (Color figure online)

N^{4x35}. Considering a bus of real size, commonly characterized by 13 rows and 52 seats, the total number of columns that compose the matrix B are 2271.

The main goal is to maximize the total revenue obtained from the accepted requests of a group of size k for the product p on the booking horizon, trying to fully satisfy, if possible, the level of service.

3   Dynamic Programming Formulation

This section is devoted to the description of a dynamic programming formulation of the problem under study. To this aim, we introduce the matrix $X(t)$ to represent the capacity of the system at the time t, reported below:

$$X(t) = \begin{pmatrix} x_{11}^1 & \cdots & x_{11}^W & \cdots & x_{s^11}^1 & \cdots & x_{s^w1}^w & \cdots & x_{sW_1}^W \\ \vdots & \ddots & \vdots & \ddots & \vdots & \ddots & \vdots & \ddots & \vdots \\ x_{1m}^1 & \cdots & x_{1m}^W & \cdots & x_{s^1m}^1 & \cdots & x_{s^wm}^w & \cdots & x_{sW_m}^W \end{pmatrix}$$

The matrix $X(t)$ has been defined by taking into account that the capacity of the system depends on the number of available seats for each row $i = 1, \ldots, m$ for each line $w = 1, \ldots, W$. In particular, the generic column of the matrix $X(t)$ represents the seats available for each row i of the line w for a specific leg s.

We denote as:

- λ_p^{tk} the probability that at the time t arrives a booking request for the product p for a group of k people;

- λ_0^{tk} the probability that no request arrives at the time t. Since it is assumed that at most one booking request could arrive for each time period, the following condition is satisfied:

$$\sum_{k=1}^{K} \lambda_0^{tk} + \sum_{k=1}^{K} \sum_{t=1}^{T} \sum_{p=1}^{|(EF)|} \lambda_p^{tk} = 1;$$

- μ_{pj}^{tkw} a boolean variable which assumes a value equal to 1 if a booking request for a product p for a group of size k is satisfied at the time t with the line w and the configuration j; a value equal to 0 otherwise.

- R_p^k the revenue obtained if a request for a product p for a group of size k is satisfied.

- Ω_j a parameter used to represent the penalty related to the service level. In particular, $\Omega_j=1$ if the service level is fully respected (people belonging to the same group are seated near each other) and $\Omega_j=0.5$ when the service level is partially or not respected.

- $V_t(X)$ the maximum expected revenue obtainable from periods $t, t+1, \ldots, T$ given that, at time t, the system capacity is $X(t)$.

The dynamic model can be represented mathematically as follows:

$$V_t(X) = \sum_{k=1}^{K} \sum_{p=1}^{|(IJ)|} \sum_{j=(n_{k-1}+1),\ldots,n_k} \lambda_p^{tk} \max_{\substack{\mu_{pj}^{tkw} \in \{0,1\} \\ \{w: \gamma_p^w=1\}}} \left[\Omega_j R_p^k \mu_{pj}^{tkw} + V_{t+1}(\tilde{X}) \right] + \sum_{k=1}^{K} \lambda_0^{tk} V_{t+1}(X)$$

with boundary conditions:

$$V_t(0) = 0, \ \forall t;$$
$$V_{T+1}(X) = 0, \ \text{if } x_{si}^w \geq 0 \text{ for all } s = 1, \ldots, S^w, \text{ for all } i, \text{ for all } k$$
$$V_t(X) = -\infty, \ \text{if } x_{si}^w < 0 \text{ for some } s \in S^w, k \text{ and } i$$

where $\tilde{X}$ represents the updated system capacity, related to the following event: at time t a request for a product p for a group of size k occurs. The passenger transport company can accept or deny the current request.

If the request is accepted by using product p^w of the line w and the configuration j we need to update the leg capacities by decreasing the seats availability on all the legs involved in p^w. In other words, we subtract from the columns of the matrix X that correspond to the legs related to the product p^w the column B_j, that is the j-th column of the matrix B, as many times as the number of legs.

Due to the curse of dimensionality, the proposed model cannot be solved to the optimality, thus a linear approximation is developed and used to define a booking limit revenue management policy.

4   Linear Programming Approximation

Starting from the dynamic programming problem, in the linear programming approximation we replace stochastic quantities by their mean values and we assume that capacity and demand are continuous. A summary of the parameters and the variable of the model is presented in what follows.

The parameters are:

- d is the random cumulative future demand at time t, and $\bar{d}$ its mean. In particular, d_p^k is the aggregate number of requests for the product p for groups of size k;

- R_p^k is the revenue for the product p for a group of size k;

- q_{wi} is the capacity of the row i on the line w (seats available on the row i of the buses belonging to the line w;

- Ω_j is the parameter of penalty related to the service level;

- $\gamma_p^w, p = 1, \ldots, |(EF)|$, is equal to 1 if the product p can be delivered by using the line w and zero otherwise;

- $h_{sp^w}^w, s = 1, \ldots, S^w, p^w = 1 \ldots, |(EF)^w|$ is equal to 1 if the leg s is used in the product p^w and zero otherwise. $h_{sp^w}^w$ is an element of the matrix H introduced above;

- b_{ij}^w represents the number of seats used to allocate, on the row i of the line w, a group of k people with $j = 1, \ldots, n_K$.
  b_{ij}^w is an element of the matrix B introduced above.

The variable is:

- y_j^{pw} integer variable representing the number of satisfied requests for the product p by the line w with the configuration j.

The linear programming model is:

$$Max \sum_{p=1}^{|(EF)|} \sum_{w \in W} \sum_{k \in K} \sum_{j=(n_{k-1}+1),\ldots,n_k} \gamma_p^w R_p^k \Omega_j y_j^{pw} \tag{1}$$

$$\sum_{w=1}^{W} \sum_{j=(n_{k-1}+1),\ldots,n_k} \gamma_p^w y_j^{pw} \leq d_p^k \quad p = 1, \ldots, |(EF)| \quad, k = 1, \ldots, K \tag{2}$$

$$\sum_{p^w=1}^{|(EF)^w|} \sum_{j=1,\ldots,n_K} \sum_{p=1}^{|(EF)|} h_{sp^w}^w \gamma_p^w b_{ij}^w y_j^{pw} \leq q_{wi} \quad w = 1, \ldots, W, s = 1, \ldots, S^w \tag{3}$$

$$i = 1, \ldots, m$$

$$y_j^{pw} \geq 0, \text{integer} \quad p = 1, \ldots, |(EF)|, w = 1, \ldots, W, j = 1, \ldots, n_K \tag{4}$$

The objective function 1 represents the total revenue deriving from the accepted requests according to the service level. Equations 2 represent the demand constraints and state that the demand for a product p (transportation service) for a group of size k can be satisfied with all the products generated by all the lines $w, w = 1, \ldots, W$, that can deliver the considered transportation service. Equations 3 represent the capacity constraints and they allow the right allocation of capacity on the rows of the buses. Finally, constraints 4 define the variables domain. Since in the objective function and in the capacity and demand constraints the summation on j depends on the group size k, it worth to notice that for $k = 0$ the value of $n_0=0$.

5 Booking Limit Policy

To solve the proposed problem, we can use the primal variables obtained from the solution of the linear approximation of the dynamic model to apply a booking limit control policy. Tajima and Misono [13] divide the vehicle into zones and define the booking limit policy as a control that limits the zones capacity, i.e., the amount of seats that can be sold to a particular class at a given time. In this case the booking limit policy is partitioned, since the available capacity is divided in some defined separated blocks and the capacity of each blocks can be used only for the requests belonging to a specific class.

In our case, since in the bus transportation there are not classes, we do not set a booking limit for a particular class but we compute a booking limit for a specific request. In particular, a value of booking limit indicates how many requests can be accepted for a specific product p, for a specific configuration j and thus for a specific group size k, on a particular line w.

In other words, the optimal solutions y_j^{*pw} give partitioned booking limits that correspond to a fixed amount of capacity of each resource, that can be used to satisfy requests for a particular product offered by the bus company. The demand for each product has access only to its allocated capacity and no other product may use this capacity.

The booking limit policy follows the scheme reported in Algorithm 1.

Algorithm 1. Booking Limit Policy.

Solve the linear programming model and let y_j^{*pw} denote its optimal solutions;

FOR each $Request_{pkt}$, request for the product p for a group of size k arrived at the time t

    IF $y_j^{*pw} > 0$ and there is enough capacity $q_{wi} - b_{ij}^w \geq 0$

    THEN Accept the request;

        SET $y_j^{*pw} = y_j^{*pw} - 1$;

        UPDATE the capacity: $q_{wi} = q_{wi} - b_{ij}^w$

           UPDATE the revenue: $RevenuePolicy = RevenuePolicy + R_p^k$;

    ELSE

        Deny the request

    END IF

END FOR

Applying this policy we solve the linear programming problem only once, to determine the values of booking limit and then we process the requests according to their ascending arrival order.

6 Numerical Results

In this section we analyse the numerical results obtained by testing all the proposed strategies. All the experiment are carried out using AIMMS 4.85.2.4, with Cplex 22.1 as solver, on a Intel(R) Core(TM) i7-10610U CPU 1.80 gigahertz 16,0 gigabytes of RAM PC, under Windows 10 Pro operating system.

In the experimental phase we want to assess the solution quality of the policy and also its scalability.

Considering the real features of the buses, each instance is characterized by a certain number of lines, with a total capacity of 52 seats. In particular, the layout of the bus operating a certain line is composed of 13 rows, each of which has 4 seats. Every line is composed of a given set of origin-destination products. A different fare is associated to each OD product for a group of size k. We consider 4 possible types of groups, depending on the size, in fact, a group can be composed from a minimum of one person (a single passenger) to a maximum of 4 people.

The values of demand, arrival time an revenues are randomly generated for each test problem. In particular, demand is generated within the interval [0,10] and the arrival time is generated with a Poisson process. In addition, even if the values of revenue are randomly generated they depend on the group size. For example, the revenue associated to a group of size 2 for a specific product p is always less than the revenue associated to the same product for a group composed by 3 people. The considered instances have different sizes, in terms of number of lines, legs and products. Table 3 reports the characteristics of each considered instance, the first column report the name, the second one the number of lines, the third column the number of legs and the fourth one the number of products.

Table 3. Test Problems.

Test	Lines	Legs	Products
T_3_1	3	3	3
T_3_2	3	6	6
T_3_3	3	6	10
T_3_4	3	10	10
T_3_5	3	15	10
T_3_6	3	15	15
T_3_7	3	20	15
T_4_1	4	8	4
T_4_2	4	10	10
T_4_3	4	10	15
T_8_1	8	5	5
T_8_2	8	10	10
T_10_1	10	5	10

In the computational study we compare the results obtained using three different strategies. Firstly, we solve the proposed model supposing to have a "perfect knowledge", thus considering to know exactly all the values of the demands. We use these values in the demand constraints.

The second strategy consists in solving the problem applying the booking limit policy defined in Sect. 5.

The third strategy refers to the solution obtained applying a first came first served (FCFS) policy (also known as first input first output FIFO), in which, the demand for a product p for a group of size k is allocated, if there is enough capacity, simply following the arrival order. Table 4 summarizes the computational results obtained using the three strategies. In particular, for each instance, it provides the value of revenue obtained for each strategy.

Table 4. Computational Results on randomly generated instances.

Test	Revenue Perfect Knowledge	Revenue Booking Limit	Revenue FCFS
T_3_1	4081	3972	3925
T_3_2	5438	5268	4880
T_3_3	5542	5542	3875
T_3_4	5730	5730	4803
T_3_5	5816	5816	4080
T_3_6	5618	4415	4289
T_3_7	5816	4346	4280
T_4_1	6801	6282	6462
T_4_2	7375	7375	5394
T_4_3	7592	5625	5692
T_8_1	4810	4810	4810
T_8_2	13244	13244	11676
T_10_1	14624	14624	14078

From the results reported in the Table 4 it easy to notice that for every test problem the revenue obtained considering the perfect knowledge strategy is always higher than the values of revenue for the booking limit and the FCFS strategies. This is an expected result, since in the perfect knowledge we exactly know the values of the demands, thus, the maximum value of objective function is reached. Thus, the value of the revenue obtained with the perfect knowledge represents an upper bound for the other policies.

It is worth to notice that, albeit in few cases, it is possible that the revenue obtained with the FCFS overcomes the revenue obtained with the booking limit policy.

In particular, this happens for the T_4_1 and for the T_4_3 but, however, the booking limit policy performs on average 12,97% better in terms of revenue than the FCFS. For the instance T_8_1, instead, the revenue is the same for each of the three cases.

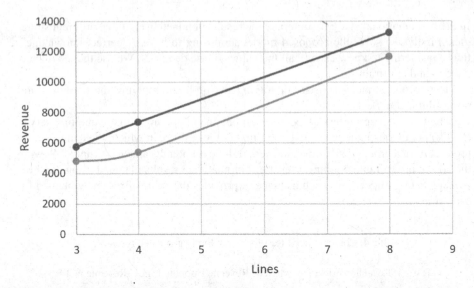

Fig. 4. Revenue as a function of the number of lines, with fixed number of products and legs: Orange line: Revenue obtained applying the booking limit policy for instances with 10 legs, and 10 products; Grey line: Revenue obtained applying the FCFS policy for instances with 10 legs, and 10 products. (Color figure online)

To summarize, for these problem tests the revenues obtained with the booking limit policy exceed the revenues obtained with the FCFS in the 76,92 % of the considered cases. The revenues obtained with the booking limit policy equal the revenues obtained with the perfect knowledge in the 53,84% of the considered cases.

In addition, looking at Fig. 4, that graphically represents the obtained revenues as a function of the number of lines, fixed number of products and legs, we can see that for the instances with 10 legs and 10 products the revenue obtained with the booking limit policy not only overcomes the revenue obtained with the FCFS but also equals the revenue derived from the resolution with the perfect knowledge. The execution times to solve the model and to apply the booking limit policy are in the orders of minutes and increase with the increasing of the instance size.

Thus, we can state that the application of the booking limit policy is an effective and efficient way to manage the allocation of group of passengers maximizing revenues.

7   Conclusions

In this paper we have presented an extension of the Bus Group Assignment Problem under Covid-19 social distancing proposed by [12].

We have readjusted and changed some of the features of the previous work, strongly related to the pandemic time, and we have been more focused on the customers' satisfaction.

In particular, we supposed that the closer the members belonging to the same group are allocated, the higher the level of service offered to the customers. We have mathematically modeled the problem, then, we have proposed a booking limit policy to solve the it.

In the computational tests we compare the performance of the booking limit policy with a first come first served strategy, and a situation of perfect knowledge. The results show that the booking limits performs better than the first come first served strategy on average.

Moreover, in many cases the revenue obtained with the booking limit policy equals the revenue obtained solving the model under perfect knowledge. This confirms that the application of the booking limit policy is an effective and efficient way to solve the proposed problem.

References

1. Yazdani, D., et al.: Real-time seat allocation for minimizing boarding/alighting time and improving quality of service and safety for passengers. Transp. Res. Part C: Emerg. Technol. **103**, 158–173 (2019). https://doi.org/10.1016/j.trc.2019.03.014. ISSN: 0968–090X
2. Sumalee, A., Tan, Z., Lam, W.H.: Dynamic stochastic transit assignment with explicit seat allocation model. Transp. Res. Part B: Methodol. **43**(8–9), 895–912 (2009). https://doi.org/10.1016/j.trb.2009.02.009. ISSN: 0191–2615
3. Yan, Z., Li, X., Zhang, Q., Han, B.: Seat allocation model for high-speed railway passenger transportation based on flexible train composition. Comput. Ind. Eng. **142**, 106383 (2020). https://doi.org/10.1016/j.cie.2020.106383. ISSN: 0360-8352
4. Milne, R.J., Salari, M.: Optimization of assigning passengers to seats on airplanes based on their carry-on luggage. J. Air Transp. Manage. **54**, 104–110 (2016). https://doi.org/10.1016/j.jairtraman.2016.03.022. ISSN: 0969–6997
5. Notomista, G., Selvaggio, M., Sbrizzi, F., Di Maio, G., Grazioso, S., Botsch, M.: A fast airplane boarding strategy using online seat assignment based on passenger classification. J. Air Transp. Manage. **53**, 140–149 (2016) https://doi.org/10.1016/j.jairtraman.2016.02.012. ISSN: 0969–6997
6. Ko, Y.D., Kwag, S.I., Oh, Y.: An efficient airline seat reallocation algorithm considering customer dissatisfaction. J. Air Transp. Manage. **85**, 101792 (2020). https://doi.org/10.1016/j.jairtraman.2020.101792. ISSN: 0969-6997
7. Akira, T., Misono, S.: Using a set packing formulation to solve airline allocation/reallocation problems. J. Oper. Res. Soc. Jpn **42**, 32–44 (1999)
8. Song, Y.S., Lee, H.Y., Yoon, M.G.: A seat allocation problem for package tour groups in airlines. Korean Manage. Sci. Rev. **25**(1), 93–106 (2008). ISSN: 1225–1100
9. Yoon, M.G., Song, Y.S., Lee, H.Y.: An integrated seat allocation model for individual and group demand in airlines. In: The 40th International Conference on Computers & Industrial Engineering, pp. 1–6 (2010). https://doi.org/10.1109/ICCIE.2010.5668417
10. Kwag, S., Lee, W.J., Ko, Y.D.: Optimal seat allocation strategy for e-sports gaming center. Int. Trans. Oper. Res. **29**(2), 783–804 (2022). https://doi.org/10.1111/itor.12809
11. Sang-Un, L.: Min-cut algorithm for arrangement problem of the seats in wedding hall. J. Inst. Internet Broadcast. Commun. **19**(1), 253–259 (2019)
12. Guerriero, F., Luzzi, M., Macrina, G.: Management of groups of passengers on buses considering the restrictions of COVID-19. In: ICORES, pp. 67–76 (2022)
13. Talluri, K.T., Van Ryzin, G.J.: The Theory and Practice of Revenue Management, vol. 1. Kluwer Academic Publishers, Boston (2004)

Reinforcement Learning Algorithms: Categorization and Structural Properties

Kenneth Schröder, Alexander Kastius, and Rainer Schlosser[✉]

Hasso Plattner Institute, University of Potsdam, Potsdam, Germany
`rainer.schlosser@hpi.de`

Abstract. Over the last years, the field of artificial intelligence (AI) has continuously evolved to great success. As a subset of AI, Reinforcement Learning (RL) has gained significant popularity as well and a variety of RL algorithms and extensions have been developed for various use cases. Although RL is applicable to a wide range of problems today, the amount of options is overwhelming and identifying the advantages and disadvantages of methods for selecting the most suitable algorithms is difficult. Sources use conflicting terminology, imply improvements to alternative algorithms without mathematical or empirical proof, or provide incomplete information. As a result, there is the chance for engineers and researchers to miss alternatives or perfect-fit algorithms for their specific problems. In this paper, we identify and explain essential properties of RL problems and algorithms. Our discussion of these concepts can be used to select, optimize, and compare RL algorithms and their extensions with respect to particular problems, as well as reason about their performance.

Keywords: Reinforcement learning · Decision support · Markov decision Processes · Conceptual comparison

1 Introduction

Most of the RL algorithms to solve Markov decision process (MDP) problems that were proposed in recent years include at least one unique feature designed to improve performance under specific challenges. These challenges usually originate from specific problem properties or from the statistical drawbacks of related optimization problems. The corresponding publications often include performance comparisons to the closely-related algorithms with respect to particularly challenging problems but mostly lack an extensive overview of the advantages and disadvantages of their solution within the broader field of research.

In addition, research articles often fail to include categorizations of their solutions concerning other RL properties and capabilities.

In [26] for example, the authors motivate in their introduction that robustness of RL algorithms with respect to hyperparameters is desirable. The evaluation section of their publication, however, does not include comparisons to other algorithms in that regard.

Introductory material often only categorizes algorithms sparsely or within the main RL families, as shown in Fig. 1, cf. [1], instead of consistently outlining and explaining the differences for various properties.

© The Author(s), under exclusive license to Springer Nature Switzerland AG 2024
F. Liberatore et al. (Eds.): ICORES 2022/2023, CCIS 1985, pp. 96–120, 2024.
https://doi.org/10.1007/978-3-031-49662-2_6

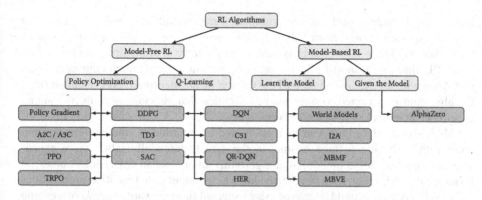

Fig. 1. Overview of modern RL algorithms as displayed in the OpenAI Spinning Up documentation. Figure adapted from [1].

This paper is an extended version of [24] in which we discussed properties of different state-of-the-art RL approaches and outlined respective implications for capabilities of RL algorithms, their variants, extensions, and families.

In this paper, we make the following contributions: First, we extend our analysis and conceptual comparison by discussing further related articles. Second, we added sections comparing discrete time vs. continuous time problems (Sect. 2.6) as well as discussing the bias-variance tradeoff (Sect. 3.3). Third, throughout the paper, we provide extended discussions and explanations. We also added several formulas and figures to better describe specifics of the RL approaches used.

This paper is organized as follows. Section 2 introduces properties which are influenced by the process definition. The following Sect. 3 discusses properties, which can be chosen freely in theory but highly influence learning dynamics. In Sect. 4, we summarize key insights and infer recommendations regarding the choice of RL algorithms for given problem characteristics. Section 5 concludes the paper.

2   Classes of MDP Problems and Suitable RL Algorithms

In this section, we distinguish different classes and properties of MDP problems. These properties can be seen as algorithm properties which cannot be chosen freely, but are predefined by the Markov process which is considered. Further, so-called value estimation and policy optimization methods are discussed.

2.1   Model-Free and Model-Based RL

When categorizing RL algorithms typically model-free and model-based algorithms are distinguished. Fig. 1 shows a categorization tree by [1] that uses this property as the first differentiator. In contrast to standard machine learning terminology, the term "model" in model-free or model-based does not refer to the trainable algorithm but to a representation of the environment with information beyond the observations. A model like this can include reward function parameters, state transition probabilities, and even optimal expected future return values for the environment at *training time*. Notably, this information does not necessarily have to be available at evaluation time for model-based RL algorithms. In self-play for example, the environment model during training is defined with a clone of the so far best version of the algorithm. At evaluation time, the algorithm is expected to have learned a policy that generalizes from its training experience and that lets it beat other opponents.

Model-based algorithms use such additional information to simulate the environment without having to take actions in the actual, potentially much more expensive environment. Some model-based algorithms do not use an environment at all. In Dynamic Programming, for example, the knowledge about state transition probabilities and rewards is used to build a table of exact expected future rewards instead of stepping through an environment and making observations. Environment models can be learned or already known through the problem definition. Another popular model-based algorithm with a given model is AlphaZero by DeepMind [28]. According to [1] model-based algorithms can improve sample efficiency but tend to be harder to implement and tune.

2.2 Policy Optimization and Value Learning

In the following, we discuss two categories of RL algorithms: policy learning and value learning methods. Algorithms like DDPG, SAC, and REINFORCE are policy learning methods, which use a parametric mapping from states to actions. The parameters are adjusted to maximize the expected discounted reward of the policy when taking an action produced by the mapping for a specific state [29].

Many of those methods also incorporate value learning, which is another option to define policies. For value learning, the goal is to develop an estimation of the expected discounted reward given a certain action is taken in a certain state. This value is called the Q-value of the state-action pair, which constitutes the foundation for Q-learning as a large group of algorithms.

The choice of policy learning versus value learning is partly defined by the process, because the majority of algorithms from the value learning category do not work in continuous action spaces, see Sect. 2.5. In other cases, both might be applied. [29] suggests, that policy learning methods might outperform value learning methods in those usecases, where the policy is easier to represent than the value function of the process.

2.3 Finite and Infinite Horizon Problems

MDP problems can have a finite or an infinite time horizon. The applicability of RL algorithms strongly depends on whether a problem has a finite or infinite horizon, as this property has implications on objective functions and learning strategies. The decisive factor behind all those implications is whether an algorithm follows a Monte Carlo (MC) or Temporal Difference (TD) approach [29]. This section introduces Monte Carlo and Temporal Difference methods. Additionally, the core ideas behind TD(n), TD(λ), and eligibility traces are examined.

The main properties of Monte Carlo methods are that they are *on-policy* by design and only work on finite horizon problems. The *current policy* plays the environment until a terminating state is reached. Then, the realized discounted returns for the full episode are calculated. Notably, the possible stochastic nature of the environment and the policy can cause drastically different realized discounted returns for the same state in-between trajectories, even under the same policy. The described Monte Carlo returns are used by some Policy Gradient variants, like REINFORCE[1].

In contrast to Monte Carlo methods, Temporal Difference (TD) approaches do not need complete trajectories or terminating states to be applied. Therefore, they can be used on infinite horizon problems as well. The core idea of TD methods is to only use a section of a trajectory between the states s_t and s_{t+n} to define target values for a value estimator $\hat{V}^\phi$. The target values can be calculated by summing the realized rewards for the observed trajectory section and adding an approximation of the expected future return of the last state s_{t+n}, given by the current version of the value-learner. Formally, the target value is calculated by:

$$G^\phi_{t:t+n} := \sum_{t'=t}^{t+n-1} \gamma^{t'-t} \mathrm{r}\left(s_{t'+1}|s_{t'}, a_{t'}\right) + \gamma^n \hat{V}_\phi(s_{t+n})$$

$$= G_{t:T} - G_{t+n:T} + \gamma^n \hat{V}_\phi(s_{t+n})$$

and the *temporal difference*, also called approximation error is given by $G^\phi_{t:t+n} - \hat{V}_\phi(s_t)$ and is used to update $\hat{V}_\phi(s_t)$.

Using the current version of a value-estimator like this for an update of the same estimator is called *bootstrapping* in RL. Algorithms that use the temporal difference over n steps are called n-step TD methods or TD(n) algorithms. Notably, TD(n) for all $n > 1$ is automatically on-policy learning, as the sequence of rewards used for the network updates depends on the policy. Therefore, the values learned by the value-learner reflect the values of states under the current policy (see Sect. 3.1). TD(1) algorithms can be trained in an off-policy or on-policy fashion, as single actions and rewards are policy-independent. The off-policy updates would be unbiased towards the current policy. DQN is an example of an off-policy TD(1) algorithm, while SARSA is an on-policy TD(1) method.

Some RL libraries like Tianshou [32] include off-policy DQN algorithms with an n-step parameter. RAINBOW uses similar updates. Different tricks are used to make this possible. RAINBOW, for example, omits off-policy exploration, see Sect. 3.2. Figure 2 categorizes methods like DP, MC, and TD according to the depth and the width of the updates calculated from the state transition tree.

Monte Carlo methods have a high variance and no bias in their network updates. This is because they look at complete trajectories, with potentially drastic differences between each other. Still, since the trajectories are complete and include no estimations, they fully reflect actual results from the environment. Temporal Difference approaches

---

[1] [25] present an overview over PG variants. They use terms like $\sum_{t=0}^{\infty} r_t$ to indicate general correctness for infinite horizon problems. Still, in practice, those sums reflect the Monte Carlo PG variants, which operate on finite horizons and break down to finite sums.

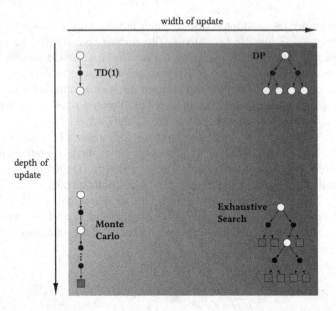

Fig. 2. Categorization of algorithms based on the depth and width of the implemented value updates in the state transition tree. Figure adapted from [29].

have a low variance in their network updates, as they only consider a few consecutive realized rewards. However, a significant bias is added towards their initial estimations of the bootstrapping term. Increasing the number of steps in TD(n) makes the updates more similar to MC updates, and bias decreases while variance increases.

Because of the bias-variance tradeoff and differences in episode lengths, for example, the best n in TD(n) is highly problem-specific. An approach called TD(λ) tries to combine the advantages of all TD(n) versions by creating a target value $G_t^{\lambda,\phi}$ that is a combination of all possible TD(n) targets. TD(λ) can be explained and calculated from a forward or a backward view, but both variants are mathematically equivalent [29].

The forward view combines all TD(n) targets $G_{t:t+n}^{\phi}$ using an exponentially weighted moving average over all possible TD(n)'s and is defined by: $G_t^{\lambda,\phi} = (1-\lambda) \cdot \sum_{n=1}^{T-t-1} \lambda^{n-1} G_{t:t+n}^{\phi} + \lambda^{T-t-1} G_{t:T}$ for the finite horizon case. For infinite horizon problems, the definition includes an infinite sum, which is approximated in practice using the truncated λ-return [29], where the sum is truncated after many steps. Notably, for $\lambda = 1$ this formula reduces to the Monte Carlo update $G_{t:T} = G_t^{1,\phi}$ and for $\lambda = 0$ it reduces to the TD(1) update $G_{t:t+1} = G_t^{0,\phi}$ for $0^0 = 1$. This forward version of TD(λ) can be seen as a Monte Carlo method because it needs to wait until the end of an episode before the updates can be computed.

The backward view was introduced to overcome this limitation. Instead of looking *forward* on the trajectory to compute the network updates, it calculates errors locally using TD(1). It then passes the results *back* to the previously visited states of the trajectory. This means that if a high estimation error is discovered at step t of a trajectory, the state-values $\hat{V}_\phi(s_{t'})$ of the previous states $s_{t'}$ with $t' < t$ are adjusted in the same direction as the local error. Notably, not all of the states visited previously are similarly responsible for the change in the value of state s_t. One way to approach this so called *credit assignment problem* [19] is to assume that the most recent decisions have the most substantial influence on achieved rewards and actions, that have been taken far in the past, are ineligible for present success. Similarly, the backward version of TD(λ) uses *eligibility traces* to quantify the eligibility of past states for discovered state-value errors. In practice, such an eligibility trace is a vector e containing decaying factors for each previously visited state of the trajectory. Different decay strategies can be applied, in which eligibility values of revisited states are increased differently. Updates in backward TD(λ) for every previous state $s \in S^{\tau_p}$ of a partial trajectory τ_p are computed as in:

$$\forall s \in S^{\tau_p} : \hat{V}_\phi(s) \leftarrow \hat{V}_\phi(s) + \alpha \delta_t e_t(s). \tag{1}$$

The learning rate is defined as α, and λ generally denotes the eligibility decay factor [27].

With backward TD(λ), it is unnecessary to wait until an episode ends. Instead, updates can be performed online, i. e. while exploring the environment. Similar to TD(n) with $n > 1$, TD(λ) is also an on-policy approach and is theoretically incompatible with off-policy techniques like replay buffers.

2.4   Discrete and Continuous State Spaces

Besides discrete state spaces also uncountable state sets, where at least one state parameter lives in a continuous space, are found in many real-world applications.

All of the table-based predecessors of RL, like Dynamic Programming or tabular Q-Learning, fail under uncountable state sets, as an infinite amount of memory and time would be necessary to compute all state-values [29]. Algorithms like this require discretization of the state space, limiting the exactness of the solution.

RL algorithms using neural networks as regression algorithms can take continuous values of states as inputs and have the ability to generalize over uncountable state sets after learning from a finite number of training samples.

2.5   Discrete and Continuous Action Sets

Similar to uncountable state space components, uncountable action sets are standard in continuous control tasks, where actions can be chosen from a range of continuous numbers.

Table-based algorithms like Q-Learning that calculate state-action values cannot be applied to problems with uncountable action sets, as an infinite amount of memory and time would be required for calculations across the whole table [29].

On-policy (see Sect. 3.1) RL algorithms can be trained under uncountable action sets by learning the parameters to a parameterized distribution over a range of actions. Training *off-policy* RL algorithms under uncountable action sets introduces new challenges. In regular off-policy RL, the policy network is trained to reflect a discrete probability distribution over the *countable* Q-values of each state. In the case of uncountable action sets, the Q-value distribution over the uncountable action set is neither fully accessible nor differentiable if the learner was trained to output single values to state-action input pairs. Generally, constructing the value-learner under uncountable action sets to output parameters to parameterized distributions is impossible, as the action values follow an unknown, possibly non-differentiable distribution. Training with arbitrary distributions would result in inaccurate state-value estimates[2]. Because of these limitations, when facing a problem with uncountable action sets, deterministic policies must be trained in an off-policy manner. Working with continuous actions, however, allows the exploitation of other properties.

DDPG for example, takes advantage of the differentiability of continuous actions output by deterministic policies [15]. The objective function of DDPG is defined as:

$$J_\pi(\theta) = E_{s \sim D}\left[\hat{Q}^\phi\left(\mu_\theta(s)|s\right)\right].$$

The policy parameters θ are trained to choose the Q-value-maximizing actions for the state $s \sim D$ of the state distribution D of the off-policy data. Gradients of the policy network parameters are backpropagated through the Q-learner $\hat{Q}^\phi$, through the continuous action policy output $\mu_\theta(s)$ and into the policy network. Gradient calculations like this would not be possible for discrete actions.

Some algorithms take one further step further and make it possible to learn *stochastic* policies under uncountable action sets. One example of those algorithms is SAC [8], which introduces some beneficial additions which improve learning performance when compared to DDPG. The policy objective of SAC is to minimize the expected KL-divergence between the distribution output by the policy and the distribution defined by the Q-values.

The objective of the SAC policy is given by:

$$\begin{aligned}
J_\pi(\theta) &= E_{s \sim D}\left[D_{KL}\left(\pi_\theta\left(\cdot|s\right) \middle|\middle| \frac{e^{\hat{Q}^\phi_{\pi_\theta}(\cdot|s)}}{Z_\phi(s)}\right)\right] \\
&= E_{s \sim D}\left[D_{KL}\left(\pi_\theta\left(\cdot|s\right)\middle|\middle| e^{\hat{Q}^\phi_{\pi_\theta}(\cdot|s) - \log Z_\phi(s)}\right)\right] \\
&= E_{s \sim D, a \sim \pi_\theta}\left[-\log\left(\frac{e^{\hat{Q}^\phi_{\pi_\theta}(a|s) - \log Z_\phi(s)}}{\pi_\theta(a|s)}\right)\right] \\
&= E_{s \sim D, a \sim \pi_\theta}\left[\log \pi_\theta(a|s) - \hat{Q}^\phi_{\pi_\theta}(a|s) + \log Z_\phi(s)\right].
\end{aligned} \tag{2}$$

It contains a KL-divergence term, initially proposed by [8], that aims at minimizing the difference between the distribution of the policy π_θ in a state s and the distribution implied by the Q-value estimations $\hat{Q}^\phi_{\pi_\theta}(\cdot|s)$. $Z_\phi(s)$ is a normalization term that does not contribute to the gradients.

The objective function can be rewritten as shown in the following lines of (2). As the Z-term does not affect the gradients of θ, it can be removed, and by additionally

---

[2] Notably, in policy optimization for uncountable action sets, the learned distribution does not have to accurately reflect the distribution of the true Q-values, as they are not used to estimate state-values.

negating the other terms, this results in the equivalent maximization objective:

$$J'_\pi(\theta) = E_{s\sim D, a\sim\pi_\theta} \left[\hat{Q}^\phi_{\pi_\theta}(a|s) - \log\pi_\theta(a|s) \right]$$
$$= E_{s\sim D, a\sim\pi_\theta} \left[\hat{Q}^\phi_{\pi_\theta}(a|s) \right] - E_{s\sim D, a\sim\pi_\theta} \left[\log\pi_\theta(a|s) \right]. \qquad (3)$$

These equations demonstrate, that SAC's policy is trained to maximize both the expected Q-values, as well as its entropy given by

$$\mathcal{H}(\pi_\theta(\cdot|s)) = -E_{a\sim\pi_\theta}[\log\pi_\theta(a|s)].$$

Notably, although the Q-values in SAC already contain an entropy coefficient, optimizing the policy to maximize the Q-values would not guarantee that the policy itself has high entropy. The policy might choose the Q-maximizing actions with 99% probability density, i.e. a very low entropy, which is why this entropy term is also necessary for the policy objective.

A problem with the formulation in (3) is the expectation over the actions $E_{s\sim D, a\sim\pi_\theta}$, which depends on the policy parameters θ, cf. [1]. Notably, with countable action sets, this is no problem at all. The expected value of the Q-values in a state s can be calculated precisely using the finite sum $\sum_{a\in A}\pi(a|s)\hat{Q}^\phi_{\pi_\theta}(a|s)$. With uncountable action sets, however, there is usually no way to calculate the expectation precisely. It is approximated using samples instead. This leads to a high variance in the resulting gradients, which can be avoided by a reformulation of the problem using the so-called reparameterization trick, which externalizes the source of randomness. Due to the reparameterization trick, the objective can be further rewritten, which exchanges the expectation over the actions with an expectation over a random sample from the standard normal distribution $\epsilon \sim \mathbb{N}$ as in:

$$J'_\pi(\theta) = E_{s\sim D, \epsilon\sim\mathbb{N}} \left[\hat{Q}^\phi_{\pi_\theta}(\tilde{a}|s) - \log\pi_\theta(\tilde{a}|s) \right], \qquad (4)$$

where $\tilde{a} = f(\epsilon, \pi_\theta(\cdot|s))$. The reparameterization function f uses the policy network outputs to transform this random sample from a standard normal distribution to the distribution defined by the policy network. This can be done by adding the mean and multiplying by the variance output by the policy network. The reparameterization trick separates sampling from the policy distribution, and in turn reduces the variance of the computed gradients.

2.6 Discrete and Continuous Time Problems

All of the previously presented RL algorithms operate on problems with discrete-time. This is a natural property of round-based problems like board games, most Atari games, or combinatorial problems. Other optimal control tasks are naturally based on continuous-time, like physics simulations or robot movement.

All of these tasks can be approached with discrete-time algorithms by discretizing the time dimension, and this is most often the case. Still, some research has been targeted at true continuous-time algorithms specifically. [2] introduced approaches to continuous-time RL based on the Hamilton Jacobi Bellman Equation [12]. In their and related work, there are no discrete state transitions. Instead, a *vector field* $\dot{x}$ (first-order derivative of the state function x), given by

$$\dot{x}(t|\phi) = f(x(t'|\phi), u_\phi(t)) \tag{5}$$

defines gradients of a state x, based on the current state x and the value of a continuous action function u. The state function itself is formally defined by:

$$x(t|\phi) = x(0) + \int_0^t f(x(t'|\phi), u_\phi(t')) \, dt'. \tag{6}$$

Both of these theoretical formulas contain circular dependencies (in $x(t|\phi)$). In practice, the state function is usually approximated [35] using functions like odeint of the scipy.integrate library which can integrate systems of ordinary differential equations as:

$$x(t|\phi) = x(0) + \sum_{i=1}^{n-1} \frac{t}{n} f\left(x\left(\frac{it}{n}|\phi\right), u_\phi\left(\frac{it}{n}\right)\right). \tag{7}$$

The continuous-time actions are modeled as functions of the continuous-time states and are usually denoted by $u_\phi(t|x) = f_\phi(x(t))$. Values of states can be calculated using integrals over the reward function r as in:

$$V_\phi(x(t)) = \int_t^\infty e^{-\frac{t'-t}{\gamma}} r(x(t'), u_\phi(t')) \, dt'. \tag{8}$$

The continuous-time RL versions approximate the value function using neural networks and iteratively improve their policy, i.e. the action function, as well as their value-estimates, similarly to their discrete-time counterparts. For this, they use special continuous-time TD updates [2].

3   Further Properties of RL Algorithms

Whereas the previously discussed aspects are all partly dictated by the problem under assessment, for others there might be multiple algorithms available which in turn vary only in their learning dynamics.

3.1   On-Policy and Off-Policy RL

In this section, we explore the differences between on-policy and off-policy for value-learning RL algorithms, like DQN. Afterward, importance sampling and its applications for off-policy learning in policy optimization methods are explored.

In value-learning, the difference between on-policy and off-policy is best explained by the type of state-action values learned by the algorithm. There are two main options for learning state-action values.

One option is to learn the expected state-action values *of the current policy*. In this case, in each learning step, the Q-values reflect expected returns under the current policy. This kind of policy is optimized by repeatedly adjusting the parameters slightly in a direction that minimizes the difference to a target value and reevaluating the expected returns. This is the main idea behind the Q-Learning variant SARSA. Algorithms like SARSA are called on-policy, as the current policy needs to be used to collect new experiences, and no outdated or unrelated data can be utilized. This dependency on policy-related actions is also reflected in the Q-value update function which uses the realized action in the Bellman expectation equation instead of an aggregation over all the possible actions. Notably, not all algorithms that learn values under the current policy are on-policy methods. The critics in SAC, for example, learn Q-values *of the current policy* in an off-policy way by inputting the succeeding states (s_{t+1}) of the off-policy experience into the current *policy network*. The result is a probability map over the actions that can be taken in s_{t+1}. The value of state s_{t+1} under the current policy can be estimated as the sum of the Q-values in s_{t+1} weighted by the calculated probability distribution.

The second possibility to approach value-learning is to learn the *optimal* state-action values achievable *in the environment in general* and independent of the current policy. This is done by Deep Q-Learning [20]. A deterministic policy is implicitly defined by maximizing actions, and stochastic policies can be formulated by applying temperature-regulated softmax over the state-action values. This is called off-policy learning, as state-action values like this can be learned using policy-unrelated experience from the environment. For off-policy learning any experience from the environment can be used to train the model. A human player for example, could deliberately collect a batch of experience to provide the algorithm with a non-random, initial training foundation. Still, in some cases, the data is generated using a greedy version of the current policy. Furthermore, all data points can be trained multiple times, as they are never outdated w. r. t. the state-action-value definition.

The optimization objectives of pure policy optimization methods are usually defined as maximizing the expected future return when following the *current policy*. Because of this, most pure policy optimization algorithms are naturally on-policy methods. During training, states and actions are sampled from the current policy, and probabilities or actions are adjusted according to their realized or expected future return. Notably, without any adjustments, using state samples from the stationary state distribution of the current policy is crucial. While it is technically possible to collect policy-independent experience from the environment and use the policy log probabilities of the chosen actions for policy updates, this optimization objective would maximize the expected value under the wrong state distribution. With this setting, the policy might visit different states when performing independently and achieve far-from-optimal expected future returns under its state distribution. Even if a lot of experience was available, covering all the state-action space, the optimization objective would be unsuitable.

One way to overcome this limitation and train a policy optimization algorithm using data that was collected from a different state distribution is *importance sampling* [29]. The importance sampling theorem, i.e.,

$$E_{s,a\sim\pi_\theta}\left[f\left(a|s\right)\right] = E_{s,a\sim D}\left[\frac{\pi_\theta\left(a|s\right)}{D\left(a|s\right)}f\left(a|s\right)\right],$$

states that the expected value of any deterministic function f when drawing from a distribution π_θ is equal to the expected value of f multiplied by the ratio of the two probabilities when drawing from a distribution D.

In the RL setting, expectations like this reflect the policy optimization objective and cannot be calculated precisely. Instead, the policy performance is updated based on a batch of samples that approximate this expectation. Similarly, the expectation can also be approximated using the importance sampling equation and samples from an arbitrary, known state distribution. As per the central limit theorem, the approximations with both methods follow a normal distribution around the actual expected value. Still, they can have significant differences in variance if the two distributions are very different [29].

Some Monte Carlo RL variants use importance sampling to reduce simulation and calculation overhead for environments with very long trajectories. For those cases, simulating and calculating the realized rewards for just a single network update is inefficient. Using importance sampling, the latest trajectories can be reused for multiple network updates if the policy does not change drastically in between a few updates.

One policy optimization approach that attempts to combine the benefits of off-policy and on-policy learning is called P3O [4]. In this work, the authors propose a combination of regular on-policy optimization, importance sampling off-policy optimization, and techniques to keep the target policy close to the behavior policy.

Actor-critic algorithms can be designed to operate on-policy or off-policy. The on-policy actor-critic variants use a policy optimization objective similar to pure policy optimization algorithms [29]. In this case, the policy is improved by altering the log probabilities of actions according to their learned Q-values or advantage-values. Off-policy actor-critic variants are possible by training the policy to reflect the action distributions implicitly defined by the learned Q-values. This is a valid objective, as the policy is trained to choose the Q-value-maximizing action in *every* state.

The choice between on-policy and off-policy algorithms primarily affects its sample efficiency and the bias-variance tradeoff of the network updates [4]. On-policy algorithms need to consistently generate new data using their current policy, making them sample inefficient. Accordingly, they are less suited for problems with high time complexity environments. On the other hand, off-policy algorithms can operate on problems with arbitrarily generated experience, even without direct access to the environment.

Off-policy algorithms usually add self-generated experience to a buffer, called the experience replay buffer, of which they can sample batches for training. Prioritized Experience Replay (PER [23]) is an extension that stores per-element training priorities, which are used for weighted sampling and could be the individual loss values of each experience. This way, elements that an agent has not performed well on yet, have a higher probability of being trained more often.

Off-policy algorithms tend to be harder to tune than on-policy alternatives because of the significant bias from old data and value-learner initializations [5]. The bias-variance tradeoff will be further discussed in Sect. 3.3.

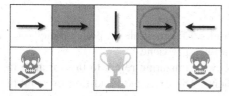 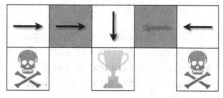

(a) A possible deterministic policy with the problem of equal decisions on the grey squares.

(b) Decisions that could be made by a stochastic policy, enabling the agent to reach the reward from both grey squares.

Fig. 3. Example of a problem with imperfect information that can only be solved with a stochastic policy. In this setting, the agent can only observe adjacent squares, making the grey squares look equivalent to the algorithm. A deterministic policy would always move left or always move right on these grey squares. With a stochastic policy and enough tries, the agent eventually always finds its way to the reward. Figures adapted from [27].

3.2 Deterministic and Stochastic Policies

Policies can be deterministic, i.e., they assign a particular action to a certain state, as well as stochastic, i.e., actions are chosen from a certain distribution. RL algorithms are designed to learn stochastic or deterministic policies. While most stochastic policies can be evaluated deterministically by selecting the actions with the highest probabilities, there is no sophisticated way of converting deterministic solutions to stochastic variants other than applying ϵ-greedy actions or temperature-controlled softmax.

It could be argued, that deterministically choosing the optimal action could outperform stochastic policies, as they tend to diverge from the optimal policy. In practice this does not as hold, for example SAC outperforms DDPG in its peak performance [7]. Also, there are problems with imperfect information that can only be reliably solved by stochastic policies, as demonstrated by [27].

Figure 3 shows a mini-game that is used by [27] as an example. In this environment, the robot is controlled by the RL algorithm, but it can only observe all adjacent squares of the playing field. This means that the observations from within the two gray squares are identical since, in both cases, the only two visible squares are both empty. A fixed deterministic policy without exploration would have to perform the same action in both of these squares, which would trap the robot on one of the two sides of the playing field, as shown in Fig. 3a. A true stochastic policy would be able to randomly move left or right when positioned on a gray square and could therefore reach the trophy eventually from every starting square, cf. Fig. 3b.

Exploration during training can be done natively with stochastic policies by randomly choosing actions according to the current probability distribution. Deterministic policies usually require ϵ-greedy exploration to allow theoretical convergence to the global optimum. For ϵ-greedy exploration, every time an action is chosen, there is a chance to choose a random action instead of the assumed optimal action, with a probability of ϵ, cf. [29]. Stochastic policies have advantages over their deterministic alternatives in multi-player games, as opponents can quickly adapt to deterministic playstyles. A straightforward example of such a game is rock-paper-scissors, see [27].

3.3 Bias-Variance Tradeoff

This section will examine the bias-variance tradeoff for multiple on-policy policy optimization versions, and for different value-learning approaches.

Variance in the context of on-policy policy optimization refers to the variance in approximations of the policy parameter gradients. The Expected Grad-Log-Probability Lemma (EGPL):

$$\mathbb{E}_{x \sim \mathbb{D}_\theta} \left[\Delta_\theta \log \mathbb{D}_\theta(x) \right] = 0$$
$$\implies \mathbb{E}_{a \sim \pi_\theta(\cdot|s)} \left[\Delta_\theta \log \pi_\theta(a|s) \right] = 0 \tag{9}$$

is essential for explaining the effects of different gradient formulations on the approximation variance. It says that when sampling values from a distribution $\mathbb{D}$ parameterized by θ, the gradients Δ of the log probabilities w.r.t. θ will be 0 in expectation. This means that those gradients of log probabilities follow some distribution with a mean of 0. The EGPL lemma can be transferred to an expectation over state-action samples from a policy. Additionally, it implies that by multiplying the gradient log probabilities with a *baseline* value $b(s)$, the result remains 0 in expectation as inferred in:

$$\mathbb{E}_{a \sim \pi_\theta(\cdot|s)} \left[\Delta_\theta \log \pi_\theta(a|s) \right] = 0$$
$$\implies \mathbb{E}_{a \sim \pi_\theta(\cdot|s)} \left[b(s) \Delta_\theta \log \pi_\theta(a|s) \right] = 0$$
$$\implies \mathbb{E}_{s \sim \pi_\theta} \left[\mathbb{E}_{a \sim \pi_\theta(\cdot|s)} \left[b(s) \Delta_\theta \log \pi_\theta(a|s) \right] \right] = 0$$
$$\implies \mathbb{E}_{(a,s) \sim \pi_\theta} \left[b(s) \Delta_\theta \log \pi_\theta(a|s) \right] = 0. \tag{10}$$

This baseline stretches or compresses the distributions of gradients of log probabilities but does not change their mean. When approximating this expectation using sampling, the baseline affects the variance of the approximations but not the mean.

For on-policy policy optimization, there are many equivalent formulas for calculating the gradients of the policy parameters θ, cf. [25,30,33]:

$$\Delta_\theta J_\pi(\theta) = \mathbb{E}_{\tau \sim \pi_\theta} \left[\sum_{t=0}^{T} \Delta_\theta \log \pi_\theta(a_t|s_t) \sum_{t'=0}^{T} \gamma^{t'} r\left(s_{t'+1}|s_{t'}^\tau, a_{t'}\right) \right] \tag{11}$$

$$= \mathbb{E}_{\tau \sim \pi_\theta} \left[\left(\sum_{t=0}^{T} \Delta_\theta \log \pi_\theta(a_t|s_t) \right) \sum_{t'=0}^{T} \gamma^{t'} r\left(s_{t'+1}|s_{t'}, a_{t'}\right) \right] \tag{12}$$

$$= \mathbb{E}_{\tau \sim \pi_\theta} \left[\sum_{t=0}^{T} \Delta_\theta \log \pi_\theta(a_t|s_t) \left(\sum_{t'=0}^{t-1} \gamma^{t'} r\left(s_{t'+1}|s_{t'}, a_{t'}\right) + \right.\right.$$
$$\left.\left. \sum_{t'=t}^{T} \gamma^{t'} r\left(s_{t'+1}|s_{t'}, a_{t'}\right) \right) \right] \tag{13}$$

$$= \mathbb{E}_{\tau \sim \pi_\theta} \left[\sum_{t=0}^{T} \Delta_\theta \log \pi_\theta(a_t|s_t) \left(\sum_{t'=0}^{T} \gamma^{t'} r\left(s_{t'+1}|s_{t'}, a_{t'}\right) - b(s_0) \right) \right] \tag{14}$$

$$= \mathbb{E}_{\tau \sim \pi_\theta} \left[\sum_{t=0}^{T} \Delta_\theta \log \pi_\theta \left(a_t | s_t \right) \sum_{t'=t}^{T} \gamma^{t'-t} \mathrm{r} \left(s_{t'+1} | s_{t'}, a_{t'} \right) \right] \tag{15}$$

$$= \mathbb{E}_{\tau \sim \pi_\theta} \left[\sum_{t=0}^{T} \Delta_\theta \log \pi_\theta \left(a_t | s_t \right) \left(\sum_{t'=t}^{T} \gamma^{t'-t} \mathrm{r} \left(s_{t'+1} | s_{t'}, a_{t'} \right) - b(s_{t'}) \right) \right] \tag{16}$$

$$= \mathbb{E}_{\tau \sim \pi_\theta} \left[\sum_{t=0}^{T} \Delta_\theta \log \pi_\theta \left(a_t | s_t \right) Q_{\pi_\theta} \left(a_t | s_t \right) \right] \tag{17}$$

$$= \mathbb{E}_{\tau \sim \pi_\theta} \left[\sum_{t=0}^{T} \Delta_\theta \log \pi_\theta \left(a_t | s_t \right) A_{\pi_\theta} \left(a_t | s_t \right) \right]. \tag{18}$$

These options originate from the different Policy Gradient variants. In the current section, the notation with an upper τ is omitted for all actions and states to improve readability (for example, s_t is used instead of s_t^τ). All of the gradient formulas from (11) to (18) can be shown to be equal in expectation [1]. This means that when approximating them using samples, they all produce the same sample mean but can have significant differences in variance. Notably, some of these definitions ((17) and (18)) include exact Q-values and advantage-values, and in practice, those are unknown and have to be estimated. With estimated Q- and advantage-values, the calculated gradient can be biased[3].

The gradient calculation for the vanilla Policy Gradient algorithm in (11) produces the worst sample variance, where each gradient log probability is scaled by the total realized reward of the whole trajectory. This creates a high sample variance for three reasons:

1. With stochastic policies and potentially stochastic environments, the *realized* rewards can differ significantly in-between episodes. Many samples are necessary to reflect the *expected* value of state-action pairs. The high variance in realized rewards directly affects the variance of the approximations.
2. For vanilla Policy Gradients, each gradient log probability is multiplied by the total trajectory return $\sum_{t'=0}^{T} \gamma^{t'} \mathrm{r} \left(s_{t'+1} | s_{t'}, a_{t'} \right)$, so it can be factored out of the first sum as in (12). Notably, the expectation of this formula is likely not zero, as the trajectory reward, unlike baselines, depends not only on the states but also on the actions. Thus the sum over the gradients of log probabilities and the total trajectory reward are likely correlated. (12) reveals the second reason for a high variance of the approximations, which are potentially high trajectory returns in expectation. Suppose the total trajectory returns for a problem are very large, strictly positive, and lie within $[1000, 1010]$. Further, suppose that most gradient log probability sums lie within the range $[-1, 1]$. Without using a baseline, their product likely lies in $[-1010, 1010]$. By subtracting an "average trajectory return"-baseline of value 1000 from the trajectory returns, the likely range of the product becomes $[-5, 5]$. As mentioned before,

---

[3] As an example, assume a Q-values estimator wrongly predicts all values as 0. This would always make the result of the gradient calculations 0 as well.

the two factors can be correlated, affecting the variance. A high correlation leads to reduced variance in both formulations. For example, assuming the strongest possible correlation in this setting, the respective ranges of samples of the products are $[-1000, 1010]$ and $[5, 5]$. This shows that sample variance can be significantly reduced by using a baseline as in (14) to shift the mean of the expected trajectory returns towards zero.

3. The third reason for the high variance in approximations of the expectation in (12) is shown in (13). By scaling each log probability by the total trajectory return, actions get reinforced based on previous rewards. These rewards do not depend on the action of a particular step. They, therefore, act as baselines that do not affect the sample mean but potentially *increase* the sample variance by shifting the mean of the expected trajectory returns away from zero. This explains why the reward-to-go Policy Gradient in (15) generally has a lower sample variance than the vanilla Policy Gradient.

Notably, the arguments of this section are only considering Monte Carlo PG variants so far, as calculating realized rewards until the end of an episode is impossible for infinite horizon problems. Only the PG variants with estimations of the rewards-to-go are eligible for infinite horizon settings, and they will be considered further below.

Multiple alternative gradient formulations have been developed to reduce the sample variance by manipulating the three presented variance-increasing factors. (16) is a popular variant that combines the reward-to-go sum with a baseline term, which empirically reduces the sample variance significantly[4] but does not resolve the problem of high variance introduced by stochastic trajectories. Theoretically-optimal baselines exist [31], but in practice, critic neural networks or exponential moving averages of the previous returns are the most frequently used options. Instead of using the *realized* returns, learned Q-values can be implemented to reinforce certain actions according to their *expected* future state-action value[5] as in (17). This version accounts for the variance of stochastic trajectories. It removes the problem of using full-trajectory returns as factors, but it lacks some form of baseline to reduce the variance originating from returns that have large absolute means. This last source of variance can be reduced by using advantage-values for reinforcing actions as in (18), because they include the missing state-value baseline ($A_{\pi_\theta} (a_t | s_t) = Q_{\pi_\theta} (a_t | s_t) - V_{\pi_\theta} (s_t)$). As the Q-values and advantage-values have to be estimated, they can introduce bias to the gradient estimates. Because of this, their supremacy compared to the Monte Carlo options depends on the quality of the value estimates.

There are many ways to estimate advantage-values. For problems with discrete action spaces, it is possible to implement a Q-value learner and estimate the advantages by subtracting all Q-values by the Q-value mean of a state. This can be computationally expensive with large action sets. For continuous action spaces, a separate state-value learner needs to be introduced to calculate the advantage by subtracting the respective state-value from the Q-value. In general, learning Q-values is considered harder than

---

[4] See the Spinning Up article: https://spinningup.openai.com/en/latest/spinningup/rl_intro3.html.

[5] See proof for exact Q-values in Spinning Up documentation: https://spinningup.openai.com/en/latest/spinningup/extra_pg_proof2.html.

learning state-values and might require many more samples or does not converge. This can lead to lousy advantage estimates, which do not have the desired benefits to the Policy Gradient computations.

Estimating the advantage-values is also possible with just a state-value learner, and the different approaches are similar to Q- and state-value estimations.

A Monte Carlo advantage estimator takes the realized rewards-to-go of a state-action pair and subtracts the corresponding state-value:

$$\hat{A}^{MC}(a|s_t) = G_{t:T} - \hat{V}_\phi(s_t). \tag{19}$$

A TD(1) advantage estimator adds the immediate reward of a state-action pair $r(s_{t'+1}|s_t, a_t)$ with the proceeding state's value estimate $\hat{V}_\phi(s_{t+1})$ and calculates the difference to the state-value estimate $\hat{V}_\phi(s)$. The formula for the TD(1) advantage estimate is given by:

$$\hat{A}^{(1)}(a_t|s_t) = r(s_{t'+1}|s_t, a_t) + \gamma\hat{V}_\phi(s_{t+1}) - \hat{V}_\phi(s_t)$$
$$= G_{t:t+1}^\phi - \hat{V}_\phi(s_t). \tag{20}$$

A TD(n) advantage estimate is calculated with the realized rewards of n steps and is defined as:

$$\hat{A}^{(n)}(a|s_t) = \left(\sum_{t'=t}^{t+n-1} \gamma^{t'-t} r(s_{t'+1}|s_{t'}, a_{t'}) + \gamma^n \hat{V}_\phi(s_{t+n}) \right) - \hat{V}_\phi(s_t)$$
$$= G_{t:t+n}^\phi - \hat{V}_\phi(s_t). \tag{21}$$

Notably, the formula for a TD(n) advantage estimate is the same as the TD(n) approximation error used for updating the state-value estimator $\hat{V}_\phi$. This is because the value estimator is trained to learn values that let this term become 0 *in expectation.* advantage-values are 0 in expectation, but their importance for policy optimization lies within their correlation to action probabilities.

Just like TD(λ), Generalized Advantage Estimation (GAE [25]) is a method to control the bias-variance tradeoff between the high bias TD(1) advantage estimates and the high variance MC advantage estimates by applying an exponentially-moving average over all estimates defined in:

$$\hat{A}^{GAE(\gamma,\lambda)}(a|s_t) = (1 - \lambda) \sum_{k=0}^{\infty} \lambda^k \hat{A}^{(k)}(a|s_t). \tag{22}$$

GAE works natively in finite horizon settings, but it can also be applied with infinite horizon problems by truncating the endless sum after many steps as γ and λ diminish the effect of rewards that lie far in the future.

Notably, bias in the learned Q-values or advantage estimates affects the policy optimization's bias and sample variance, similar to bad baselines. This is important because of the bias-variance tradeoff in *value learning*, which will be discussed next.

In value-learning, there is a bias-variance tradeoff in the target values, i. e. decreasing the variance tends to increase the bias. As mentioned in Sect. 2.3, this tradeoff is

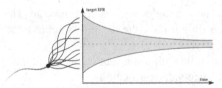 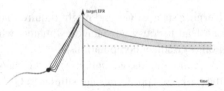

(a) Development of target expected future returns for a Monte Carlo estimator. The high variance of the target values is caused by the many different trajectories that can occur during training. Over time, the variance decreases as the algorithm converges to fewer, more promising trajectories.

(b) Development of target expected future returns for a TD(1) estimator. The early bias is caused by the initialization of the Q-values. The variance is low as only a single realized reward is considered. Over time the Q-value estimations get more accurate, and bias decreases.

Fig. 4. Illustration of the development of target expected future returns (target EFR) for an arbitrary but fixed state over time for a Monte Carlo estimator and a TD(1) estimator.

determined by the depth of the value updates. Monte Carlo calculations have high variance but no bias. In expectation, the sum of realized returns after a state correctly reflects the state-value, but stochastic policies or environments might realize many different trajectories and rewards. In the case of on-policy value-learning, this variance decreases over time as the policy becomes less random and narrows its options to the most promising trajectories. Both of these properties are visualized in Fig. 4a. Temporal Difference calculations of single steps have very low variance, as they only use a single realized reward with bootstrapping in their updates. This means that the sum of the immediate reward after a state and a bootstrapped value varies much less than the realized rewards of Monte Carlo trajectories. The benefits in variance come at the cost of bias towards the initializations of the value estimators. At the beginning of the learning phase of state-action values, the estimated Q-values are still far from the actual state-action values, but those wrong estimations are used for bootstrapping. Off-policy Deep Q-Learning suffers from an additional *overestimation bias*. This is caused by choosing the maximum Q-value of the following state in bootstrapping, which significantly slows down convergence in combination with bad initializations. Algorithms like Double DQN (DDQN) have been developed to reduce this overestimation bias but are known to introduce some underestimation bias [9,22]. Over time, the learned Q-value estimations tend to converge to the correct values, but there are no guarantees[6]. Figure 4b visualizes the composition of TD(1) target values for a particular state and their development over training.

The larger the number of steps in Temporal Difference calculations, the closer the TD target values get to the Monte Carlo targets and variance increases as bias decreases. TD(n) and TD(λ) can therefore be used to tune the bias-variance tradeoff in value-learning.

---

[6] See this discussion on Stack Exchange: https://ai.stackexchange.com/questions/11679.

During training, the variance in approximations of the gradients can be lowered further by increasing the number of samples used for the approximation, i. e. by increasing the batch size.

3.4 Exploration-Exploitation Tradeoff

Next, we discuss the tradeoff between exploration and exploitation during training. The exploration-exploitation tradeoff is a dilemma that not only occurs in RL but also in many decisions in real life. When selecting a restaurant for dinner, the two main options are to revisit the favorite restaurant or try a new one. By choosing the favorite restaurant, there is a high likelihood of achieving the known satisfaction, but without *exploring* other options, one will never know whether there are even better alternatives. By only exploring new restaurants, the pleasure will generally be lower in expectation than with the favorite restaurant. The same concept applies to RL as well. The agent needs to balance exploration for finding better solutions with exploitation to direct exploration and motivate convergence towards optimal solutions.

Many different exploration-exploitation strategies exist, including no exploration, ϵ-greedy exploration, upper confidence bounds exploration, Boltzmann exploration, maximum entropy exploration, and noise-based exploration [34]. ϵ-greedy exploration is often applied when learning deterministic policies to produce new trajectories. Otherwise, the deterministic policy would not include any exploration, which would prevent the agent from discovering better policies than those already present in the data.

Upper confidence bounds exploration introduces a notion of confidence to Q-value estimations. Actions are chosen based on the sum of each Q-value with an individual uncertainty value. The uncertainty value is defined to be inversely proportional to the number of times an action was taken. The sum of a Q-value with its uncertainty value represents the upper confidence bound of that Q-value, i. e., Q-values of actions that have rarely been trained have high uncertainty and could be much larger in reality. Upper confidence bounds encourage exploration of rarely visited actions while considering their current Q-value estimation. Figure 5 displays exemplary Q-values with their respective confidence intervals.

Boltzmann exploration applies the softmax operation over the Q-values to define a stochastic policy and regulates the strength of exploration by the softmax temperature parameter. In maximum entropy exploration, the agents' objective functions are extended by an entropy term that penalizes the certainty of the learned policy. Algorithms like SAC use this type of exploration. Noise-based exploration adds random noise to observations, actions, or even model parameters to introduce variation.

Some settings are especially challenging for exploration to find better solutions consistently. For example, very sparse or deceptive rewards can be problematic, which is called the hard exploration problem. [34] and [18] provide further information on exploration challenges and possible solutions.

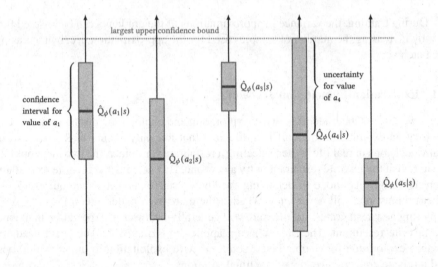

Fig. 5. Visualization of the confidence intervals according to five Q-value estimations with their respective uncertainty values. Upper confidence bounds exploration greedily chooses the action with the largest upper confidence bound.

3.5 Hyperparameter Tuning and Robustness

In the context of RL, hyperparameter sensitivity characterizes how much an algorithm's performance depends on carefully tuned hyperparameters and how much the hyperparameters need to be adjusted between different problems. In other words, it describes the size of the hyperparameter space that generally produces good results. Brittleness can be used as a synonym for hyperparameter sensitivity in some literature [7] but also to convey a neural network's ability to generalize to unseen inputs. Many publications of new RL algorithms mention hyperparameter sensitivity and claim or imply improvements compared to previous work. [7,8], for example, note that their algorithm SAC is less sensitive than DDPG. [26] also suggest in their publication of PPO that it is less sensitive than other algorithms. These papers fail to include concrete sensitivity analysis to support their claims.

Other researchers have published work on hyperparameter tuning and sensitivity comparisons on specific RL tasks [10,11,14,16]. Most emphasize the high sensitivity of all compared RL algorithms and the lack of generally well-performing configurations. These articles suggest a need for less sensitive algorithms to be developed and for more research on best-performing hyperparameters in different settings.

Robustness is used as an antonym to hyperparameter sensitivity in some articles [26] to describe how successful an algorithm is on various problems without hyperparameter tuning. Most of the literature uses robustness as a measure of how well a learned algorithm can handle differences between its training and test environment [3,17]. The second definition of robustness is specifically important for real-world applications that incorporate a shift between training and testing environments.

There are several approaches to increase the robustness of RL algorithms. One idea is to compose a distribution of environments and optimize the average performance of an agent on multiple environment samples from this distribution. The properties of the distribution highly influence the success of this procedure. One the other hand, an optimized average performance can result in the agent being exceptionally good in a few environments and much worse in the others.

A second option for robustness in RL is to create an adversarial setup, where an adversary is trained to adjust the environment such that the agent's performance drops. Hence, the agent constantly trains on different environments. Such setups are promising but not easy to implement. [3] show that maximum entropy RL algorithms like SAC are robust to environment changes, as the respective agents learn to recover from disturbances introduced by "surprising" actions.

3.6 Learning Stability

A crucial feature of an RL algorithm is the so-called learning stability. This particularly involves the tendency to forget intermediate best-performing policies throughout training. This phenomenon is sometimes referred to as *catastrophic forgetting* in the literature [6], although catastrophic forgetting is also used in the context of sequentially training a model on different tasks [13].

Learning instability in policy optimization algorithms is mainly caused by noise in the gradient estimations, producing destructive network updates. Such noise is usually created by a high variance in the gradient estimates as discussed in Sect. 3.3. Therefore, learning stability can be increased by choosing RL algorithms with low-variance gradient estimators. Another option to prevent destructive updates to the policy during training is to limit the change of the policy in-between updates, as done by PPO. A less sophisticated alternative with similar effects is to clip the gradient norms before using them in network updates. This is called gradient norm clipping.

In [21], the authors propose to transfer stochastic weight averaging (SWA) to the RL setting to increase learning stability. SWA has improved generalization in supervised and unsupervised learning and is based on averaging the weights of the models collected during training [21].

4 Summary

In this section, we summarize the key takeaways for the different RL properties discussed in the Sect. 2 and Sect. 3.

(i) Model-Free and Model-Based RL. The categorization of model-free and model-based RL algorithms indicates whether an algorithm learns or is given additional knowledge of the environment beyond the observations. This allows simulations of the environment and wider updates and exploration compared to following one trajectory at a time. It can be especially powerful if acting in the real environment is expensive or an environment model is available anyway. Model-free algorithms are more popular, and model-based implementations are not supported by as many frameworks because of the problem-specifics.

(ii) Policy Optimization and Value Learning Methods. For many problems, there is only one alternative. For all others, it depends on the complexity of a problem's value function, if it is easier to learn a value function or directly search for the optimal policy.

(iii) Bias-Variance Tradeoff. Many different formulations of the reinforcing term in the objectives are available in policy optimization, which differ significantly in approximation variance. The order of the various options in decreasing variance is roughly given by:

- vanilla, Monte Carlo (entire trajectory) return
- Monte Carlo reward-to-go
- Monte Carlo calculations with baselines
- true Q-Values
- true advantage values.

Using estimators for the Q-Values or advantage values introduces bias to the gradient estimates. Advantage values can be estimated in many ways, similar to the different TD options. Generalized Advantage Estimation (GAE) is the equivalent of TD(λ) for advantage estimation. In value learning, the bias-variance tradeoff is mostly defined by the choice between Monte Carlo, TD(1), TD(n), and TD(λ) targets. Monte Carlo calculations have high variance, TD(1) has low variance but suffers from strong initialization bias and potentially significant overestimation bias, especially when trained off-policy. TD(n) and TD(λ) provide options to interpolate between the MC and TD(1) extremes.

(iv) Finite and Infinite Horizon Problems. Whether a problem has a finite or infinite horizon has implications on which options for value estimations are applicable. Monte Carlo methods only work with finite horizons, as whole trajectories are necessary. MC targets for value estimators have no bias but high variance, especially with extended episodes. Temporal Difference methods are applicable to both finite and infinite horizon methods and allow tuning of the problem-specific tradeoff between bias and variance. TD(λ) combines all TD(n) updates and ideally their problem-specific benefits and is technically an on-policy method, as it includes Monte Carlo estimates. Truncating the infinite sum is possible for applying TD(λ) to infinite horizon problems. Forward and backward views can be used to calculate TD(λ) estimates. Usually, the backward view is applied with eligibility traces, as it allows online updates of the estimator.

(v) Discrete and Continuous State Spaces. Traditional table-based methods can only operate on countable state sets, as only finite memory is available on a computer. Neural networks can generalize to infinitely-many inputs with limited memory, making them suitable for problems with uncountable state sets.

(vi) Discrete and Continuous Action Sets. Table-based algorithms are in most cases inapplicable for uncountable action sets, but RL with neural networks can be used. Most on-policy RL algorithms can handle both countable and uncountable action sets. Off-policy learning of uncountable action sets includes additional challenges because the distribution type of the action values of a state is unknown. TD methods' state

values cannot be calculated easily, and adding another expected value over these distributions is required in the objectives. Using action samples and the reparameterization trick, these expectations can be approximated and gradients can be backpropagated through the continuous actions into the policy network. Learning stochastic policies under uncountable action sets is also possible, as done by SAC.

(vii) Discrete and Continuous Time Problems. Continuous-time problems are often discretized and approached with traditional RL algorithms, but formulations for actual continuous-time algorithms exist. These use the Hamilton Jacobi Bellman Equation and incorporate ideas from discrete problems, like TD updates and value estimations by neural networks.

(viii) On-Policy and Off-Policy RL. On-policy algorithms need to be trained with experience of the latest version of the policy. Off-policy alternatives can use and re-use any experience acquired in an environment. A significant advantage of off-policy algorithms is sample efficiency, but they tend to be harder to tune because of the bias-variance implications between the respective learning strategies. Learning *values* of states or state-action pairs *under the current policy* is usually done *on-policy*, in which case, all Monte Carlo, TD(n), and TD(λ) methods can be used. SARSA is an example that uses TD(1) calculations. Only the TD(1) targets can be used if values under the current policy are learned *off-policy*. The bootstrapped value of the succeeding state can then be estimated by the current policy's probability distribution of that state and the respective Q-values, as done by SAC. Actual values within the environment are learned *off-policy*, and the maximizing action is chosen for updates. One example of such an algorithm is Deep Q-Learning.

Most policy optimization objectives are defined over the stationary state distribution of the current policy and therefore require on-policy training. Importance sampling still allows for off-policy gradient calculations in policy optimization but can significantly increase gradient approximation variance. Consequently, it is only used in special cases, for example, to execute a few consecutive network updates with just one batch of data. Prioritized Experience Replay is an extension that can only be applied to off-policy algorithms.

(ix) Deterministic and Stochastic Policies. Some problems with imperfect information can not be solved by deterministic policies, in which case stochastic alternatives are more powerful. Deterministic policies or deterministic evaluations of stochastic policies tend to have higher total expected returns as only the best known actions are chosen. Stochastic policies have benefits in exploration and in multi-player games, where unpredictability is beneficial.

(x) Exploration-Exploitation Tradeoff. The exploration-exploitation tradeoff is a natural dilemma when not knowing whether an optimum is reached in a decision process. Exploration is necessary to discover better solutions, but exploitation is needed for directed learning instead of random wandering. Many different exploration strategies exist, such as ϵ-greedy, stochastic exploration, maximum entropy exploration, upper confidence bounds exploration, etc. Some problems are especially challenging

for exploration. The hard exploration problem, for example, occurs in environments with very sparse rewards.

(xi) Hyperparameter Tuning and Robustness. The sensitivity of algorithms indicates how narrow the usable ranges of hyperparameters are and how well they can be trained on multiple problems with the same configuration. Sensitivity is often mentioned in the literature but rarely analyzed in detail. It is highly problem-specific, which makes studying sensitivity on multiple problems valuable. Algorithms like PPO can make algorithms less sensitive by limiting the incentives of drastic adjustments to the policy. The ability of an algorithm to cope with changes between training and test environment is referred to as robustness. It can be improved by training in multiple environments, creating adversarial setups, or applying maximum entropy RL.

(xii) Learning Stability. The learning stability of an algorithm is an indicator for its tendency to forget intermediate best-performing policies throughout training. Low variance gradient estimators help improve algorithms' stability, and methods like PPO's objective function clipping can help prevent destructive updates. Additional options include gradient norm clipping and stochastic weight averaging.

5   Conclusion

To solve real-world problems with incomplete information, RL is a promising approach as it only requires a suitable reward function and no optimal or complete data. Over the training process, the model incrementally builds better solutions by observing the consequences of consecutive decisions. The practicality of RL is impaired by the opaque selection of algorithms and their extensions as well as the more complicated tuning compared to supervised learning. In this context, we reviewed multiple properties of RL problems and algorithms and inferred helpful guidelines to decide under which circumstances to apply which algorithm. As a complementing future research direction, we will verify the presented conceptual insights by evaluating numerical results, e.g., for specific combinatorial problem applications.

References

1. Achiam, J.: Spinning up in deep reinforcement learning (2018). https://spinningup.openai.com/
2. Doya, K.: Reinforcement learning in continuous time and space. Neural Comput. **12**(1), 219–245 (2000)
3. Eysenbach, B., Levine, S.: Maximum entropy RL (provably) solves some robust RL problems. arXiv preprint arXiv:2103.06257 (2021)
4. Fakoor, R., Chaudhari, P., Smola, A.J.: P3O: policy-on policy-off policy optimization. In: Uncertainty in Artificial Intelligence, pp. 1017–1027. PMLR (2020)
5. Fujimoto, S., et al.: Addressing function approximation error in actor-critic methods. In: ICML, pp. 1587–1596. PMLR (2018)

6. Géron, A.: Hands-on Machine Learning with Scikit-Learn, Keras, and TensorFlow: Concepts, Tools, and Techniques to Build Intelligent Systems. O'Reilly Media, Inc., Sebastopol (2019)
7. Haarnoja, T., et al.: Soft actor-critic algorithms and applications. arXiv preprint arXiv:1812.05905 (2018)
8. Haarnoja, T., et al.: Soft actor-critic: off-policy maximum entropy deep reinforcement learning with a stochastic actor. In: ICML, pp. 1861–1870. PMLR (2018)
9. Hasselt, H.: Double q-learning. Adv. Neural Inf. Process. Syst. 23 (2010)
10. Henderson, P., et al.: Deep reinforcement learning that matters. In: Proceedings of the AAAI Conference on Artificial Intelligence, vol. 32, no. 1 (2018). https://doi.org/10.1609/aaai. v32i1.11694, https://ojs.aaai.org/index.php/AAAI/article/view/11694
11. Hussenot, L., et al.: Hyperparameter selection for imitation learning. In: ICML, pp. 4511–4522. PMLR (2021)
12. Kirk, D.E.: Optimal Control Theory: An Introduction. Courier Corporation, Chelmsford (2004)
13. Kirkpatrick, J., et al.: Overcoming catastrophic forgetting in neural networks. Proc. Natl. Acad. Sci. 114(13), 3521–3526 (2017)
14. Liessner, R., et al.: Hyperparameter optimization for deep reinforcement learning in vehicle energy management. In: ICAART (2), pp. 134–144 (2019)
15. Lillicrap, T.P., et al.: Continuous control with deep reinforcement learning. arXiv preprint arXiv:1509.02971 (2015)
16. Mahmood, A.R., et al.: Benchmarking reinforcement learning algorithms on real-world robots. In: Conference on Robot Learning, pp. 561–591. PMLR (2018)
17. Mankowitz, D.J., et al.: Robust reinforcement learning for continuous control with model misspecification. arXiv preprint arXiv:1906.07516 (2019)
18. McFarlane, R.: A survey of exploration strategies in reinforcement learning. McGill University (2018)
19. Minsky, M.: Steps toward artificial intelligence. Proc. IRE 49(1), 8–30 (1961)
20. Mnih, V., et al.: Playing atari with deep reinforcement learning. arXiv preprint arXiv:1312.5602 (2013)
21. Nikishin, E., et al.: Improving stability in deep reinforcement learning with weight averaging. In: Uncertainty in Artificial Intelligence Workshop on Uncertainty in Deep Learning (2018)
22. Ren, Z., Zhu, G., Hu, H., Han, B., Chen, J., Zhang, C.: On the estimation bias in double q-learning. Adv. Neural Inf. Process. Syst. 34 (2021)
23. Schaul, T., et al.: Prioritized experience replay. arXiv preprint arXiv:1511.05952 (2015)
24. Schröder, K., Kastius, A., Schlosser, R.: Welcome to the jungle: a conceptual comparison of reinforcement learning algorithms. In: Proceedings of the 12th International Conference on Operations Research and Enterprise Systems - Volume 1: ICORES, pp. 143–150 (2023)
25. Schulman, J., et al.: High-dimensional continuous control using generalized advantage estimation. arXiv preprint arXiv:1506.02438 (2015)
26. Schulman, J., et al.: Proximal policy optimization algorithms. arXiv preprint arXiv:1707.06347 (2017)
27. Silver, D.: Lectures on reinforcement learning (2015). https://www.davidsilver.uk/teaching/
28. Silver, D., et al.: Mastering chess and shogi by self-play with a general reinforcement learning algorithm. arXiv preprint arXiv:1712.01815 (2017)
29. Sutton, R.S., Barto, A.G.: Reinforcement Learning: An Introduction. MIT Press, Cambridge (2018). https://www.andrew.cmu.edu/course/10-703/textbook/BartoSutton.pdf
30. Sutton, R.S., McAllester, D., Singh, S., Mansour, Y.: Policy gradient methods for reinforcement learning with function approximation. Adv. Neural Inf. Process. Syst. 12 (1999)
31. Weaver, L., Tao, N.: The optimal reward baseline for gradient-based reinforcement learning. arXiv preprint arXiv:1301.2315 (2013)

32. Weng, J., et al.: Tianshou: a highly modularized deep reinforcement learning library. arXiv preprint arXiv:2107.14171 (2021)

33. Weng, L.: Policy gradient algorithms (2018). lilianweng.github.io, https://lilianweng.github.io/posts/2018-04-08-policy-gradient/

34. Weng, L.: Exploration strategies in deep reinforcement learning (2020). https://lilianweng.github.io/

35. Yildiz, C., et al.: Continuous-time model-based reinforcement learning. In: ICML, pp. 12009–12018. PMLR (2021). https://youtu.be/PIouASLg_-g

Comparison of Multi-objective Linear Programming Solutions Using Performance Metrics Based on Data Envelopment Analysis Models

Javier E. Gómez-Lagos[1] , Marcela C. González-Araya[2]([✉]) ,
and Luis G. Acosta Espejo[3]

[1] Doctorado en Sistemas de Ingeniería, Faculty of Engineering, Universidad de Talca, Campus Curicó, Camino a Los Niches, km 1, Curicó, Chile
javier.gomez@utalca.cl

[2] Departament of Industrial Engineering, Faculty of Engineering, Universidad de Talca, Campus Curicó, Camino a Los Niches km 1, Curicó, Chile
mgonzalez@utalca.cl

[3] Departamento de Ingeniería Comercial, Universidad Técnica Federico Santa María, Avenida Santa María 6400, Vitacura, Santiago, Chile
luis.acosta@usm.cl

Abstract. In this study, three performance metrics based on data envelopment analysis (DEA) are proposed, aiming to evaluate and compare solution methods for solving multi-objective linear programming (MOLP) models. Particularly, the proposed metrics are based on the slack-based measure (SBM) model and the super-efficiency DEA model (SE-DEA). For the SBM model, an integer version is formulated (INT-SBM model), which offers the advantage of evaluating non-dominated solutions in the non-convex region of the Pareto frontier. In this study, this case is demonstrated through an example. The SE-DEA model has the advantage of identifying the non-dominated solutions that define the Pareto frontier. Furthermore, each metric is associated to one of the cardinality, accuracy, and diversity categories, and are classified as unary or binary. The cardinality and accuracy metrics are estimated by using a procedure where the INT-SBM model is applied. On the other hand, the diversity metric is calculated in a procedure by using the SE-DEA model. In order to compare two sets of solutions obtained for a MOLP tactical harvest planning model based on two strategies of the multi-objective greedy randomized adaptive search procedure (MO-GRASP), the proposed metrics are applied. The results indicate that the metrics effectively discriminate between the MOLP solution methods and can support the selection of a suitable method for solving a MOLP model.

Keywords: Multi-objective linear programming · Performance metrics · MOLP · Solution methods · Data envelopment analysis · Slack-based measure · Model · Super-efficiency model

F. Liberatore et al. (Eds.): ICORES 2022/2023, CCIS 1985, pp. 121–137, 2024.
https://doi.org/10.1007/978-3-031-49662-2_7

1 Introduction

In operations research, many practical problems have been addressed through multi-objective linear programming (MOLP). In this way, as mentioned by (Gómez-Lagos et al., 2023), MOLP models have been proposed in transport (Demir et al., 2014), agriculture (Varas et al., 2020), manufacturing (Mirzapour Al-E-Hashem et al., 2011), location (Karataş & Yakıcı, 2018), among others. Solving these MOLP models is usually challenging (Deb, 2014), and, as a result, both exact and heuristic methods have been proposed for their solution. However, selecting an appropriate solution method is not a straightforward task. Consequently, various performance metrics have been proposed to analyze the solutions obtained by these methods. Regarding this matter, Riquelme et al. (2015) presented a literature review on performance metrics for evaluating MOLP solution methods, classifying them into three categories: cardinality, accuracy, and diversity. Cardinality represents the number of non-dominated solutions found by a MOLP solution method. Accuracy refers to the convergence of the non-dominated solutions to the Pareto frontier, measuring the distance between each non-dominated solution and the theoretical Pareto frontier (Riquelme et al., 2015). Diversity considers the distribution and spread of non-dominated solutions. Distribution considers the relative distance among the non-dominated solutions, while spread corresponds to the range of objective function values covered by the non-dominated solutions. It is worth mentioning that different metrics have been proposed within each category (Audet et al., 2021; Riquelme et al., 2015). Furthermore Riquelme et al. (2015) also classified the performance metrics as unary or binary. A metric is unary if the non-dominated solutions are obtained using only one solution method. Conversely, a metric is binary if the non-dominated solutions are obtained by using two solution methods. In Audet et al. (2021), a literature review analyzing 63 performance metrics was carried out, classifying them into cardinality, accuracy, spread, and distribution categories. In another study, Halim et al. (2021) critically analyzed the advantages and disadvantages of 100 performance metrics. In addition, they defined the desirable attributes and properties that a performance metric must have.

In the literature, data envelopment analysis (DEA) models have been used to estimate MOLP performance metrics. Bal & Satoglu (2019) used the BCC model (Banker et al., 1984) as a metric to evaluate the performance of Pareto optimal solutions obtained by the augmented epsilon constraint method 2. This solution method was applied to a MOLP model with four objective functions, aiming to enhance the coordination of an appliance supply chain. Hong & Jeong (2019) applied the CCR model (Charnes et al., 1978) to evaluate solutions obtained through the weighting method. The method was used to solve a MOLP model with five objective functions, which aimed to determine strategic decisions for a facility location–allocation problem. Recently, Gómez-Lagos et al. (2023) proposed three performance metrics based on DEA models for evaluating solutions obtained by MOLP solution methods.

In this study, three performance metrics based on DEA models to evaluate MOLP solution methods are proposed. Each metric is associated with a category of cardinality, accuracy, or diversity, and can be classified as unary or binary. The current study differs from Gómez-Lagos et al. (2023) in the development of a non-linear model to a linear model based on the Slack Based Measure (SBM) model (Tone, 2001), and in the demonstration that the proposed integer SBM model (INT-SBM) can identify and evaluate non-dominated solutions in the non-convex region of a Pareto frontier.

This study is divided as follows: Sect. 2 describes the applied DEA models. In Sect. 3, the performance metrics based on DEA models are proposed and the procedure for calculating each metric is described. Section 4 presents the results of this study, while Sect. 5 summarizes the conclusions.

2 Applied Data Envelopment Analysis Models

The DEA models used for assessing different performance metrics of MOLP solution methods, as the associated procedure for applying them, are presented in this section. As mentioned previously, the categories considered in this analysis are cardinality, accuracy, and diversity. The proposed cardinality and accuracy metrics are estimated using a DEA model based on the slacks-based measure proposed by Tone (2001). On the other hand, the proposed diversity metric is calculated applying the super efficiency DEA model developed by Andersen & Petersen (1993). The characteristics for selecting these DEA models and the way that they can be used for estimating every metric are detailed in the following sub-sections.

The nomenclature of parameters and decision variables used in the DEA models are defined in Table 1 and Table 2, respectively. In this definition, the MOLP nature of the DEA assessment is considered.

Table 1. Parameters of the DEA models (Gómez-Lagos et al., 2023).

Parameter	Definition
m	Number of objective functions to be minimized in the MOLP model
s	Number of objective functions to be maximized in the MOLP model
n	Number of solutions obtained by a MOLP solution method
x_{ij}	Value of the minimized objective function i obtained by solution j, where $i = 1, \ldots, m, j = 1, \ldots, n$
y_{rj}	Value of the maximized objective function r obtained by solution j, where $r = 1, \ldots, s, j = 1, \ldots, n$
j_0	Evaluated solution in every execution of a DEA model

In the following sub-sections, the applied DEA models are described.

Table 2. Decision variables of the DEA models (Gómez-Lagos et al., 2023).

Variable	Definition
S_i^-	Slack of the minimized objective function i, where $i = 1, \ldots, m$
S_r^+	Slack of the maximized objective function r, where $r = 1, \ldots, s$
λ_j	Intensity of the solution j for establishing the target in the Pareto frontier of the evaluated solution $j_0, j = 1, \ldots, n$
θ	Proportional reduction of the objective functions for the evaluated solution j_0

2.1 Integer Slack-Based Measure Model (INT-SBM)

In this sub-section, a DEA model based on the slacks-based measure model (SBM) developed by Tone (2001) is proposed in order to obtain a measure related to the domination degree of every solution. In this way, efficient solutions will correspond to non-dominated solutions. Furthermore, the SBM does not require inputs in an output-oriented model, nor does it require outputs in an input-oriented model. In this way, it can evaluate MOLP solution methods that solve models that have only maximization objective functions or only minimization objective functions.

The SBM proposed by Tone (2001) is a non-linear model, which is linearized applying the linear transformation proposed by Charnes & Cooper (1962). In this study, the proposed DEA model is based on the model presented by Tone (2001), but it uses binary variables. The binary variables allow us to identify non-dominated solutions located in the non-convex region of the Pareto frontier. On the other hand, the SBM proposed by Tone (2001) does not allow us to identify this kind of non-dominated solutions. For this reason, the SBM model proposed in this study is called integer-SBM (INT-SBM). Figure 1 illustrates an example where some non-dominated solutions are in the non-convex region of the Pareto frontier (highlighted through the red ellipse). In the abscissa axis (x), a FO to be minimized is considered, and in the ordinates axis (y), a FO to be maximized is considered.

It is important to notice that the INT-SBM model must be executed for every solution j obtained by a MOLP method, where j_0 is the evaluated solution in a specific model execution. The non-linear INT-SBM model (Model 1) for every evaluated solution j_0 is formulated as follows.

(*Model 1*)

$$Minimize\ \xi_{j0} = \frac{1 - \frac{1}{m} \sum_{i=1}^{m} S_i^- / x_{ij_0}}{1 - \frac{1}{s} \sum_{r=1}^{s} S_r^+ / y_{rj_0}} \tag{1}$$

Subject to

$$x_{ij_0} = \sum_{j=1}^{n} x_{ij} \lambda_j + S_i^-, i = 1, \ldots, m, \tag{2}$$

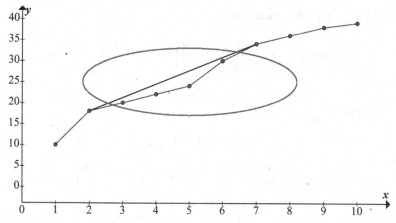

Fig. 1. Non-dominated solutions located in the non-convex region of the Pareto frontier.

$$y_{rj_0} = \sum_{j=1}^{n} y_{rj} \lambda_j - S_r^+, r = 1, \ldots, s, \tag{3}$$

$$\sum_{j=1}^{n} \lambda_j = 1, \tag{4}$$

$$\lambda_j \in \{0, 1\}, j = 1, \ldots, n, \tag{5}$$

$$S_i^- \geq 0, i = 1, \ldots, m, \tag{6}$$

$$S_r^+ \geq 0, r = 1, \ldots, s, \tag{7}$$

As can be observed in Model 1, the objective function (1) is non-linear. This objective function estimates the efficiency score of the solution j_0 based on the maximation of slacks associated to their values of the MOLP objective functions. Constraints (2) and (3) calculate the slacks associated to the values of the minimized and maximized MOLP objective functions of the solution j_0, respectively. Constraint (4) corresponds to the convexity constraint of the efficient frontier, and it is associated to variable returns to scale. Finally, constraints (5) to (7) establish the nature of the decision variables.

In the Model 1, ξ_{j_0} corresponds to the efficiency score of the evaluated solution j_0, which can take values greater than zero, and equal or lower than one. Thus, $\xi_{j_0} = 1$ represents a non-dominated solution, and $\xi_{j_0} < 1$ represents a dominated solution. In addition, this score represents the closeness to the Pareto frontier.

For linearizing the objective function (1), a decision variable t is introduced according to the transformation proposed by Charnes & Cooper (1962). The decision variable t is defined as follows:

$$t = \frac{1}{1 + \frac{1}{s} \sum_{r=1}^{s} \frac{S_r^+}{y_{rj_0}}} \tag{8}$$

Therefore, the objective function (1) is:

$$\text{Minimize} \xi_{j_0} = \left(1 - \frac{1}{m} \sum_{i=1}^{m} \frac{S_i^-}{x_{ij_0}}\right) \times t = t - \frac{1}{m} \sum_{i=1}^{m} \frac{S_i^- \times t}{x_{ij_0}} \tag{9}$$

This multiplication by t requires to linearize the decision variables using the following equations:

$$\zeta_i^- = S_i^- \times t, i = 1, \ldots, m, \tag{10}$$

$$\zeta_r^+ = S_r^+ \times t, r = 1, \ldots, s, \tag{11}$$

$$\Upsilon_j = \lambda_j \times t, j = 1, \ldots, n. \tag{12}$$

In this way, a new linearized INT-SBM model (Model 2) can be formulated as: (*Model 2*)

$$\text{Minimize} \xi_{j_0} = t - \frac{1}{m} \sum_{i=1}^{m} \frac{\zeta_i^-}{x_{ij_0}} \tag{13}$$

Subject to

$$1 = t + \frac{1}{s} \sum_{r=1}^{s} \frac{\zeta_r^+}{y_{rj_0}}, \tag{14}$$

$$x_{ij_0} t = \sum_{j=1}^{n} x_{ij} \Upsilon_j + \zeta_i^-, i = 1, \ldots, m, \tag{15}$$

$$y_{rj_0} t = \sum_{j=1}^{n} y_{rj} \Upsilon_j - \zeta_r^+, r = 1, \ldots, s, \tag{16}$$

$$\sum_{j=1}^{n} \Upsilon_j = t, \tag{17}$$

$$\Upsilon_j \in \{0, t\}, j = 1, \ldots, n, \tag{18}$$

$$\zeta_i^- \geq 0, i = 1, \ldots, m, \tag{19}$$

$$\zeta_r^+ \geq 0, r = 1, \ldots, s, \tag{20}$$

$$t \geq 0. \tag{21}$$

It is important to notice that Eq. (14) corresponds to Eq. (8) after performing algebraic operations on both sides of the equation. As can be observed in constraint (18), the

variable Υ_j only can take values equals to zero or t. For this reason, the big-M constraints are used in the formulation for transforming Υ_j in a continuous variable. Consequently, the linearized INT-SBM formulation used for estimating the performance metrics is:
(*Model 3*)

$$\text{Minimize } \xi_{j_0} = t - \frac{1}{m}\sum_{i=1}^{m} \frac{\zeta_i^-}{x_{ij_0}} \tag{13}$$

Subject to

$$1 = t + \frac{1}{s}\sum_{r=1}^{s} \frac{\zeta_r^+}{y_{rj_0}}, \tag{14}$$

$$x_{ij_0}t = \sum_{j=1}^{n} x_{ij}\Upsilon_j + \zeta_i^-, i = 1, \ldots, m, \tag{15}$$

$$y_{rj_0}t = \sum_{j=1}^{n} y_{rj}\Upsilon_j - \zeta_r^+, r = 1, \ldots, s, \tag{16}$$

$$\sum_{j=1}^{n} \Upsilon_j = t, \tag{17}$$

$$\sum_{j=1}^{n} \lambda_j = 1, \tag{22}$$

$$\Upsilon_j \geq t - (1 - \lambda_j), j = 1, \ldots, n, \tag{23}$$

$$\Upsilon_j \leq t + (1 - \lambda_j), j = 1, \ldots, n, \tag{24}$$

$$\lambda_j \in \{0, 1\}, j = 1, \ldots, n, \tag{25}$$

$$\Upsilon_j \geq 0, j = 1, \ldots, n, \tag{26}$$

$$\zeta_i^- \geq 0, i = 1, \ldots, m, \tag{19}$$

$$\zeta_r^+ \geq 0, r = 1, \ldots, s, \tag{20}$$

$$t \geq 0. \tag{21}$$

In Model 3, the new constraints are from constraint (22) until constraint (26). Equation (22) is incorporated for ensuring that Υ_j only takes values equal to zero or t. On the other hand, constraints (23) and (24) establish that Υ_j is equal to t when λ_j is equal to one. Finally, constraints (25) and (26) correspond to the nature of the decision variables λ_j and Υ_j. The other constraints of Model 3 were explained in Model 2 description.

2.2 Super-Efficiency DEA Model (SE-DEA)

Andersen & Petersen (1993) proposed the super-efficiency DEA model for ranking all the evaluated units according to their efficiency score. This efficiency score could be greater than one, for an input-oriented model, or lower than one, for an output-oriented model. These values are possible because the data of every evaluated solution j_0 are not considered in the observed data of the DEA model for determining the DEA efficient frontier. Figure 2 presents an example where the Pareto frontier is modified when an extreme solution is withdrawn from the observed data. In this figure, the dashed line represents the frontier when the extreme solution highlighted in yellow is withdrawn. As mentioned previously, the efficiency score for this extreme solution will be greater than one, for an input-oriented DEA model, and lower than one, for an output-oriented DEA model, because it defines the Pareto frontier. Additionally, as it is explained later in this sub-section, for variable returns to scale, the super-efficiency DEA model could be infeasible.

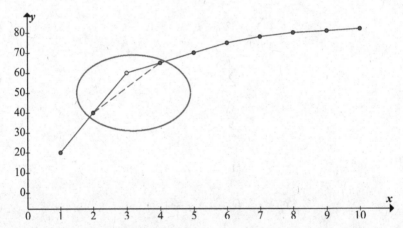

Fig. 2. Pareto frontier modification when an extreme solution is withdrawn from the observed data set.

In this study, an input-oriented super-efficiency DEA model (SE-DEA) was used, aiming to improve the discrimination among the non-dominated solutions. It is important to mention that the model orientation does not vary the identification of the super-efficient solutions' set. In addition, the SE-DEA model must be executed for every solution j, where j_0 corresponds to the evaluated solution in a specific model execution.

The SE-DEA formulation is:

$$\text{Minimize } \delta_{j_0} = \theta \tag{27}$$

Subject to

$$\sum_{\substack{j=1 \\ j \neq j_0}}^{n} \lambda_j x_{ij} \leq \theta x_{ij_0}, i = 1, \ldots, m, \tag{28}$$

$$\sum_{\substack{j=1 \\ j \neq j_0}}^{n} \lambda_j y_{rj} \geq y_{rj_0}, r = 1, \ldots, s, \tag{29}$$

$$\sum_{\substack{j=1 \\ j \neq j_0}}^{n} \lambda_j = 1, \tag{30}$$

$$\lambda_j \geq 0, j = 1, \ldots, n, \tag{31}$$

$$\theta \text{ free} \tag{32}$$

The objective function (27) minimizes the proportional reduction of the minimized objective functions obtained by solution j_0. Constraint (28) establishes that the proportional reduction of the minimized objective functions obtained by solution j_0 must be greater or equal than the composed target in the efficient frontier (left hand of the constraint). Constraint (29) estimates that the maximized objective functions obtained by solution j_0 must be lower or equal than the composed target in the efficient frontier (left hand of the constraint). Constraint (30) imposes the convexity of the efficient frontier, which is associated to variable returns to scale (Banker et al., 1984). Constraints (31) and (32) establish the nature of the decision variables.

In the SE-DEA model, δ_{j_0} corresponds to the efficiency score of the evaluated solution j_0. This efficiency score, differently from a traditional BCC input-oriented model (Banker et al., 1984), could achieve values greater than one or even the model could be infeasible. The necessary and sufficient conditions for infeasibility of SE-DEA models when variable returns to scale are considered (constraint 30), are presented in the study of Seiford & Zhu (1999). Consequently, a solution j_0 that is an extreme point of the Pareto efficient frontier will have a δ_{j_0} value greater than one or the associated model could be infeasible.

In the following sub-section, the performance metrics for evaluating MOLP solution methods and the steps for implementing them using the formulated DEA models are described.

3 Performance Metrics Based on DEA Models

As mentioned previously, the considered categories for evaluating MOLP solution methods are cardinality, accuracy, and diversity. The proposed metrics in every category and the steps for calculating them are presented as follows.

3.1 Cardinality Metric (*CM*)

The cardinality metric (*CM*) represents the domination degree of the solutions obtained by a MOLP method. For this reason, it is a unary metric, using the information of a unique solution set. In this study, it is calculated using the INT-SBM model, where data of all the solutions (S) obtained by a solution method are evaluated. It is important to highlight that the dominated and non-dominated solutions obtained by a solution method are considered as observed data of the model. The following steps must be carried out for obtaining the cardinality metric *CM*.

Step 1: Execute the INT-SBM model for every solution of set S. In this step, a vector Ξ is obtained, which corresponds to the vector of ξ_i, the efficient measure of the INT-SBM model for every solution i of the set S.

Step 2: Calculate the efficiency average of vector Ξ. This value will correspond to the cardinality metric *CM*.

The cardinality metric *CM* is greater than zero, and lower than or equal to one. A value equal to one means that any solution dominated by other does not exist in the set S. On the other hand, a value close to zero means that few non-dominated solutions exist in the set S.

3.2 Accuracy Metric (*AC*)

The accuracy metric (*AC*) represents the domination degree of one MOLP solution method over other MOLP solution method. Furthermore, it is a binary metric because it needs two sets of non-dominated solutions for making the comparison. In this study, for estimating the accuracy metric *AC*, the INT-SBM model and the metafrontier approach, proposed by O'Donnell et al. (2008), are used together. The metafrontier approach allows us to classify the non-dominated solutions into different groups. In this way, two sets of non-dominated solutions, S_1 and S_2, obtained by two different solution methods, are compared. The following steps must be carried out for estimating the proposed accuracy metric *AC*.

Step 1: Execute the INT-SBM model for every non-dominated solution of set S_1. In this step, a vector Ξ^1 is obtained, which corresponds to the vector of ξ_i, the efficient measure of the INT-SBM model for every non-dominated solution i of the set S_1.

Step 2: Execute the INT-SBM model for every non-dominated solution of set S_2. In this step, a vector Ξ^2 is obtained, which corresponds to the vector of ξ_i, the efficient measure of the INT-SBM model for every non-dominated solution i of the set S_2.

Step 3: Execute the INT-SBM model for every solution belonging to the union of sets S_1 and S_2. In this step, a vector Ξ^3 is obtained, which corresponds to the vector of ξ_i, the efficient measure of the INT-SBM model for solution i belonging to the union of sets S_1 and S_2.

Step 4: Separate the efficiency vector Ξ^3 in two sets, efficiency scores of solutions from set S_1 (Ξ_1^3), and efficiency scores of solutions from set S_2 (Ξ_2^3).

Step 5: Calculate the efficiency averages of vectors Ξ^1, Ξ^2, Ξ_1^3, and Ξ_2^3, individually.

Step 6: Make the difference between the efficiency averages of vectors Ξ^1 and Ξ_1^3, which corresponds to AC_1, and between Ξ^2 and Ξ_2^3, which corresponds to AC_2.

Step 7: Calculate the minimum value between AC_1 and AC_2. This value will correspond to the accuracy metric AC.

It is important to notice that AC is greater than or equal to zero, and lower than one. Moreover, the solution method with the minimum value AC will be the best method, meaning that this method obtains a lower number of dominated solutions than the other solution method.

3.3 Diversity Metric (*DM*)

The diversity metric DM evaluates a change in the Pareto frontier when a new solution is added. This is a unary metric because it uses the information of a unique solution set. In this study, the diversity metric DM is calculated using the SE-DEA model. In this model, the non-dominated solutions (*NS*) obtained by a MOLP method are evaluated. The following steps must be carried out for obtaining the DM.

Step 1: Execute the SE-DEA model for every solution of the set *NS*. A vector Δ is obtained, which corresponds to the vector of δ_i, that is, a vector of the efficiency score obtained by the SE-DEA model for every solution i of the set *NS*.

Step 2: Identify the subset of *NS* that corresponds to extreme solutions. These solutions are those that in step 1, obtained a δ_i value greater than one or the respective SE-DEA model is infeasible. This subset denominated *ES*, defines the Pareto frontier.

Step 3: Calculate DM using Eq. (33).

$$DM = \frac{|ES|}{|NS|} \tag{33}$$

The diversity metric DM is greater than zero, and lower than or equal to one. A value close to zero means that most of the solutions are a linear combination of extreme solutions in ES. A value equal to one means that all the solutions are not a linear combination of other extreme solutions in ES. In this way, the best value for the DM is one.

4 Analysis of MOLP Solutions Applying the Proposed DEA Metrics

In this section, the proposed metrics are analyzed in order to demonstrate their adequacy for evaluating different Pareto frontiers' structures. For this reason, in the first subsection, the proposed INT-SBM is applied to show its performance when non-convex regions in the Pareto frontier exist Furthermore, in the second sub-section, two Pareto frontiers obtained by two MOLP metaheuristics are compared using the proposed metrics. The Pareto frontiers were estimated for a real case study (Gómez-Lagos et al., 2021).

4.1 Analysis of the Non-Convex Region of a Pareto Frontier

In this sub-section, an analysis of the non-convex region of a Pareto frontier using the INT-SBM and the SBM models is shown. For this purpose, two Pareto frontiers are

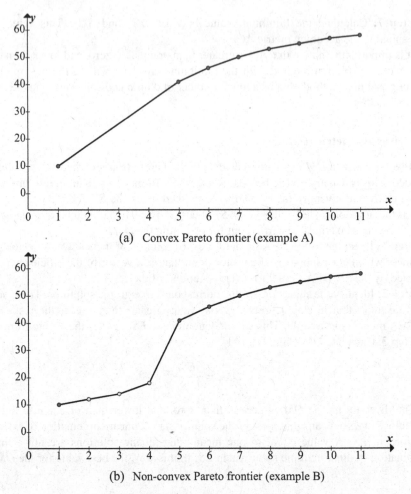

(a)   Convex Pareto frontier (example A)

(b)   Non-convex Pareto frontier (example B)

Fig. 3. Analyzed Pareto frontiers.

compared according to the accuracy metric (*AC*) proposed in Sect. 3. Figure 3 presents the two analyzed Pareto frontiers.

In Fig. 3.a, a convex Pareto frontier is shown (example A), and, in Fig. 3.b, a non-convex Pareto frontier is illustrated (example B). The non-dominated solutions of example b include the non-dominated solutions of example A. Moreover, three non-dominated solutions are included in the Pareto frontier of example B for establishing the non-convex region. The non-dominated solutions of both frontiers are evaluated using the INT-SBM and SBM models. The purpose of this comparison is to demonstrate that the SBM model (Tone, 2001) is not able to recognize and assess the non-convex region of a Pareto frontier. The results of both models are summarized in Table 3.

In Table 3, the first eight non-dominated solutions, which are in the convex region of the Pareto frontiers, are highlighted in red in Figs. 3.a and 3.b. The last three non-dominated solutions are in the non-convex region of the Pareto frontier depicted in

Table 3. Comparison of efficiency scores obtained by INT-SBM and SBM models.

Non-dominated Solutions		Efficiency Score Average for Example A		Efficiency Score Average for Example B	
		INT-SBM	SBM	INT-SBM	SBM
Located in the Convex Region	1	1.00	1.00	1.00	1.00
	2	1.00	1.00	1.00	1.00
	3	1.00	1.00	1.00	1.00
	4	1.00	1.00	1.00	1.00
	5	1.00	1.00	1.00	1.00
	6	1.00	1.00	1.00	1.00
	7	1.00	1.00	1.00	1.00
	8	1.00	1.00	1.00	1.00
Located in the Non-convex Region	9	–	–	1.00	0.63
	10	-	–	1.00	0.51
	11	–	–	1.00	0.51
Average		1.00	1.00	1.00	0.88

Fig. 3.b and are highlighted in yellow. The efficiency scores obtained by both models for solutions in the convex region of the Pareto frontier are equal to one. On the other hand, the efficiency scores obtained by the models for the solutions in the non-convex region of the Pareto frontier differ. Using the INT-SBM model, the efficiency scores for these solutions are one. However, using the SBM model, the efficiency scores for these solutions are lower than one. Therefore, the SBM model considers the solutions in the non-convex region of a Pareto frontier as inefficient, without recognizing their non-dominance characteristic. Comparing the average scores presented in Table 3, the SBM model evaluates the Pareto frontier with non-convex regions worse than the Pareto frontier without non-convex regions. Consequently, the SBM model is not suitable for assessing Pareto frontiers, because it does not consider the possibility of non-convex regions.

4.2 Performance Analysis of Pareto Frontiers for a Real Case Study

In this sub-section, the solutions obtained by two MOLP methods are used for calculating the proposed metrics. The solutions were obtained for a MOLP model based on the tactical harvest planning model proposed by Gómez-Lagos et al. (2021), where the same case study used in this study was analyzed. In this MOLP model, the first objective corresponds to the harvest costs' minimization (Z_1); the second objective corresponds to the harvest days' minimization (Z_2); and the third objective corresponds to the harvest fruit in the optimal conditions' maximization (Z_3). The two applied MOLP methods are two solution strategies of the MO-GRASP algorithm (algorithms a and b), that is, two solution strategies of the multi-objective greedy randomized adaptive search procedure (Martí et al., 2015).

After executing every MO-GRASP algorithm 1000 times for solving the MOLP model, two sets of solutions were obtained, S_a and S_b; one set of 1000 solutions for every algorithm. In Fig. 4, the trade-off between the objective function values obtained by the set S_a are represented. The first trade-off corresponds to Z_1 and Z_2; the second, Z_1 and Z_3; and the third, Z_2 and Z_3.

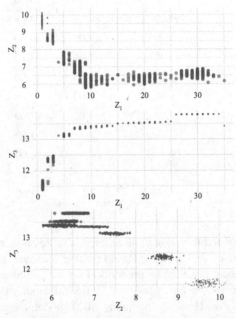

Fig. 4. Trade-offs between objective function values of set S_a (Gómez-Lagos et al., 2023).

A conflict between the objective functions can be observed in Fig. 4 because when an objective function improves, the other deteriorates.

Figure 5 represents the trade-off between the objective function values obtained by the set S_b. The conflict between the objective functions is also observed in Fig. 5, where an objective improves the other objective function worsens.

Table 4 summarizes the metrics calculated for both sets of solutions, S_a and S_b. In this way, it can be observed that the set S_a obtains the best value for the cardinality metric CM (0.980). Regarding the accuracy metric AC, the set S_a again achieves the best value (0.999), meaning that main of solutions of S_a are not dominated by the solutions of S_b. Finally, for the diversity metric DM, both sets obtain low values. However, the set S_b obtains the best value (0.131), meaning that around 13% are extreme-efficient solutions, that is, define the Pareto frontier.

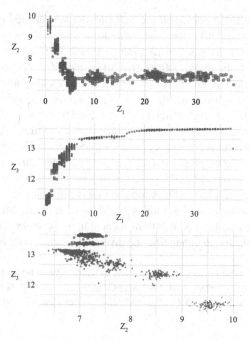

Fig. 5. Trade-offs between objective function values of set S_b (Gómez-Lagos et al., 2023).

Table 4. Values of the performance metrics for S_a and S_b (Gómez-Lagos et al., 2023).

Set of Solutions	CM	AC	DM
S_a	0.980	0.999	0.101
S_b	0.977	0.962	0.131

From the values presented in Table 4, it could be suggested selecting the algorithm for solving the MOLP model because it shows the best performance in the binary metric *AC*, and in the unary metric *CM*. Furthermore, for the unary metric *DM*, around 10% the solutions obtained by the algorithm a are extreme-efficient, close to *DM* obtained by algorithm b.

5   Conclusions

In this study, three performance metrics associated with cardinality, accuracy, and diversity categories, and based on DEA models were proposed for evaluating MOLP solution methods. A procedure was presented for calculating each metric through the application of the proposed DEA models, which are the INT-SBM model, an integer version of the SBM model (Tone, 2001), and the SE-DEA model developed by Andersen & Petersen (1993). The INT-SBM model was developed to properly evaluate non-dominated solution in the non-convex region of the Pareto frontier. In this way, an illustrative example comparing the INT-SBM model with the SBM model was carried out, demonstrating that the INT-SBM can identify and assess the non-convex region of a Pareto frontier. The proposed metrics were applied to analyze Pareto frontiers in a real case study, where the Pareto frontiers obtained from two MO-GRASP algorithms were compared. The INT-SBM efficiency score, used to calculate the cardinality and accuracy metrics, enabled discrimination among the non-dominated solutions regardless their location in the Pareto frontier. The SE-DEA efficiency score, used to estimate the diversity metric, facilitated the identification of the set of non-dominated solutions that determine each Pareto frontier. Therefore, the proposed metrics allow for discrimination by considering desirable characteristics of the Pareto frontiers obtained through different MOLP solution methods, and even aid in the selection of a suitable solution method.

For future research, exploring DEA models that incorporate zero or negative values within the set of non-dominated solutions could be of interest. Moreover, new DEA models to identify non-dominated solutions located in non-convex regions of the Pareto frontier, with an emphasis on enhancing the diversity metric, could be developed.

Acknowledgements. . DSc Marcela González-Araya would like to thank FONDECYT Project 1191764 (Chile) for its financial support. Ms Javier Gómez-Lagos would like to thank CONICYT PFCHA/BECA DE DOCTORADO NACIONAL/2019 under Grant 21191364 for its financial support.

References

Andersen, P., Petersen, N.: A procedure for ranking efficient units in data envelopment analysis. Manage. Sci. **39**(10), 1261–1264 (1993). https://about.jstor.org/terms

Audet, C., Bigeon, J., Cartier, D., Le Digabel, S., Salomon, L.: Performance indicators in multiobjective optimization. Eur. J. Oper. Res. **292**(2), 397–422 (2021). https://doi.org/10.1016/j.ejor.2020.11.016

Bal, A., Satoglu, S.I.: The use of data envelopment analysis in evaluating Pareto optimal solutions of the sustainable supply chain models. Procedia Manuf. **33**, 485–492 (2019). https://doi.org/10.1016/j.promfg.2019.04.060

Banker, R.D., Charnes, A., Cooper, W.W.: Some models for estimating technical and scale inefficiencies in data envelopment analysis. Manage. Sci. **30**(9), 1078–1092 (1984). https://doi.org/10.1287/mnsc.30.9.1078

Charnes, A., Cooper, W., Rhodes, E.: Measuring the efficiency of decision-making units. Company Eur. J. Oper. Res. **2**, 429–444 (1978)

Charnes, A., Cooper, W.W.: Programming with linear fractional functionals. Naval Res. Logistics Quart. **9**, 181–186 (1962). https://doi.org/10.1002/nav.3800090303

Deb, K.: Multi-objective optimization. In: Burke, E., Kendall, G. (eds.) Search Methodologies: Introductory Tutorials in Optimization and Decision Support Techniques, Second Edition, pp. 403–450. Springer, Boston (2014). https://doi.org/10.1007/978-1-4614-6940-7_15

Demir, E., Bektaş, T., Laporte, G.: The bi-objective pollution-routing problem. Eur. J. Oper. Res. **232**(3), 464–478 (2014). https://doi.org/10.1016/j.ejor.2013.08.002

Gómez-Lagos, J.E., González-Araya, M.C., Soto-Silva, W.E., Rivera-Moraga, M.M.: Optimizing tactical harvest planning for multiple fruit orchards using a metaheuristic modeling approach. Eur. J. Oper. Res. **290**(1), 297–312 (2021). https://doi.org/10.1016/j.ejor.2020.08.015

Gómez-Lagos, J., González-Araya, M., & Acosta Espejo, L. (2023). Performance Metrics Based on Data Envelopment Analysis for Evaluating Multi-Objective Linear Programming Solution Methods. 151–157. https://doi.org/10.5220/0011660200003396

Halim, A.H., Ismail, I., Das, S.: Performance assessment of the metaheuristic optimization algorithms: an exhaustive review. Artif. Intell. Rev. **54**(3), 2323–2409 (2021). https://doi.org/10.1007/s10462-020-09906-6

Hong, J.D., Jeong, K.Y.: Combining data envelopment analysis and multi-objective model for the efficient facility location–allocation decision. J. Ind. Eng. Int. **15**(2), 315–331 (2019). https://doi.org/10.1007/s40092-018-0294-2

Karatas, M., Yakıcı, E.: An iterative solution approach to a multi-objective facility location problem. Appl. Soft Comput. J. **62**, 272–287 (2018). https://doi.org/10.1016/j.asoc.2017.10.035

Martí, R., Campos, V., Resende, M.G.C., Duarte, A.: Multiobjective GRASP with path relinking. Eur. J. Oper. Res. **240**(1), 54–71 (2015). https://doi.org/10.1016/j.ejor.2014.06.042

Mirzapour Al-E-Hashem, S.M.J., Malekly, H., Aryanezhad, M.B.: A multi-objective robust optimization model for multi-product multi-site aggregate production planning in a supply chain under uncertainty. Int. J. Prod. Econ. **134**(1), 28–42 (2011). https://doi.org/10.1016/j.ijpe.2011.01.027

O'Donnell, C.J., Rao, D.S.P., Battese, G.E.: Metafrontier frameworks for the study of firm-level efficiencies and technology ratios. Empirical Econ. **34**(2), 231–255 (2008). https://doi.org/10.1007/s00181-007-0119-4

Riquelme, N., Von Lücken, C., Barán, B.: Performance metrics in multi-objective optimization. In: Proceedings - 2015 41st Latin American Computing Conference, CLEI 2015 (2015). https://doi.org/10.1109/CLEI.2015.7360024

Seiford, L.M., Zhu, J.: Infeasibility of super-efficiency data envelopment analysis models. INFOR J. **37**(2), 174–187 (1999). https://doi.org/10.1080/03155986.1999.11732379

Tone, K.: A slacks-based measure of efficiency in data envelopment analysis. Eur. J. Oper. Res. **130**(3), 498–509 (2001). https://doi.org/10.1016/S0377-2217(99)00407-5

Varas, M., Basso, F., Maturana, S., Osorio, D., Pezoa, R.: A multi-objective approach for supporting wine grape harvest operations. Comput. Ind. Eng. **145**, 106497 (2020). https://doi.org/10.1016/j.cie.2020.106497

Comparing Power Flow Models in Tree Networks with Stochastic Load Demands

M. H. M. Christianen[1]($\boxtimes$)(iD), M. Vlasiou[1,2](iD), and B. Zwart[1,3](iD)

[1] Eindhoven University of Technology, Eindhoven, The Netherlands
m.h.m.christianen@tue.nl
[2] University of Twente, Enschede, The Netherlands
[3] Centrum Wiskunde & Informatica, Amsterdam, The Netherlands

Abstract. The process of charging electric vehicles (EVs) within an electricity network is a complex stochastic process. Various factors contribute to this complexity, including the stochastic arrivals and demands of users at charging stations, the nonlinear nature of power flow in the network, and the need to uphold reliability constraints for the network's proper functioning. While nonlinear power flow equations can be approximated by computationally simpler linear equations, the consequences of linearizing the physics in such a complex stochastic process require careful examination. In this study, we apply a blend of analytical and simulation techniques to compare the performance of the nonlinear *Distflow* model with the linear *Linearized Distflow* model in the context of EV charging. The results demonstrate that across various parameter settings, the network's performance is comparable when using either of the power flow models. Specifically, in terms of the mean number of EVs and mean charging time, there is a relative difference of less than 5% between the two models. These findings suggest that the Linearized Distflow model can be effectively employed as a simplified approximation for the Distflow model, providing a faster yet efficient analysis of network performance.

Keywords: Electric vehicle charging · Power flow models · Bandwidth-sharing networks

1 Introduction

In recent years, congestion has become a recurring issue in electrical distribution networks. The combination of increasing electricity consumption, the rapid adoption of renewable energy generation, the ongoing energy transition, and the slow expansion of network infrastructure has pushed these networks to their limits. The consequences of congestion are far-reaching, with potential network damage and the risk of blackouts. The significance of congestion and its potential consequences are highlighted in a study by [11], where they evaluate the impact of the energy transition on an actual electricity network in the Netherlands. The authors show that charging a small number of EVs is enough to cause a blackout in a neighborhood.

In the coming years, the integration of EVs leads to increased congestion in distribution networks. The increasing demand for EVs is driven by factors such as decreasing

F. Liberatore et al. (Eds.): ICORES 2022/2023, CCIS 1985, pp. 138–167, 2024.
https://doi.org/10.1007/978-3-031-49662-2_8

battery costs, advancements in charging technologies, and the implementation of various decision policies [12]. As a result, there is expected to be a rapid and substantial integration of EV load into the power grid.

This paper considers the stochastic process of charging EVs in a neighborhood such that the voltage losses in the distribution network, a small part of the electricity network, stay small. The stochastic nature of this process arises from the random arrivals of EVs at charging stations and the energy demands of EV owners. The voltage losses, also known as voltage drops, on the cables within the network arise from the physical properties of the network and are directly related to power losses in the network.

In a distribution network, the total power generated by the generator is shared among all EVs connected to the network. However, due to power losses that occur during the transportation of electricity, not all EVs charging at different stations can receive an equal amount of power. These losses arise due to resistance in network cables. As a result, the power allocation for each EV in the network depends on the total number of EVs present and their respective charging locations within the network.

While the problem setup allows for a more general problem description, i.e., considering the process of controlling stochastic loads in a neighborhood such that the voltage losses in the distribution network stay small, our specific focus is on EV charging. The model description is appropriate for EVs due to two key elements: neglecting reactive power and considering a maximum number of parking spaces.

In general, power can be divided into two components: active power and reactive power. Active power, also known as real power, is the portion of power that performs useful work, such as generating mechanical energy or producing heat. Reactive power, on the other hand, represents the power consumed or produced by inductive or capacitive elements in an electrical system. The exclusion of reactive power is suitable for EVs as they typically do not introduce a significant reactive power component. This characteristic aligns with resistive loads such as incandescent light bulbs, electric heaters, toasters, and electric stoves, which primarily consume active power without generating substantial reactive power.

Additionally, the consideration of a maximum parking lot capacity makes our setup specific to EVs. It is common for parking lots or charging stations to have a restricted number of charging points available, which naturally establishes a finite capacity for accommodating EVs. At the same time, parking lots often have limited physical space. Consequently, this imposes a maximum threshold on the number of EVs that can be accommodated simultaneously within the given parking lot.

In our modeling approach, we represent the process as a queuing system. Within this system, EVs are treated as individual "jobs", while the charging stations are considered the "servers" responsible for delivering the service, which is the power supplied to the EVs. The power allocation is subject to physical laws and network constraints. At each charging station, all EVs are served simultaneously, and the service begins immediately upon arrival. Furthermore, each EV receives an equal proportion of the power allocated to that particular charging station.

The queuing model we utilize belongs to the broader category of queuing networks known as *bandwidth-sharing networks*. In these networks, customers (in our setting: EVs) necessitate simultaneous service, which entails the parallel usage of multiple "servers" (in our setting: all distribution lines between the EV's location to the generator within the distribution network).

An early paper in the field of queuing literature related to EV charging is [21], where the authors focus on maximizing utility while minimizing the probability of overloads. They emphasize the importance of avoiding synchronization in EV-charging start times to prevent overloads. Next, [4] and [28] delve into queuing models specifically for fast charging. In these studies, the authors propose control mechanisms that result in improved outcomes, such as reduced blocking of EVs, more efficient utilization of power resources, and minimal queuing. In [1], the authors develop a stochastic resource-sharing network for EV charging. They also introduce a fluid approximation to estimate the number of uncharged EVs within a distribution network.

Besides general literature on charging electric vehicles in a queuing framework, there is related literature on the stability of queuing networks that are used in the context of EV charging, although very limited. In [6], the authors investigate the stability of queuing networks that can be used in the context of EV charging through simulations. They find that there exists a threshold on the arrival rates of EVs, beyond which some cars experience increasingly long waiting times to complete their charging. This finding highlights the importance of managing arrival rates to ensure efficient charging processes. The first analytical study in [9] explores the stability of queuing networks for stochastic load demands on a line, which fits the framework of EV charging. The authors compare the critical arrival rates under different power flow models and explicitly compute the differences between these rates as the number of nodes in the network approaches infinity.

In order to model the physical laws and network constraints, we utilize two different approximations of the alternating current (AC) power flow equations [17]. The first model we examine is the nonlinear *Distflow* model, where the nonlinear terms explicitly account for the electric power losses within the system. This model captures the complex nature of power flow in the network. The second model we consider is a simplified version of the Distflow model known as the *Linearized Distflow* model, as proposed in [2,3]. The Linearized Distflow model approximates the power flow equations by disregarding the nonlinear terms that represent the electric power losses in the original Distflow model.

The practical utilization of the Linearized Distflow model as a substitute for the Distflow model is based on the assumption that power losses on cables are generally small. Experimental studies have demonstrated that this approximation introduces only a minor relative error, typically on the order of 1% [10]. However, it is important to investigate the potential magnification of these small relative errors when applied in complex stochastic processes, which is the focus of our research in this paper. Several other numerical studies have also been conducted to validate the accuracy and effectiveness of the Linearized Distflow model [3,5,7,15,19,23,24,26,27]. These studies provide further evidence supporting the practical utility of the Linearized Distflow model in various scenarios.

Regardless of the chosen power flow model, it is crucial to adhere to the physical laws and constraints of the network. One significant constraint in distribution networks is the regulation of voltage losses, commonly referred to as voltage drop, across network cables. These losses arise due to the resistance of the network's cables. Maintaining control over voltage losses is essential to ensure that all network users receive electricity at a safe and reliable voltage level within specified standards, which may vary across different countries [13]. For instance, in accordance with Dutch legislation, the voltage drop within a distribution network must not exceed 4.5% [25].

In this study, we assess the accuracy and effectiveness of the Linearized Distflow model compared to the Distflow model. To achieve this, we utilize simulation techniques along with analytical results that establish the relationship between the power flow equations of both models. Our goal is to compare the performance of different power flow models in the stochastic process of EV charging, specifically in terms of the mean number of EVs and the average charging time for each EV within the network. Furthermore, we focus on analyzing the *critical* arrival rates under both power models. These critical arrival rates represent specific rates at which the mean number of EVs reach their maximum capacity. By studying these critical arrival rates, we gain valuable insights into the difference between the two power flow models. Moreover, to capture more realistic scenarios, we introduce heterogeneity into the network. This entails incorporating variability in the distribution of arrival rates across different nodes within the network. By considering this heterogeneity, we can explore how variations in arrival rates impact the overall network performance.

This paper builds upon our previous work presented in [8], expanding it in two significant directions. First, we consider a more general network topology compared to the previous study. While the network topology in [8] was limited to a line, this paper focuses on a radial, i.e., a tree, topology. This extension allows us to explore the implications of the power flow models in a more general and realistic network structure, as distribution networks often have a tree topology. Second, in contrast to our previous work in [8], this paper provides a comprehensive comparison of both power flow models, allowing for a more informed evaluation of their respective performances and characteristics. By considering a more general network topology and introducing analytical comparisons between the power flow models, we deepen our understanding of the effects of different network topologies and power flow approximations on the overall performance of the system.

This paper yields several notable contributions, which can be summarized as follows. First, we observe that the performance of the Linearized Distflow model is comparable to that of the Distflow model. Specifically, the mean number of EVs and the average charging time of EVs under both power flow models exhibit similar behavior. Moreover, the relative difference in critical arrival rates is below 5%. These findings provide further evidence that the Linearized Distflow model serves as a valid and accurate approximation of the Distflow model, even in scenarios with high heterogeneity and tree network topologies. Second, regarding the heterogeneity within the network, our numerical examples consider cases where one particular node has almost all the load of the system. Even in such highly heterogeneous scenarios, the performance of the network remains unchanged when employing either the Linearized Distflow or the Distflow model. However, it is important to note that the network's performance is not symmetric under the same loads. If the load of an individual node is way larger than the other loads of the other nodes, the performance of the network is different from the performance of the network if the same largest load is placed on a different node.

The paper is structured as follows. We introduce the queuing network and outline the constraints and assumptions of the electrical distribution network. Additionally, we present the power flow models that we consider for analysis. In Sect. 3, we present two analytical results that allow us to compare the feasible regions of the optimization problem governing the dynamics of the queuing model under both power flow models. These results provide valuable insights into the differences between the two models. Section 4 presents a series of numerical experiments that demonstrate the accuracy and effectiveness of the Linearized Distflow model. We also investigate the impact of variability in the distribution of arrival rates on the network's performance. A summary of the paper's results is provided in Sect. 5, where we discuss the main findings and their implications. Finally, in Sect. 6, we gather some proofs of results in this paper.

2   Model Description

This section provides a detailed model description. We discuss the main components of the EV-charging model, i.e.; we describe the characteristics of the queuing, the distribution network, and the power flow models.

2.1   Queuing Model of EV-Charging

We use a queuing model to study the process of charging EVs in a distribution network. In this model, EVs, referred to as *jobs* require service, which is provided by charging stations acting as *servers*. The service being delivered is the power supplied to EVs. Each parking lot is equipped with a single charging station comprising multiple parking spaces, with each space having its own EV charger. The capacity of each queue corresponds to the number of parking spaces available in a charging station, denoted as K, where K is a positive value. It represents the maximum number of EVs that can be accommodated at each charging station simultaneously.

Thus, in the queuing system, we consider N single-server queues, where each queue represents a parking lot with its own arrival stream of jobs. We denote the state of the system at time t as $\mathbf{X}(t) = (X_1(t), \ldots, X_N(t))$ where $X_i(t)$ represents the number of jobs in queue i at time t. At each queue, jobs arrive independently according to Poisson processes with arrival rates λ_i, where i ranges from 1 to N. The service requirements of the jobs are independent exponentially distributed random variables with a mean of 1. If the capacity of a queue is already reached when a new job arrives, the job does not enter the system but is assumed to leave immediately. In other words, if all parking spaces in a charging station are occupied, a newly arriving EV cannot be accommodated and is considered to depart without receiving service.

At each queue, all jobs are served simultaneously and there is no queuing. Each job starts receiving service immediately upon arrival. The service capacity, which represents the charging rate, is allocated equally among all jobs in the queue. We denote the vector of service capacities allocated to each queue at time t as $\mathbf{p}(t) = (p_1(t), \ldots, p_N(t))$. From now on, for simplicity, we drop the dependence on time t from the notation. Therefore, we write X_j and p_j instead of $X_j(t)$ and $p_j(t)$, respectively.

The service capacities, represented by the vector $\mathbf{p}$, are dependent on the current state of the system, characterized by the vector $\mathbf{X}$, which represents the number of jobs at each parking lot. For each state of the system, i.e. a given number of EVs charging at each parking lot, we assume that the charging rates $\mathbf{p}$ are determined as the unique solution of the optimization problem:

$$\max_{\mathbf{p}} \sum_j X_j \log \left(\frac{p_j}{X_j} \right), \tag{1}$$

which are called proportional fair allocations. The feasible region of this optimization problem can take many forms and depends heavily on the power flow model that is used. In Sect. 2.2, we discuss the feasible regions for both power flow models in more detail.

Under our assumption of independent arrivals of EVs according to Poisson processes and independent service requirements that are exponentially distributed, we can model the number of EVs charging at each station as an N-dimensional continuous-time Markov process. The evolution of the queue at node j is given by

$$X_j(t) \rightarrow X_j(t) + 1 \text{ at rate } \lambda_j,$$

and

$$X_j(t) \rightarrow X_j(t) - 1 \text{ at rate } p_j.$$

2.2 Distribution Network Model

As distribution networks are typically radial [1], we model the distribution network as a directed graph $\mathcal{G} = (\mathcal{N}^+, \mathcal{E})$. Here, $\mathcal{N}^+ = \{0\} \cup \mathcal{N}$ represents the set of nodes, with node 0 being the root node. The number of nodes in the network, excluding the root node, is denoted as N, and $\mathcal{E}$ represents the set of directed edges. We assume that $\mathcal{G}$ has a tree topology. Each edge $(i, j) \in \mathcal{E}$ corresponds to a line connecting nodes i and j, where node j is located farther away from the root node than node i. The impedance of each edge $(i, j) \in \mathcal{E}$ is characterized by the values $z_{ij} = r_{ij} + \mathrm{i} x_{ij}$, where r_{ij} and x_{ij} are the resistance and reactance along the lines, respectively. We make the following natural assumption, given that $r_{ij} >> x_{ij}$ in distribution networks [14, 20].

Assumption 1 . Each edge $\epsilon_{i,j} \in \mathcal{E}$ has a resistance value $r_{ij} > 0$ and reactance value $x_{ij} = 0$.

Furthermore, introduce the complex power consumption at node j as $s_j = p_j + \mathrm{i} q_j$, where p_j and q_j denote the active and reactive power consumption at node j, respectively. By convention, a positive active (reactive) power term corresponds to consuming active (reactive) power, while a negative active (reactive) power term corresponds to generating active (reactive) power. Since EVs can only consume active power [6], it is natural to make the following assumption:

Assumption 2. Each node $j \in \mathcal{N}$ has a non-negative active power $p_j \geq 0$ and a reactive power $q_j = 0$.

We denote the voltage at node j as V_j. Given Assumptions 1 and 2, the voltages at each node j, V_j, can be chosen to have zero imaginary components [1,6].

Voltage Drop Constraint. The distribution network constraints, i.e., in our case only the voltage drop constraint, represent the feasible region of (1) and are described by a set $\mathcal{C}$. The set $\mathcal{C}$ is contained in an N-dimensional vector space and represents feasible power allocations. In our setting, a power allocation is feasible if the maximal voltage drop; i.e., the relative difference between the root voltage V_0 and the minimum voltage among all nodes between the root node and any other node, is bounded by a value $\Delta \in (0, \frac{1}{2}]$. Thus, the distribution network constraints can be described as follows:

$$\mathcal{C} := \left\{ \mathbf{p} : \frac{V_0 - \min_{j \in \mathcal{N}^+} V_j}{V_0} \leq \Delta \right\}, \tag{2}$$

where $0 < \Delta \leq \frac{1}{2}$.

2.3   Power Flow Equations

We introduce two commonly used models to represent the power flow that are used for radial systems; i.e., systems where all charging stations have only one (and the same) source of supply. These models are called the *Distflow* model and *Linearized Distflow* model [3,16]. Both power flow models yield a relation in three complex variables (as derived in [16]), namely the magnitude of the squared voltages $|V_j|^2$ for all nodes $j \in \mathcal{N}^+$, the complex power S_{ij} and the complex current I_{ij} for all edges $\epsilon_{ij} \in \mathcal{E}$.

In the following sections, we discuss each power flow model, which provides us with expressions for the (squared) voltages. These voltage expressions make the voltage drop constraint in Eq. (2) more concrete for each power flow model.

Linearized Distflow Model. For the Linearized Distflow model, we follow the approach in [16]. Using Assumptions 1 and 2 yields the following explicit expression for the squared voltage $(V_j^L)^2$ at every node $j \in \mathcal{N}^+$:

$$(V_j^L)^2 = V_0^2 - \sum_{(h,i) \in \mathcal{P}(0,j)} 2 \left(r_{hi} \sum_{l \in \mathcal{T}(i)} p_l \right). \tag{3}$$

Distflow Model. For the Distflow model, we derive a relation in a single complex variable, namely the complex voltage at each node. However, given Assumptions 1 and 2, this relation simplifies to a relation in the real voltage at each node.

This power flow model is based on three principles. The first principle is Kirchhoff's current law (4); the injected current at each bus is equal to the sum of outward flowing current from that bus, i.e.,

$$I_j = -I_{ij} + \sum_{k : j \to k} I_{jk}, \quad (i,j), (j,k) \in \mathcal{E}. \tag{4}$$

Second, by Ohm's law, the current flowing from bus i to bus j is proportional to the voltage across points i and j, where the constant of proportionality is given by the resistance r_{ij}, i.e.,

$$I_{ij} = \frac{V_i^D - V_j^D}{r_{ij}}, \quad (i,j) \in \mathcal{E}.$$

Third, by the complex power formula, the injected current at bus j is given by

$$I_j = \frac{p_j}{V_j^D}, \quad j \in \mathcal{N}^+. \tag{5}$$

Before we proceed with the alternative derivation, we introduce some notation and conventions. We denote $\mathcal{L}$ as the set of leaf nodes in the tree $\mathcal{G}$ and $\mathcal{T}(j)$ as the subtree rooted at node j, including node j itself. To indicate that node k belongs to subtree $\mathcal{T}(j)$, we write "$k \in \mathcal{T}(j)$". Similarly, $(k,l) \in \mathcal{T}(j)$ denotes that the edge (k,l) belongs to subtree $\mathcal{T}(j)$. Additionally, $\mathcal{P}(j,k)$ represents the set of edges on the unique path from node j to node k.

It is worth noting that there are different labeling schemes for nodes in a network. However, for our purposes, the specific labeling is not crucial. The only conventions we adhere to are that the label 0 corresponds to the root of the tree, and $\nu \in \mathcal{L}$ refers to one of the leaf nodes of the tree.

We continue with the alternative derivation. Using Kirchhoff's current law, Ohm's law, and the definition of complex power, we obtain the following set of equations governing the voltages differences in the network;

$$V_h^D - V_i^D = r_{hi} \frac{p_i}{V_i^D}, \quad i \in \mathcal{L}, \tag{6}$$

$$V_h^D - V_i^D = \sum_{k:i \to k} \frac{r_{hi}}{r_{ik}}(V_i^D - V_k^D) + r_{hi} \frac{p_i}{V_i^D}, \quad i \notin \{0, \mathcal{L}\}, \tag{7}$$

$$0 = \sum_{k:0 \to k} \frac{1}{r_{0k}}(V_0^D - V_k^D) + \frac{p_0}{V_0^D}. \tag{8}$$

From (7), we observe that the voltage differences between neighboring nodes form a telescoping series. As a result, we can express the voltage differences between subsequent nodes in a simple closed-form expression. The expression is given in Lemma 1.

Lemma 1. *Let $(h,i) \in \mathcal{E}$, the voltage differences $V_h^D - V_i^D$ is given by*

$$V_h^D - V_i^D = r_{hi} \sum_{\ell \in \mathcal{T}(i)} \frac{p_\ell}{V_\ell^D}. \tag{9}$$

The proof of Lemma 1 can be found in Appendix 6.2.

Furthermore, having the Lemma 1 at our disposal, we recognize another telescoping sum in the left-hand side of (9) for the voltage difference between neighboring nodes when summing over a set of edges on the unique path from node j to node k. This allows us to write an alternative expression for the voltage at every node $j \in \mathcal{N}^+$ in terms of voltages located deeper in the network. The alternative expression is presented in Lemma 2.

Lemma 2. *Let* $j \in \mathcal{N}^+$, *the voltage* V_j^D *is given by*

$$V_j^D = V_k^D + \sum_{(h,i) \in \mathcal{P}(j,k)} \left(r_{hi} \sum_{\ell \in \mathcal{T}(i)} \frac{p_\ell}{V_\ell^D} \right), \tag{10}$$

with the convention that node k *is further away from the root node than node* j.

The proof of Lemma 2 can also be found in Appendix 6.2.

2.4 Summary

In Sects. 2.1–2.3, we covered all individual components of the modeling of the process of charging EVs within a distribution network, employing two different power flow models. In this section, we provide a complete description of this process under both power flow models.

Despite the difference in the power flow model, the processes of charging EVs exhibit the same characteristics. Both processes are defined by:

1. The network topology $\mathcal{G}$, including the set of nodes $\mathcal{N}^+$ that represent parking lots with charging stations and a set of edges $\mathcal{E}$ that represent distribution lines, along with the characteristics of the network, given by the resistances r_{ij} on each edge.
2. The behavior of an individual parking lot by the arrivals and departures of EVs, as well as the current number of EVs being charged at that particular location. If k denotes the number of EVs at a parking lot, then an arrival at this parking lot increases k by 1 and a departure decreases k by 1 up to the maximum number of parking spaces of a parking lot, given by parameter K.
3. The transitions between values of k that occur according to the arrival rates λ_j and the charging rates p_j for parking lot j. In our setting, the charging rates p_j depend on the total number of EVs currently in the network and their location. We assume that the charging rates p_j are the unique solution of the optimization problem in (1), constrained by the voltage drop constraint in (2) for a given voltage drop parameter Δ.

Besides the similarities described above, the difference in both processes can be found in the voltage drop constraint. The calculation of voltages required in the voltage drop constraint depends on the specific power flow model employed. If the Distflow model is utilized, these voltages are computed using Eq. (10). On the other hand, if the Linearized Distflow model is employed, these voltages are computed using Eq. (3).

To distinguish between the two processes, we assign the label "model D" to the process defined by characteristics (1)–((3)) above and Eq. (10) when utilizing the Distflow model. Similarly, we assign the label "model L" to the process defined by characteristics (1)–(3) above and Eq. (3) when employing the Linearized Distflow model.

3   Analytical Results

In this section, we derive two analytical results to compare both power flow models, specifically their feasible regions for optimization problem (1). First, we develop a *duality* in tree networks for the Distflow model. Second, we construct an *inner region* of the feasible region for the Distflow model in terms of the Linearized Distflow model equations.

The first result provides a way to characterize the feasible region of optimization problem (1) for the Distflow model. Observe that the feasible region consists of the implicit expression for the voltages for the Distflow model (cf. Eq. (10)) and the voltage drop constraint (cf. Sect. 2.2). In other words, the first result provides a way to check if there exist voltages that satisfy the non-linear power flow Eqs. (10) and the voltage drop constraint (2) for any given set of active power $\mathbf{p}$.

The reason to develop such a result is simple. The active powers $\mathbf{p}$ are the solution to the optimization problem in (1). If we are not able to check whether the powers $\mathbf{p}$ are a feasible solution to (1), it is impossible to find the powers $\mathbf{p}$ that maximize the expression in (1).

Therefore, in Sect. 3.1, we first establish an equivalence between two cases, namely case I and II. Case I involves fixing the voltage at the root node and verifying the existence of voltages that satisfy the power flow equations and voltage drop constraint. Case II involves setting the voltages at the leaf nodes and checking if there exist voltages such that the maximum voltage difference in the network is smaller than a predefined voltage drop parameter Δ (cf. Eq. (2)). This latter case is computationally favorable given the recursive structure of the voltages starting at the leaf nodes. However, even in this case, it remains nontrivial to check if there exist voltages that satisfy the power flow equations for the Distflow model and the voltage drop constraint. Hence, as a second step, we provide a method for the computations of voltages in case II and present an example involving a small network.

The second result is a direct way to compare the feasible regions of the optimization problems in (1) for both the Distflow model and the Linearized Distflow model. In our previous work [8], we observed that the feasible region under the Linearized Distflow is slightly larger than the feasible region under the Distflow model. This is due to the overestimation of voltages under the Linearized Distflow model. However, with the second result, we construct a feasible region in terms of the Linearized Distflow model equations that is slightly smaller than the feasible region under the Distflow model.

Having an inner region in terms of the Linearized Distflow model offers at least two advantages:

1. The inner region is not only smaller but easy to compute as well. This is due to the availability of the explicit expressions for the squared voltages in the network under this power flow model (cf. Eq. (3)).
2. The inner region provides a lower (or upper) bound on the optimal value when embedded in a maximization (or minimization) problem. For example, consider the optimization problem in (1).

In combination with the slightly larger feasible region under the Linearized Distflow model, the last advantage enables us to bound the solution to such optimization problems for the Distflow model between solutions of computationally easier optimization problems.

Therefore, we construct in Sect. 3.2 an inner region of the feasible region under the Distflow model in terms of the Linearized Distflow model.

3.1 Duality in Tree Networks

In this section, we derive a duality result in tree networks for the Distflow model. The duality result enables us to verify the feasibility of any given power allocation. To establish this duality, we follow the approach described in [22], which provides the proof of the duality result for a line network. In our study, we extend this result from a line network to a more complex tree network structure.

In Corollary 1, we show that a sequence of voltages $V_j^D, \ldots, V_\nu^D$ from any node j in the network to any leaf node $\nu \in \mathcal{L}$ forms a decreasing sequence. This is a technical result that allows us to show that the voltage at node j, as a function of the voltage at a leaf node $\nu \in \mathcal{L}$, is an increasing function. This has been established in Lemma 3. This allows us to prove the desired equivalence between cases I and II.

We first start with the decreasing sequence of voltages.

Corollary 1. *Consider any leaf node* $\nu \in \mathcal{L}$. *Suppose the voltage* V_ν^D *is positive* $(V_\nu^D > 0)$. *Then, for every path from node* j *to node* ν *in the network, the voltages* $V_j^D, \ldots, V_\nu^D$ *form a decreasing sequence.*

Proof. This is an immediate consequence of Lemma 2. We prove it by induction. From Lemma 2, we have for the leaf node ν, i.e. $\nu \in \mathcal{L}$, and its unique parent node M:

$$V_M^D = V_N^D + r_{MN} \frac{p_N}{V_N^D} \geq V_N^D,$$

since r_{MN}, p_N and V_N^D are positive by assumption. Now, suppose that we have a decreasing sequence of voltages for some node k in the unique path from node j to ν. In other words, suppose that $V_k^D \geq V_i^D$ for all edges (k, i) on the unique path from node k to node N $(i.e., (k, i) \in \mathcal{P}(k, N))$. By the induction hypothesis, we have for the unique parent node K of node k,

$$V_K^D = V_k^D + r_{Kk} \sum_{\ell \in \mathcal{T}(k)} \frac{p_\ell}{V_\ell^D} \geq V_k^D + r_{Kk} \frac{\sum_{\ell \in \mathcal{T}(k)} p_\ell}{V_N^D} \geq V_k^D + 0 = V_k^D,$$

since the resistance r_{Kk}, all powers p_ℓ for $\ell \in \mathcal{T}(k)$, and V_N^D are positive.

Before we continue to Lemma 3, we introduce some new notation. To establish the equivalence between case I and case II, we consider the voltage at each node j as a function of the voltage at a leaf node $\nu \in \mathcal{L}$. Thus, let the voltage V_ν^D be positive and define $v_j : \mathbb{R}_+ \to \mathbb{R}_+$ to be the function defined by the equation $v_j(V_\nu^D) = V_j^D$. This is another formulation of the voltage at node j under the Distflow model, as shown in Eq. (10), expressed as a function of the voltage at a leaf node $\nu \in \mathcal{L}$.

In Lemma 3, we show that the function v_j is an increasing function.

Lemma 3. *Consider any leaf node* $\nu \in \mathcal{L}$. *Suppose the voltage* V_ν^D *is positive* $(V_\nu^D > 0)$. *If* $V_0^D \leq 2V_\nu^D$, *then*

$$0 \leq \frac{dv_j}{dV_\nu^D} \leq 1.$$

Proof. This is true for $j = \nu$, since $\frac{dv_\nu}{dV_\nu^D} = \frac{dV_\nu^D}{dV_\nu^D} = 1$. Suppose that there is some $J \in \mathcal{N}^+$ such that $0 \le \frac{dv_j}{dV_\nu^D} \le 1$ for all $j \in \mathcal{T}(J)$, then we have for the (unique) parent node K, by Lemma 2 and the definition of the function v_j,

$$
\frac{dv_K}{dV_\nu^D} = \frac{d}{dV_\nu^D} \left(V_\nu^D + \sum_{(i,k)\in\mathcal{P}(K,\nu)} \left(r_{ik} \sum_{\ell\in\mathcal{T}(k)} \frac{p_\ell}{V_\ell} \right) \right)
$$

$$
= 1 - \sum_{(i,k)\in\mathcal{P}(K,\nu)} \left(r_{ik} \sum_{\ell\in\mathcal{T}(k)} \frac{p_\ell}{v_\ell^2(V_\nu^D)} \frac{dv_\ell}{dV_\nu^D} \right)
$$

$$
= 1 - \sum_{(i,k)\in\mathcal{P}(K,\nu)} \left(r_{ik} \sum_{\ell\in\mathcal{T}(k)} \frac{p_\ell}{V_\ell^2} \frac{dv_\ell}{dV_\nu^D} \right). \tag{11}
$$

First, we show that $\frac{dv_K}{dV_\nu^D} \ge 0$. By Corollary 1, the sequence of voltages $V_K^D, \dots, V_\nu^D$ on the unique path from node K to node ν is a decreasing sequence. This implies that $-\frac{1}{V_\ell^D} \ge -\frac{1}{V_\nu^D}$ for any node ℓ on the path from node K to node ν. In combination with the induction hypothesis this yields,

$$
\frac{dv_K}{dV_\nu^D} \ge 1 - \frac{1}{V_\nu^D} \sum_{(i,k)\in\mathcal{P}(K,\nu)} \left(r_{ik} \sum_{\ell\in\mathcal{T}(k)} \frac{p_\ell}{V_\ell} \right)
$$

$$
= 1 - \frac{1}{V_\nu^D} \left(V_K^D - V_\nu^D \right),
$$

where the last equality follows from Lemma 2. Combining the fact that the sequence of voltages $V_K^D, \dots, V_\nu^D$ on the unique path from node K to node ν is a decreasing sequence and the assumption that $V_0^D \le 2V_\nu^D$, yields,

$$
\frac{dv_K}{dV_\nu^D} = 2 - \frac{V_K^D}{V_\nu^D} \ge 2 - \frac{2V_\nu^D}{V_\nu^D} = 0.
$$

Now, we show that $\frac{dv_K}{dV_\nu^D} \le 1$. By the induction hypothesis, we have $-\frac{dv_\ell}{dV_\nu^D} \le 0$ for all nodes ℓ on the unique path from node K to node ν. Starting from (11), we immediately get,

$$
\frac{dv_K}{dV_\nu^D} = 1 - \sum_{(i,k)\in\mathcal{P}(K,\nu)} \left(r_{ik} \sum_{\ell\in\mathcal{T}(k)} \frac{p_\ell}{V_\ell^2} \frac{dv_\ell}{dV_\nu^D} \right) \le 1,
$$

which concludes the induction step.

Note that the assumption $V_0^D \le 2V_\nu^D$ in Lemma 3 is not restrictive due to the voltage drop constraint stated in Eq. (2). The voltage drop constraint ensures that voltages cannot go below 50% of the voltage at the root node. Therefore, the assumption

$V_0^D \leq 2V_\nu^D$ imposes the same limitation on the voltages in the network as imposed by the voltage drop constraint.

Now, we have all the preliminary results to prove the desired equivalence between the two cases. Specifically, we consider the case where the voltage at the root node is fixed and check if there exist voltages that satisfy the power flow equations and voltage drop constraint. We compare this case to the scenario where the voltages at the leaf nodes are set, and we verify if there exist voltages such that the largest voltage difference in the network is smaller than a predefined voltage drop parameter. The formal statement of this equivalence is summarized in Proposition 1.

Proposition 1. *Let the resistances r_{ik} for all edges $(i, k) \in \mathcal{E}$ and power consumptions p_ℓ for all nodes $\ell \in \mathcal{N}^+$ be given. Then, the following equivalence, for a voltage drop parameter, $0 < \Delta \leq \frac{1}{2}$, for the voltages (under the Distflow model) at the root node and any leaf node ($\nu \in \mathcal{L}$) of the network holds:*

$$\left(\exists x \geq (1 - \Delta)c \; s.t. \; V_\nu^D = x \; and \; v_0(V_\nu^D) = c \right) \; if \; and \; only \; if$$
$$\left(V_\nu^D = (1 - \Delta)c \; and \; v_0(V_\nu^D) \leq c \right), \tag{12}$$

where $c > 0$.

Proof. In order to prove the implication from left to right, we take the negation of the right-hand side of (12). Suppose $v_0(V_\nu^D) > c$. By Lemma 3, v_0 is a increasing function in V_ν^D, so $c < v_0((1 - \Delta)c) \leq v_0(x)$ for $x \geq (1 - \Delta)c$. Hence, there exists no $x \geq (1 - \Delta)c$ such that $v_0(x) = c$.

For the order implication, we first observe that v_0 is a continuous function, because it is a composition of continuous functions itself. To prove the implication from the right-hand side of (12) to the left-hand side of (12), we assume that at $V_\nu^D = (1 - \Delta)c$, we have $v_0(V_\nu^D) \leq c$. Then, by the intermediate value theorem, we know that there exists $(1 - \Delta)c \leq x \leq c$ such that $v_0(x) = c$.

In Proposition 1, the equivalence between cases I and II holds for any leaf node $\nu \in \mathcal{L}$. However, in order to satisfy the condition of maintaining the maximum voltage difference in the network below a predefined threshold Δ, we employ the equivalence between cases I and II specifically for the leaf node where the minimal voltage in the network is attained. This choice ensures that the voltage constraints are met throughout the complete network. The result is summarized in Theorem 1.

Theorem 1. *Let the resistances r_{ik} for all edges $(i, k) \in \mathcal{E}$ and power consumptions p_ℓ for all nodes $\ell \in \mathcal{N}^+$ be given. Denote by H the node with the minimal voltage in the network. If $V_H^D = (1 - \Delta)c$ and $v_0(V_H^D) \leq c$, then the power allocation $\mathbf{p}$ is feasible (given the resistances r_{ik} for all edges $(i, k) \in \mathcal{E}$).*

Proof. Let the resistances r_{ik} for all edges $(i, k) \in \mathcal{E}$ and power consumptions p_ℓ for all nodes $\ell \in \mathcal{N}^+$ be given. Denote by H the node with the minimal voltage in the network. If $V_H^D = (1 - \Delta)c$ and $v_0(V_H^D) \leq c$, then, by Proposition 1, there exists a voltage $V_0^D = c$ and $V_H^D \geq (1 - \Delta)c$. Hence, the relative maximal voltage drop (cf.

Eq. (2)) in the network is

$$\frac{V_0^D - V_{.H}^D}{V_0^D} \leq \frac{c - (1 - \Delta)c}{c} = \Delta,$$

which implies that the power allocation $\mathbf{p}$ is feasible.

It is important to notice that finding the node with the minimal voltage in the network influences the practical use of Theorem 1. Therefore, we conjecture that the minimum voltage is attained at the node that is *heaviest loaded*. Here, we define the heaviest loaded node as the leaf node that yields the largest weighted sum of loads along the path from the root node to that leaf node. The weighted sum takes into account the resistances by multiplying the load on each node by the sum of resistances along the unique path from the root node to that particular node. The conjecture can be summarized as follows:

Conjecture 1. Let the resistances r_{ik} for all edges $(i, k) \in \mathcal{E}$ and power consumptions p_ℓ for all nodes $\ell \in \mathcal{N}^+$ be given. Denote by H the node with the heaviest load, i.e.,

$$H := \arg\max_{\nu \in \mathcal{L}} \left\{ \sum_{(i,k) \in \mathcal{P}(0,\nu)} r_{ik} \sum_{\ell \in \mathcal{T}(k)} p_\ell \right\}. \tag{13}$$

Then, the minimum voltage in the network is located at node H, i.e., $V_\nu^D \leq V_\ell^D$ for all $\ell \in \mathcal{N}^+$.

Conjecture 1 has been confirmed to be true based on our numerical experiments for different small distribution networks across a wide range of parameter settings, providing evidence for its validity.

Notice that it is possible to find the heaviest loaded node H without explicitly the voltages based on the Distflow model. The determination of node H can be done based on network characteristics and a specific power allocation (cf. Eq. (13)).

Furthermore, observe that this conjecture is not crucial for the applicability of Theorem 1, it provides a means to streamline the process of determining feasible power allocations. In cases where the specific node with the minimum voltage is unknown, one can iterate the procedure for each leaf node in the network, verifying if there exists a scenario where the maximum voltage difference is below the predefined threshold Δ. However, if the node at which the minimum voltage is attained is known, the procedure only needs to be executed once, specifically for node H, reducing computational overhead.

Procedure for the Computation of Voltages for the Distflow Model. Until now, we have established a method to determine the feasibility of a given power allocation $\mathbf{p}$ according to Theorem 1. However, it is necessary to develop a procedure for computing the voltages in case II, starting from the leaf nodes and working our way up in the network.

Before we start with the procedure, it is helpful to introduce some new notation. For all remaining leaf nodes, we define the *total load* as the weighted sum of loads along the path from the root node to each specific leaf node. Similar to the case of the heaviest loaded node, the weighted sum accounts for the resistances by multiplying the load at each node by the sum of resistances along the unique path from the root node to that particular node.

We propose to use a forward-backward type algorithm to compute the voltage at the root node for the Distflow model. The algorithm follows these steps:

1. Initialization: compute the total load for all leaf nodes, sort the voltages in ascending order based on their total load, and set the voltages at all leaf nodes equal to the minimum allowed voltage.
2. Backward computation: knowing all voltages in a subtree rooted at node j, allows for the computation of the voltage at its unique parent node i, i.e., we compute in a backward manner,

$$V_i^D = V_j^D + r_{ij} \sum_{\ell \in \mathcal{T}(j)} \frac{p_\ell}{V_\ell}. \tag{14}$$

3. Forward recomputation: when we encounter a node with multiple children, we need to recompute the voltages along all paths leading back to their respective leaf nodes, excluding the path with the highest total load. This recomputation is performed in a forward manner. Given the voltage at a junction node i and the voltages in a subtree rooted in one of the children nodes j of the junction node, excluding the child node j itself, we can compute the voltage at this child node j. This step ensures that all voltage values in the subtree rooted at node i are appropriately updated.

To compute the voltage at node j during the forward step, we follow these steps. First, we rewrite Eq. (14) in terms of the voltage at node j. Then, we multiply both sides of the equation by the voltage at node j and rearrange the terms, resulting in a second-order equation in the voltage at node j:

$$(V_j^D)^2 = \left(V_i - r_{ij} \sum_{\ell \in \mathcal{T}(j), \ell \neq j} \frac{p_\ell}{V_\ell} \right) V_j^D - r_{ij} p_j. \tag{15}$$

Equation (15) provides a relationship between the voltage at node j and the voltages and power allocations of its parent node i and children nodes $\ell \in \mathcal{T}(j)$, excluding node j itself. By solving (15), we can determine the voltage at node j in terms of the network's characteristics, the power allocations and all voltages in the subtree of node j, excluding node j itself:

$$V_j^D = \frac{1}{2} \left(V_i^D - r_{ij} \sum_{\ell \in \mathcal{T}(j), \ell \neq j} \frac{p_\ell}{V_\ell^D} \right) \cdot$$
$$\cdot \left(1 \pm \sqrt{1 - \frac{4 r_{ij} p_j}{\left(V_i^D - r_{ij} \sum_{\ell \in \mathcal{T}(j), \ell \neq j} \frac{p_\ell}{V_\ell^D} \right)^2}} \right). \tag{16}$$

In Sect. 1, it was observed that the voltages at each node j, denoted as V_j^D, can be chosen to have zero imaginary components. Moreover, to ensure that the voltage drop is not violated, we have to use the plus-sign in (16). Using the minus-sign in (16) would result in $V_i^D \geq 2V_j^D$ for any edge $(i, j) \in \mathcal{E}$, indicating that the voltage drop on this edge is at least equal to the maximum allowed voltage drop in the network. Therefore, when computing voltages in a forward manner, we use the expression in (16) with the plus-sign. This allows us to calculate the voltage at node j as follows:

$$
V_j^D = \frac{1}{2} \left(V_i^D - r_{ij} \sum_{\ell \in T(j), \ell \neq j} \frac{p_\ell}{V_\ell^D} \right) \cdot
$$

$$
\cdot \left(1 + \sqrt{1 - \frac{4 r_{ij} p_j}{\left(V_i^D - r_{ij} \sum_{\ell \in T(j), \ell \neq j} \frac{p_\ell}{V_\ell^D} \right)^2}} \right) . \quad (17)
$$

Starting at the leaf node that is heaviest loaded, denoted as node H. By iteratively performing the backward computation and forward recomputation steps as described earlier, we can compute the voltages for all nodes in the network.

Example 1. As an illustrative example, consider a tree network with a fixed number of parking lots, specifically $N = 3$. The network consists of a root node (node 0) connected to node 1, which serves as a parent node to two children, nodes 2 and 3. In this setting, we set the voltage at the root node as $V_0 = 1$. For the cables connecting th nodes, we assume equal resistance values. Specifically, the resistance of each cable is set to $r_{01} = r_{12} = r_{13} = 0.1$. Next, we allocate power to the parking lots. In this example, we allocate equal power to nodes 1 and 2, specifically $p_1 = p_2 = 0.1$, while node 3 is allocated a higher power, specifically $p_3 = 0.2$. To control the voltage drop across the network, we introduce a parameter $\Delta = 0.05$.

The node with the heaviest load is node 3. This can be determined by comparing the total load that is placed on the path from the root node to node 3 with the total load that is placed on the path from the root node to node 2. Specifically, we have:

$$
r_{01} (p_1 + p_2 + p_3) + r_{13} (p_3) = 0.06 > 0.05 = r_{01} (p_1 + p_2 + p_3) + r_{12} (p_2) .
$$

Since the total load on the first path is greater than the total load on the second path, we conclude that node 3 is heaviest loaded. Consequently, we set the voltage at the third node equal to the minimum allowed voltage in the network, which is $1 - \Delta = 0.95$. Thus, we set $V_3 = 0.95$.

By moving backward in the tree, we compute the voltage at node 1:

$$
V_1^D = V_3^D + r_{13} \frac{p_3}{V_3^D} = 0.95 + 0.1 \frac{0.2}{0.95} = 0.9711.
$$

Then, moving forward in the tree, solving for the voltage at node 2, gives

$$
V_2^D = \frac{1}{2} V_1^D \left(1 + \sqrt{1 - \frac{r_{12} p_2}{\left(\frac{1}{2} V_1^D \right)^2}} \right) = 0.9606,
$$

and finally, moving backward in the tree, we find for the voltage at the root node that

$$V_0^D = V_1^D + r_{01} \left(\frac{p_1}{V_1^D} + \frac{p_2}{V_2^D} + \frac{p_3}{V_3^D} \right) \tag{18}$$

$$= 0.9711 + 0.1 \left(\frac{0.1}{0.9711} + \frac{0.1}{0.9606} + \frac{0.1}{0.95} \right) = 1.0023.$$

According to Theorem 1, based on the previous voltage calculations, we can conclude that the power allocation $\mathbf{p} = (0.1, 0.1, 0.2)$ is not feasible for the given network configuration.

However, by making a slight change in the power allocation, we can obtain a feasible power allocation. Let the power allocation for nodes 1 and 2 be the same as before, i.e., let $p_1 = p_2 = 0.1$, but change the power allocation at node 3 to $p_3 = 0.11$. In this case, the node with the heaviest load remains node 3, as we have:

$$r_{01}(p_1 + p_2 + p_3) + r_{13}(p_3) = 0.042 > 0.041 = r_{01}(p_1 + p_2 + p_3) + r_{12}(p_2).$$

By following the same steps as in the previous case, we compute consecutively

$$V_1^D = 0.9618, \ V_2^D = 0.9511, \ \text{and} \ V_0^D = 0.9932.$$

According to Theorem 1, the power allocation $\mathbf{p} = (0.1, 0.1, 0.11)$ is indeed feasible for the given network configuration.

3.2 Inner Region of Feasible Region

In this section, we construct an inner region of the feasible region of the optimization problem in (1) for the Distflow model. An inner region, by definition, restricts the feasible region by excluding certain feasible points. In our case, we express an inner region of the feasible region under the Distflow model in terms of the Linearized Distflow model.

To show the desired inner region of the feasible region, we first obtain the feasible set of power allocations for both power flow models. This set consists of all power allocation vectors, denoted as $\mathbf{p}$, that satisfy the voltage drop constraint set in Eq. (2). In other words, the set that consists of all power allocations vectors $\mathbf{p}$ such that the voltage at each node j satisfies the constraint $V_j(\mathbf{p}) \geq (1 - \Delta)V_0$.

Next, we obtain an inner region of the feasible region of the optimization problem in (1) for the Distflow model for a given voltage drop parameter Δ. Additionally, we carefully select another voltage drop parameter η in such a way that the inner region of the feasible region of the optimization problem in (1) for the Distflow model is equal to the feasible region of the optimization problem in (1) for the Linearized Distflow model that uses the voltage drop parameter η. This inner region corresponds to the set of power allocation vectors $\mathbf{p}$ for which the voltage at each node j satisfies the constraint $V_j^L(\mathbf{p}) \geq (1 - \eta)V_0$, where $V_j^L(\mathbf{p})$ is the voltage at node j obtained using the Linearized Distflow model equations. The result is summarized in Theorem 2.

Theorem 2. *Given* $\Delta \in (0, \frac{1}{2}]$, *we define* $\eta = 1 - \sqrt{1 - 2\Delta(1 - \Delta)}$. *Then, the feasible set of the optimization problem in* (1) *for the Linearized Distflow model, using the constraint* $V_j^L \geq (1 - \eta)V_0$ *for each node* j, *is contained within the feasible set of the optimization problem in* (1) *of the Distflow model, using the constraint* $V_j^D \geq (1 - \Delta)V_0$ *for each node* j, *i.e.,*

$$\left\{ \mathbf{p} : V_j^L(\mathbf{p}) \geq 1 - \eta, \, j \in \mathcal{N}^+ \right\} \subseteq \left\{ \mathbf{p} : V_j^D(\mathbf{p}) \geq 1 - \Delta, \, j \in \mathcal{N}^+ \right\}. \quad (19)$$

Proof. We first derive the feasible set of power allocations for the Distflow model, we then obtain an inner region of this feasible set and lastly, we construct a feasible set of power allocations in terms of the Linearized Distflow model equations that is equal to this inner region.

The feasible set of power allocations for the Distflow model is given by,

$$\left\{ \mathbf{p} : V_j^D(\mathbf{p}) \geq (1 - \Delta)V_0, \, j \in \mathcal{N}^+ \right\} = \left\{ \mathbf{p} : V_H^D(\mathbf{p}) \geq (1 - \Delta)V_0 \right\}, \quad (20)$$

where H denotes the heaviest loaded node at which we conjecture that the minimum voltage in the network is attained. Then, by the expression of the voltages for the Distflow model in Eq. (10), we find

$$\left\{ \mathbf{p} : V_H^D(\mathbf{p}) \geq (1 - \Delta)V_0 \right\} = \left\{ \mathbf{p} : \sum_{(i,k) \in \mathcal{P}(0,H)} \left(r_{ik} \sum_{l \in \mathcal{T}(k)} \frac{p_l}{V_l^D} \right) \leq \Delta V_0 \right\}. \quad (21)$$

However, since we have a minimum allowed voltage in the network, i.e., $V_j^D(\mathbf{p}) \geq (1 - \Delta)V_0$ for all $j \in \mathcal{N}^+$, we obtain an inner region for the feasible set of power allocations for the Distflow model equations, described by the following set:

$$\left\{ \mathbf{p} : \sum_{(i,k) \in \mathcal{P}(0,H)} \left(r_{ik} \sum_{l \in \mathcal{T}(k)} p_l \right) \leq \Delta(1 - \Delta)V_0^2 \right\}. \quad (22)$$

On the other hand, the feasible set of power allocations for the Linearized Distflow model is given by,

$$\left\{ \mathbf{p} : V_j^L \geq (1 - \eta)V_0, \, j \in \mathcal{N}^+ \right\} =$$

$$\left\{ \mathbf{p} : \sum_{(i,k) \in \mathcal{P}(0,H)} \left(r_{ik} \sum_{l \in \mathcal{T}(k)} p_l \right) \leq \eta \left(1 - \frac{1}{2}\eta \right) V_0^2 \right\}, \quad (23)$$

since the minimal voltage under the Linearized Distflow model is attained at the heaviest loaded node H. Now, we choose η in such a way that the inner region of the feasible region of the Distflow model in (22) is equal to the feasible region of the Linearized Distflow model in (23). Therefore, we choose η such that

$$\eta \left(1 - \frac{1}{2}\eta \right) V_0^2 = \Delta(1 - \Delta)V_0^2,$$

or in other words, such that,

$$1 - \eta = \sqrt{1 - 2\Delta(1 - \Delta)}. \tag{24}$$

The construction of the Linearized Distflow model ensures that the feasible set of the Distflow model (under the constraint $V_j^D \geq (1 - \Delta)V_0$ for each node j) is contained within the feasible set of the Linearized Distflow model (under the constraint $V_j^L \geq (1 - \Delta)V_0$ for each node j). In other words, we have an outer relaxation of the feasible set of the Distflow model equations in terms of the Linearized Distflow model equations. By combining this outer relaxation with the inner region obtained in Theorem 2, we have the following sequence of increasing sets:

$$\left\{ \mathbf{p} : V_j^L(\mathbf{p}) \geq (1 - \eta)V_0, \ j \in \mathcal{N}^+ \right\} \subseteq \left\{ \mathbf{p} : V_j^D(\mathbf{p}) \geq (1 - \Delta)V_0, \ j \in \mathcal{N}^+ \right\}$$

$$\subseteq \left\{ \mathbf{p} : V_j^L(\mathbf{p}) \geq (1 - \Delta)V_0, \ j \in \mathcal{N}^+ \right\}, \tag{25}$$

where $1 - \eta$ is defined as in (24). Therefore, we do not only have a lower (upper) bound on the optimal value if we embed the inner region in a maximization (minimization) problem, but we also have an upper (lower) bound in a maximization (minimization) problem. In other words, we can bound the optimal solution of an optimization problem under the Distflow model, using computationally simpler optimization problems under the Linearized Distflow model.

4   Numerical Results

In Sect. 2, we presented our model for the EV-charging process for two power flow models, which we labeled as model D and model LD, representing the Distflow and the Linearized Distflow models, respectively. Both descriptions involve solving constrained optimization problems in (1) to allocate charging rates to each parking lot. The constraints heavily depend on the power flow model that is used. Therefore, in Sect. 3, we obtained analytical results concerning the feasible regions of the optimization problems in (1) for the Distflow and Linearized Distflow model. In this section, we obtain general insights into the performance of models D and LD by simulation over a wide range of parameter settings. This allows us to compare the behavior of model D and model LD. Here, we investigate the impact of varying the total arrival rate to the network as well as the distribution of the arrival rate among different parking lots. Our focus is primarily on two key performance metrics: the mean number of EVs in the network and the mean charging time of an EV. These metrics provide insights into the overall state of the system and the efficiency of the charging process. By examining the behavior of these metrics under different arrival rate scenarios, we gain a valuable understanding of the network's performance.

4.1   Line Network

In this section, we obtain these general insights using a small distribution network with a line topology. We set the number of parking lots to $N = 2$, the voltage at the root node at $V_0 = 1$, the resistance of each cable to $r = 0.1$, and the parameter to control the voltage drop to $\Delta = 0.05$. Initially, we assume that all EVs arrive independently at each parking lot, following Poisson processes with equal rates of $\lambda_1 = \frac{1}{2}\lambda$ and $\lambda_2 = \frac{1}{2}\lambda$. Here, we vary the total arrival rate to the network, λ, between 0.05 and 1. Later, we relax the assumption of equal arrival rates and vary the fraction of EVs that arrive at each parking lot. The model is depicted in Fig. 1.

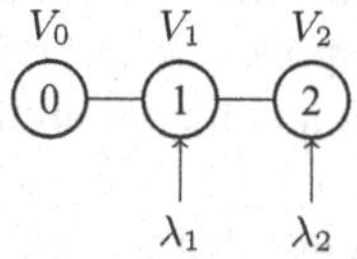

Fig. 1. Line network with $N = 2$ charging stations and arriving vehicles at rate λ_i, $i \in \mathcal{N}$.

In order to effectively manage and control the network, it is important to identify a critical arrival rate, denoted as λ_c. This critical arrival rate serves as a threshold at which the mean number of EVs grows to its maximum capacity. This occurs when the actual arrival rate surpasses the critical arrival rate. See Fig. 2a, where the mean number of EVs is plotted versus the individual arrival rate to each parking lot. The figure displays the mean number of EVs for both power flow models, specifically the Distflow model (dashed) and the Linearized Distflow model (solid). We observe the mean number of EVs at parking lot 1 (blue) and parking lot 2 (red). When examining the variation in the total arrival rate, we notice that the solid curves are close to the dashed lines for the two parking lots. Additionally, the solid curves are consistently below the dashed curves. This observation is not surprising given the findings in [9], where it was noted that the Linearized Distflow power flow model allows for too optimistic arrival rates. This is due to the fact that the Linearized Distflow model overestimates the voltages compared to the Distflow model. Consequently, the power allocated to each parking lot is higher under the Linearized Distflow compared to the Distflow model. Higher allocated power means faster charging. As a result, EVs leave the parking lots faster and the mean number of EVs charging is lower.

Besides the observation that there are critical arrival rates under both power flow models, using the constructed inner region of the feasible region of Theorem 2 and the observation of the outer relaxation in (25) allows us to compute the critical arrival rates under the Linearized Distflow model that bound the critical arrival rate under the Distflow model. As discussed in Sect. 1, there are only K parking spaces at each parking lot. Therefore, from a theoretical perspective, the queuing model that we consider is always stable. However, the critical arrival rate λ_c that we consider is the arrival rate that corresponds to the maximum *stable* arrival rate such that the Markov process $\mathbf{X}$ is positive recurrent given that there is no maximum capacity at each parking lot.

According to [18, Theorem 11], the Markov process $\mathbf{X}$ is positive recurrent for an arrival rate λ if there exists a critical arrival rate $\lambda_c \in \mathcal{C}$ such that $\lambda < \lambda_c$. Here, the constraint set $\mathcal{C}$ depends on the power flow model. For the Linearized Distflow model (independent of the voltage drop parameter), the explicit form of the constraint set $\mathcal{C}$ allows for a simple computation of the maximum stable arrival rate (under the

Linearized Distflow model) that yields a positive recurrent Markov process $\mathbf{X}$. Indeed, given the assumption that the arrival rates for all parking lots are equal, the expression for the squared voltage at the end of the line under the Linearized Distflow model (cf. Eq. (3)) simplifies to,

$$(V_2^L)^2 = 1 - 0.3\lambda_c^L,$$

where λ_c^L denotes the critical arrival rate under the Linearized Distflow model. The maximum stable arrival then corresponds to the arrival rate such that the difference between the voltage at the root node and the minimum voltage in the network is equal to the maximal voltage drop.

For the outer relaxation, this implies that the maximum stable arrival rate under the Linearized Distflow model $\lambda_c^L(\Delta)$ is the solution to the equation $(V_2^L)^2 = (1 - \Delta)^2$. Hence, we denote and compute the critical arrival rate under the Linearized Distflow model using the voltage drop parameter $\Delta = 0.05$, as follows:

$$\lambda_c^L(\Delta) = \frac{1 - (1 - 0.05)^2}{0.3} = 0.325.$$

Similarly, for the inner region, we need for the maximum stable arrival rate under the Linearized Distflow model $\lambda_c^L(\eta)$ that $V_2^L = 1 - \eta$. Thus, we denote and compute the critical arrival rate under the Linearized Distflow mode using the voltage drop parameter η, as follows:

$$\lambda_c(\eta) = \frac{2 \cdot 0.05 \cdot (1 - 0.05)}{0.3} = 0.317.$$

Given the critical arrival rates for both the inner region and outer relaxation, under the Linearized Distflow model, we conclude that the maximum feasible arrival rate under the Distflow model $\lambda_c^D(\Delta)$ is bounded by,

$$0.317 \leq \lambda_c^D(\Delta) \leq 0.325. \tag{26}$$

The critical arrival rates for each parking lot are then equal to half the total critical arrival rates as in (26). The critical arrival rates per parking lot are visualized in Fig. 2b. In this figure, we zoomed in on the mean number of EVs versus the individual arrival rate to each parking lot for a small range of arrival rates as plotted in Fig. 2a. Besides the mean number of EVs, we see the critical arrival rate under the Linearized Distflow model for the inner region (dashed vertical black line) and the outer relaxation (solid vertical black line). For arrival rates that are smaller than these critical arrival rates, we see that the mean number of EVs does not grow to the maximum capacity of the network. On the contrary, for arrival rates that are bigger than these critical arrival rates, it seems that the number of EVs in the network quickly grows to the maximum capacity of the network.

Up to this point, our analysis assumed equal arrival rates for all parking lots. However, in the subsequent analysis, we introduce variability to the distribution of arrival rates by considering different arrival rates for each parking lot. We vary both the sum of the arrival rates and the fraction of EVs that arrive at each parking lot. For each

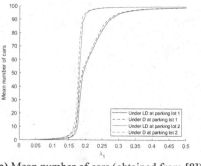

(a) Mean number of cars (obtained from [8])

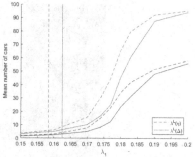

(b) Zoom in on mean number of cars

Fig. 2. Mean number of cars vs. the individual arrival rate per parking lot for the Distflow model (dashed) and the Linearized Distflow model (solid) at parking lot 1 (blue) and at parking lot 2 (red). (Color figure online)

combination of these parameters, we examine the resulting mean number of EVs in the network. This allows us to gain insights into the impact of varying arrival of the network's performance.

The heat map of the mean number of cars exhibits an interesting structure for all combinations of the sum of all arrival rates and fractions of EVs that arrive at each parking lot. In Fig. 3, we present the mean number of EVs in the network as a function of the sum of all arrival rates and the fractions of EVs that arrive at each parking lot. The heat map reveals a non-symmetric structure in the mean number of EVs in the network.

When we increase the fraction of EVs that arrive at parking lot 1 (and thus decrease the fraction of EVs that arrive at parking lot 2) compared to the situation with equal arrival rates, we observe a faster decline in the total mean number of EVs in the network than when we increase the fraction of EVs that arrive at parking lot 2. This is natural given the total available power that can be allocated to each parking lot. This outcome aligns with the total available power that can be allocated to each parking lot. Due to the power loss on the cables, the available power for charging at parking lot 1 is roughly twice that of parking lot 2.

To illustrate this, we compare two scenarios: one where a given number of EVs are charging at parking lot 1 with no EVs charging at parking lot 2 (representing a situation where the fraction of EVs that arrive at parking lot 1 is high), and another scenario where the same number of EVs are charging at parking lot 2 with no EVs charging at parking lot 1 (representing a situation where the fraction of EVs that arrive to parking lot 2 is high). Since the allocated power to parking lot 1 in the first situation is greater than the allocated power to parking lot 2 in the second scenario, the mean number of EVs at parking lot 1 tends to be smaller than the number of EVs at parking lot 2.

Moreover, if we focus on a fixed sum of arrival rates, such as $\lambda_1 + \lambda_2 = 0.8$, the variability in the distribution of the total arrival rate has a minor influence on the mean number of EVs across a wide range of arrival rate ratios. Specifically, we observe that the relative difference remains below 5% when the fraction of EVs arriving at parking lot 1 varies between 20% and 60%.

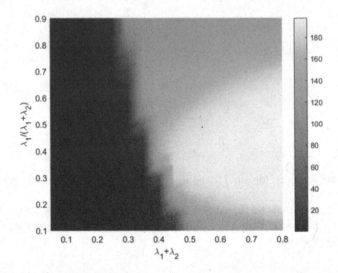

Fig. 3. Mean number of cars (obtained from [8]). (Color figure online)

Another observation from the heat map of the mean number of cars is that there is a clear distinction between networks that have reached their capacity and those that have not. This distinction is visually evident in the color-coded regions.

In the blue region of the heat map, the mean number of EVs is relatively low, indicating that the network is operating below its maximum capacity. However, as we move towards the green and yellow regions, the mean number of EVs increases significantly and approaches its maximum capcity of 200 EVs.

Specifically, the green region on the heat map indicates a network where the number of EVs charging at either one of the parking lots has reached its capacity. This suggests that one of the parking lots is fully occupied with charging EVs, while the other parking lot may still have available spaces.

In contrast, the yellow region indicates a network where both parking lots have reached their capacity, meaning that the number of EVs at both parking lots has reached its maximum. In this scenario, the network is operating at its full charging capacity, with all parking spaces occupied by charging EVs.

4.2   Tree Network

In the previous section, we conducted a series of experiments on a small line distribution network. In this section, we extend our investigations to a small tree distribution network. Specifically, we consider a tree network with the number of parking lots equal to 3, i.e., $N = 3$.

In the tree distribution network, the root node (node 0) is connected to node 1, which has two children, nodes 2 and 3. Furthermore, we set the voltage at the root node to $V_0 = 1$, a uniform resistance value $r = 0.1$ for each cable. Additionally, we use the parameter, $\Delta = 0.05$, to control the voltage drop.

Similar to our previous experiments, we assume that at each parking lot, EVs arrive independently according to Poisson processes with rates $\lambda_1 = \frac{2}{5}\lambda, \lambda_2 = \frac{3}{10}\lambda, \lambda_3 = \frac{3}{10}\lambda$. Here, λ represents the total arrival rate, which we vary between the range of 0.05 and 0.8. The model is illustrated in Fig. 1.

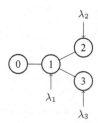

Fig. 4. Tree network with $|\mathcal{N}| = 3$ charging stations and arriving vehicles at rate λ_i, $i \in \mathcal{N}$.

In Fig. 5, the mean number of cars and mean charging time of an EV are visualized against the total arrival rate to the network. The plot showcases the mean number of EVs for both the Distflow model (dashed lines) and the Linearized Distflow model (solid lines) at each parking lot, represented by distinct colors: parking lot 1 (blue), parking lot 2 (red), and parking lot 3 (green).

For every parking lot, the solid curves are close to, and always below, the dashed lines. This behavior stems from the fact that the Linearized Distflow model overestimates voltages compared to the Distflow model, as discussed earlier in Sect. 4.1.

From Fig. 5, it becomes clear that at every parking lot, as the actual total arrival rate to the network surpasses the critical arrival rate, the mean number of EVs at each parking lot rapidly approaches its maximum capacity. Simultaneously, the mean charging time experiences a substantial increase across all parking lots when the actual total arrival rate exceeds the critical arrival rate.

Although the allocated power to parking lot 1 is not enough to lower the mean number of cars in the system if the total arrival rate to the network is greater than the critical arrival rate, the allocated power to parking lot 1 lowers the mean charging time of an EV. This effect can be attributed to the power loss that occurs on the cables. The power allocated to parking lot 1 is higher than the power allocated to parking lots 2 and 3, which leads to shorter mean charging times for EVs at parking lot 1. This phenomenon is clearly illustrated in Fig. 5.

We extend our analysis to the tree network depicted in Fig. 4 and compute bounds on the critical arrival rate for the Distflow model using the inner region approximation from Theorem 2 and the outer relaxation from (25), similar to what was done for the line network in Sect. 4.1. Assuming arrival rates of $\lambda_1 = \frac{2}{5}\lambda$, $\lambda_2 = \frac{3}{10}\lambda$, and $\lambda_3 = \frac{3}{10}\lambda$ for each parking lot, and equal cable resistances, the total load on leaf nodes 2 and 3 is identical. Consequently, the expression for the squared voltage at the heaviest loaded node under the Linearized Distflow model simplifies to:

$$(V_2^L)^2 = 1 - 0.26\lambda_c^L,$$

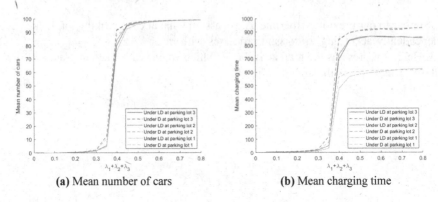

(a) Mean number of cars   **(b)** Mean charging time

Fig. 5. Performance measures vs. total arrival rate per parking lot for the Distflow model (dashed) and the Linearized Distflow model (solid) at parking lot 1 (blue), at parking lot 2 (red), and parking lot 3 (green). (Color figure online)

where λ_c^L represents the critical arrival rate under the Linearized Distflow model. By computing the critical arrival rates for both the inner region and outer relaxation under the Linearized Distflow model, we obtain the following bounds on the maximum feasible arrival rate under the Distflow model:

$$0.365 \leq \lambda_c^D(\Delta) \leq 0.375.$$

The critical arrival rates are visually represented in Fig. 5. Here, we see that for arrival rates smaller than these critical values, the mean number of EVs does not reach the maximum capacity of the network, and the mean charging time of an EV remains low. However, as the arrival rates exceed these critical rates, both the number of EVs in the network and the mean charging time of an EV increase significantly, indicating that the network's capacity is being quickly reached.

Instead of assuming fixed arrival rates for all parking lots, we consider different arrival rates for each parking lot. Following the experiments conducted in the case of the line network discussed in Sect. 4.1, we explore various combinations of the sum of the total arrival rate to the network and the fraction of EVs arriving at each parking lot. Our objective is to examine the mean number of EVs in the network for these different combinations. To visualize the relationship between the mean number of EVs and the sum of the total arrival rate along with the fractions of EVs arriving at each parking lot, we adopt a similar approach as shown in Fig. 3. Specifically, we fix the fraction of EVs arriving at parking lot 2 and analyze three distinct scenarios: one where this fraction corresponds to 10% of the total arrival rate, another with 30%, and a third with 60%.

Figure 6 reveals that the observations made in the case of the line network also apply to the tree network. Specifically, we observe the following consistent patterns:

1. The mean number of EVs exhibits a non-symmetric structure, indicating that the distribution of EVs across parking lots significantly impacts the overall network performance.

2. When considering a fixed total arrival rate, the variability in the distribution of arrival rates among parking lots has minimal influence on the mean number of EVs. This finding holds true across a wide range of fraction combinations.
3. A clear distinction persists between networks that have reached their capacity and those that have not. In the former, where the mean number of EVs approaches the maximum network capacity, the heatmap exhibits higher values, while in the latter, the mean number of EVs remains relatively low.

These consistent patterns across both the line and tree network demonstrate the robustness of our findings and provide valuable insights into the behavior of EV charging networks under varying arrival rate distributions.

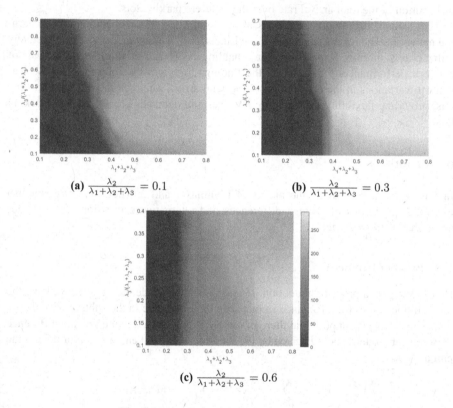

(a) $\frac{\lambda_2}{\lambda_1+\lambda_2+\lambda_3} = 0.1$ (b) $\frac{\lambda_2}{\lambda_1+\lambda_2+\lambda_3} = 0.3$

(c) $\frac{\lambda_2}{\lambda_1+\lambda_2+\lambda_3} = 0.6$

Fig. 6. Mean number of cars in tree network.

5   Summary

In summary, our analysis reveals that the performance of both the Distflow model and the Linearized Distflow model in the process of charging EVs is comparable. Simulation results indicates that, across a wide range of total arrival rates, the mean number of EVs and mean charging time exhibit a relative difference of less than 5% between the two models. This suggests that the Linearized Distflow model provides a reasonable approximation of the more complex Distflow model.

Furthermore, the critical arrival rates obtained from both power flow models demonstrate close alignment, indicating consistent behavior and agreement between the models. This observation is reinforced by the computation of critical arrival rates under the Linearized Distflow model, which provides valuable bounds for the critical arrival rate under the Distflow model.

Additionally, we find that the variability in the distribution of the total arrival rates to parking lots has minimal influence on network performance. As long as the fraction of EVs arriving at each parking lot falls within the range of 20% to 60%, the network's performance remains relatively the same, with negligible impact from variations in the distribution of the total arrival rate over the different parking lots.

These insights have practical implications for the design and operation of EV charging networks. The results suggest that the Linearized Distflow model can be effectively utilized as a simplified approximation, enabling faster and more efficient analysis of network performance. Furthermore, the findings highlight the robustness of the network to variations in the distribution of the total arrival rate over the different parking lots, providing flexibility in managing EV charging demands across different parking lots.

6 Proofs

In this section, we provide the proofs of Lemmas 1 and 2. These lemmas establish important properties of voltage differences and alternative expressions for voltages under the Distflow model.

6.1 Proof of Lemma 1

Proof. We give a proof by induction. If node i is a leaf node, i.e., $i \in \mathcal{L}$, we have the result immediately from (6) since in that case the only node in the subtree of node i is node i itself. Now, suppose that there is some $j \in \mathcal{N}^+$ such that the voltage difference between subsequent nodes is given by (9) for all $k \in \mathcal{T}(j)$, then we have for the unique parent node i,

$$
V_h^D - V_i^D = \sum_{j:i \to j} \frac{r_{hi}}{r_{ij}} (V_i^D - V_j^D) + r_{hi} \frac{p_i}{V_i^D}
$$

$$
= \sum_{j:i \to j} \frac{r_{hi}}{r_{ij}} \left(r_{ij} \sum_{\ell \in \mathcal{T}(j)} \frac{p_\ell}{V_\ell^D} \right) + r_{hi} \frac{p_i}{V_i^D}, \tag{27}
$$

where we used the induction hypothesis in (27). Notice that iterating over all child nodes j of node i, and then summing over all nodes in the subtrees of nodes j is the same as summing over all nodes in the subtree of node i excluding node i itself. Hence, after simplifying the first term in (27), we find for the voltage difference,

$$
\begin{aligned}
V_h^D - V_i^D &= r_{hi} \sum_{j:i \to j} \sum_{\ell \in \mathcal{T}(j)} \frac{p_\ell}{V_\ell^D} + r_{hi} \frac{p_i}{V_i^D} \\
&= r_{hi} \sum_{\ell \in \mathcal{T}(i), \ell \neq i} \frac{p_\ell}{V_\ell^D} + r_{hi} \frac{p_i}{V_i^D}.
\end{aligned}
\tag{28}
$$

Combining the two terms on the right-hand side of (28) yield

$$
V_h^D - V_i^D = r_{hi} \sum_{\ell \in \mathcal{T}(i)} \frac{p_\ell}{V_\ell^D},
$$

as desired.

6.2 Proof of Lemma 2

The result is an immediate consequence of Lemma 1. Summing over all edges on the unique path from node j to node k yields a telescoping sum on the left-hand side of (9), and gives

$$
V_j^D - V_k^D = \sum_{(h,i) \in \mathcal{P}(j,k)} \left(r_{hi} \sum_{\ell \in \mathcal{T}(i)} \frac{p_\ell}{V_\ell^D} \right),
\tag{29}
$$

which is, after bringing the voltage V_k^D to the right-hand side of (29), equal to the desired result.

References

1. Aveklouris, A., Vlasiou, M., Zwart, B.: A stochastic resource-sharing network for electric vehicle charging. IEEE Trans. Control Netw. Syst. 6(3), 1050–1061 (2019). https://doi.org/10.1109/TCNS.2019.2915651
2. Baran, M.E., Wu, F.F.: Optimal capacitor placement on radial distribution systems. IEEE Trans. Power Deliv. 4(1), 725–734 (1989). https://doi.org/10.1109/61.19265
3. Baran, M.E., Wu, F.F.: Optimal sizing of capacitors placed on a radial distribution system. IEEE Trans. Control Netw. Syst. 4(1), 735–743 (1989)
4. Bayram, I.S., Michailidis, G., Papapanagiotou, I., Devetsikiotis, M.: Decentralized control of electric vehicles in a network of fast charging stations. In: GLOBECOM - IEEE Global Telecommunications Conference, pp. 2785–2790 (2013). https://doi.org/10.1109/GLOCOM.2013.6831496
5. Cao, Y., Wei, W., Wang, J., Mei, S., Shafie-Khah, M., Catalao, J.P.: Capacity planning of energy hub in multi-carrier energy networks: a data-driven robust stochastic programming approach. IEEE Power Energy Soc. General Meet. 2019 11(1), 3–14 (2019). https://doi.org/10.1109/PESGM40551.2019.8973417
6. Carvalho, R., Buzna, L., Gibbens, R., Kelly, F.: Critical behaviour in charging of electric vehicles. New J. Phys. 17(9), 95001 (2015). https://doi.org/10.1088/1367-2630/17/9/095001

7. Chen, C., Wang, J., Qiu, F., Zhao, D.: Resilient distribution system by microgrids formation after natural disasters. IEEE Trans. Smart Grid 7(2), 958–966 (2016). https://doi.org/10.1109/TSG.2015.2429653
8. Christianen., M., Vlasiou., M., Zwart., B.: Simulation study for the comparison of power flow models for a line distribution network with stochastic load demands. In: Proceedings of the 12th International Conference on Operations Research and Enterprise Systems - ICORES, pp. 167–174. INSTICC, SciTePress (2023). https://doi.org/10.5220/0011670600003396
9. Christianen, M., Cruise, J., Janssen, A., Shneer, 'S., Vlasiou, M., Zwart, B.: Comparison of stability regions for a line distribution network with stochastic load demands (2021)
10. Farivar, M., Chen, L., Low, S.: Equilibrium and dynamics of local voltage control in distribution systems. In: Proceedings of the IEEE Conference on Decision and Control, pp. 4329–4334 (2013). https://doi.org/10.1109/CDC.2013.6760555
11. Hoogsteen, G., Molderink, A., Hurink, J.L., Smit, G.J., Kootstra, B., Schuring, F.: Charging electric vehicles, baking pizzas, and melting a fuse in Lochem. CIRED - Open Access Proc. J. 2017(1), 1629–1633 (2017). https://doi.org/10.1049/oap-cired.2017.0340
12. IEA: Global EV Outlook 2021 - Accelerating ambitions despite the pandemic (2021). https://iea.blob.core.windows.net/assets/ed5f4484-f556-4110-8c5c-4ede8bcba637/GlobalEVOutlook2021.pdf
13. Kersting, W.: Distribution System Modeling and Analysis, 4th edn. CRC Press, Boca Raton (2018)
14. Khatod, D.K., Pant, V., Sharma, J.: A novel approach for sensitivity calculations in the radial distribution system. IEEE Trans. Power Deliv. 21(4), 2048–2057 (2006). https://doi.org/10.1109/TPWRD.2006.874651
15. Li, R., Wei, W., Mei, S., Hu, Q., Wu, Q.: Participation of an energy hub in electricity and heat distribution markets: an MPEC approach. IEEE Trans. Smart Grid 10(4), 3641–3653 (2019). https://doi.org/10.1109/TSG.2018.2833279
16. Low, S.H.: Convex relaxation of optimal power flow - Part i: formulations and equivalence. IEEE Trans. Control Netw. Syst. 1(1), 15–27 (2014). https://doi.org/10.1109/TCNS.2014.2309732
17. Molzahn, D., Hiskens, I.: A survey of relaxations and approximations of the power flow equations. Found. Trends Electric Energy Syst. 4(1), 1–221 (2019). https://doi.org/10.1561/9781680835410
18. Shneer, S., Stolyar, A.: Stability and moment bounds under utility-maximising service allocations: finite and infinite networks, pp. 1–26 (2018). http://arxiv.org/abs/1812.01435
19. Tan, S., Xu, J.X., Panda, S.K.: Optimization of distribution network incorporating distributed generators: an integrated approach. IEEE Trans. Power Syst. 28(3), 2421–2432 (2013). https://doi.org/10.1109/TPWRS.2013.2253564
20. Tonso, M., Morren, J., De Haan, S.W., Ferreira, J.A.: Variable inductor for voltage control in distribution networks. In: 2005 European Conference on Power Electronics and Applications 2005 (2005). https://doi.org/10.1109/epe.2005.219707
21. Turitsyn, K., Sinitsyn, N., Backhaus, S., Chertkov, M.: Robust broadcast-communication control of electric vehicle charging, pp. 203–207 (2010). https://doi.org/10.1109/smartgrid.2010.5622044
22. Vasmel, N.: Electrical grid failures. Master thesis, Leiden University (2019)
23. Wang, Z., Chen, B., Wang, J., Kim, J., Begovic, M.M.: Robust optimization based optimal DG placement in microgrids. IEEE Trans. Smart Grid 5(5), 2173–2182 (2014). https://doi.org/10.1109/TSG.2014.2321748
24. Wang, Z., Chen, H., Wang, J., Begovic, M.: Inverter-less hybrid voltage/var control for distribution circuits with photovoltaic generators. IEEE Trans. Smart Grid 5(6), 2718–2728 (2014). https://doi.org/10.1109/TSG.2014.2324569

25. van Westering, W., Hellendoorn, H.: Low voltage power grid congestion reduction using a community battery: design principles, control and experimental validation. Int. J. Electr. Power Energy Syst. **114**, 105349 (2020). https://doi.org/10.1016/j.ijepes.2019.06.007

26. Yeh, H.G., Gayme, D.F., Low, S.H.: Adaptive VAR control for distribution circuits with photovoltaic generators. IEEE Trans. Power Syst. **27**(3), 1656–1663 (2012). https://doi.org/10.1109/TPWRS.2012.2183151

27. Yuan, W., Wang, J., Qiu, F., Chen, C., Kang, C., Zeng, B.: Robust optimization-based resilient distribution network planning against natural disasters. IEEE Trans. Smart Grid **7**(6), 2817–2826 (2016). https://doi.org/10.1109/TSG.2015.2513048

28. Yudovina, E., Michailidis, G.: Socially optimal charging strategies for electric vehicles. IEEE Trans. Autom. Control **60**(3), 837–842 (2015). https://doi.org/10.1109/TAC.2014.2346089

Robust Optimization for Operating Room Scheduling with Uncertain Surgical Durations: Impact of Risk-Aversion on Delay

Mari Ito[1]([✉])[iD], Yoshito Namba[2], and Ryuta Takashima[2][iD]

[1] Center for Mathematical and Data Sciences, Kobe University, 1-1, Rokkodai-cho, Nada-ku, Kobe, Hyogo 657-8501, Japan
mariito@opal.kobe-u.ac.jp
[2] Department of Industrial and Systems Engineering, Tokyo University of Science, 2641 Yamazaki, Noda, Chiba 278-8510, Japan
7421521@ed.tus.ac.jp, takashima@rs.tus.ac.jp

Abstract. We introduce a robust optimization model for scheduling operating rooms with uncertain surgical durations. The model addresses multiple operating rooms and surgical procedures. In the numerical analysis, we verify the influence of the risk-averse tendency on the schedule. The schedules created by the robust optimization are compared with those of stochastic programming. The results suggest that robust optimization avoids long delays, and obtains a solution faster than stochastic programming. In specific control conservative, robust optimization exhibits the same performance as stochastic programming. The robust optimization model is more effective for operating room managers who desire to obtain an accurate solution quickly.

Keywords: Operations research in health service · Operating room scheduling · Robust optimization · Stochastic programming

1 Introduction

Hospital management is critical to improving the quality of service to patients [19]. Surgeries account for most of the hospital revenue and expenditure [15,18]. Efficient surgical management is required to achieve optimal hospital management. By clarifying the cost structure underlying operating room time, Dexter and Macario revealed that improved operating room scheduling can effectively reduce costs [8]. Creating a robust operating room schedule is effective in managing the operating room.

In the flow of the operating room scheduling, the surgeon and patient decide the surgery date through mutual agreement. The surgeon then reports the

Supported by JSPS KAKENHI (Japan) Grant Number JP21K14371.

surgery date and estimated duration of surgery to the operating room manager. The manager decides when and in which operating room to perform the surgery based on information such as the estimated duration of surgery and department.

It is the issue of operating room management that surgery is often not performed according to the schedule. The quality of service to patients is affected because of the waiting time occurrence owing to the delay from the scheduled end time of surgery. Surgical duration is uncertain, influenced by the patient's condition, lack of information on the preoperative diagnosis, and the surgeon's skill. The challenge is to cope with the uncertainty of the surgical duration.

The manager desires to avoid a risk of delay, with surgery being delayed significantly from the scheduled end time. Long delays lead to increased overtime for surgical staff, not only increasing costs, but also reducing staff satisfaction. In operating room scheduling, it is necessary to consider decision-making to avoid the risk of delay.

In this study, we propose a robust optimization model that considers surgical procedures and minimizes delay from the regular opening time of the operating room. After calculating the delay for uncertain surgical duration parameter sets in the numerical analysis and comparing the performance of the proposed model to those of the stochastic programming model, we verify whether risk avoidance tendencies are reflected in schedules. In the proposed model, we consider multiple operating rooms that could not be considered in Namba et al. [14].

From the numerical results, three important points are obtained as follows:

- Robust optimization tends to avoid long delays.
- Robust optimization, which has only information on the 10th and 90th percentile duration of the scenario, exhibits the same performance as stochastic programming, which has complete information on the scenario in specific control conservative.
- Robust optimization obtains a solution faster than stochastic programming because the number of variables and constraints in robust optimization is smaller than in stochastic programming.

The remainder of the paper is organized as follows: Sect. 2 provides a literature review. Section 3 introduces the problem setting and robust optimization model of operating room scheduling. In Sect. 4, we describe the data used in the study and report the results of our numerical experiments. We conclude the paper in Sect. 5 with comments regarding matters for future exploration.

2 Literature Review

Operating room scheduling has been studied extensively [5,10,20]. A few studies have proposed a stochastic model for operating room scheduling [2,9,17]. Batun et al. [4] presented a two-stage stochastic mixed-integer programming model with uncertain surgical duration. Addis et al. [1] proposed the operating room rescheduling with uncertain patient arrival and surgical duration.

Kamran et al. [16] approached the operating room scheduling problem with different formulations of stochastic programming. Ito et al. [11,12] proposed a stochastic programming model for scheduling an operating room using the conditional value-at-risk (CVaR). The CVaR expresses the risk-aversion of the manager towards the risk that the surgical duration estimated by the surgeon could be significantly delayed. Another technique that reflects delayed risk aversion is robust optimization. Bandi and Gupta [3] developed a new criterion and a robust optimization approach for staffing and operating room scheduling problems under uncertain case mix and case lengths. Denton et al. [6] proposed an operating room scheduling model with robust optimization to address the uncertain surgical duration.

Our work is somewhat related to Denton et al. [6], but is particularly different in that their study did not address the sequence of surgeries in the operating room. It is important to consider the sequence of surgeries within the operating room [7]. The manager is making efforts regarding the order of surgeries, e.g., surgeries belonging to the same department consecutively perform when arranging surgical equipment and adjusting schedules.

3   Operating Room Scheduling

3.1   Robust Optimization

We propose a robust optimization model for the operating room scheduling problem under uncertain parameter sets; the worst-case that results in maximum total surgical duration. The operating room scheduling determines the allocations to an operating room and surgery procedures.

Surgeries are limited to elective surgeries with prior consent between the patient and surgeon; thus, we did not consider the interruption of emergency surgery. In addition, all operating rooms are treated with the same function. We define surgical duration as the difference between a patient's entry times and when the patient leaves the operating room.

The formulation of the maximum surgical duration problem, which is considered the main problem, is presented in Sect. 3.2. The formulation of the operating room scheduling problem is shown as follows:

Notation
Index sets.

J: Set of surgeries.
D: Set of departments.
E_d $(d \in D)$: Set of surgeries belonging to the department d.
M: Set of operating rooms.

Parameters.

$d_m(m \in M)$: Regular opening time of the operating room.
$n_d(d \in D)$: Number of surgeries in department d, $n_d = |E_d|$.

$\overline{p_j}, \underline{p_j}(j \in J)$: Upper and lower bounds on the duration for surgery j.
τ: Control conservative. Set how conservatively you want to control the worst-case scenario from the decision-maker's perspective. This represents the number of surgeries for which the upper bound of the surgical duration is reached.

Variables.

$p_j(j \in J)$: Duration of surgery j.
$c_j(j \in J)$: Finishing time of surgery j.
$t_m(m \in M)$: Delay from the regular opening time of the operating room.
$z_{ij}(i, j \in J, i \neq j)$: Binary variable for surgery precedence, where $z_{ij} = 1$ if surgery i is processed before surgery j, $z_{ij} = 0$ otherwise.
$x_{mj}(m \in M, j \in J)$: Binary variable for surgery assignment to the operating room, where $x_{mj} = 1$ if in operating room m, surgery j is assigned, $x_{mj} = 0$ otherwise.
$\gamma_{mij}(m \in M, i, j \in J, i \neq j)$: Linearized binary variables, where $\gamma_{mij} = 1$ if in operating room m, surgery i precedes surgery j, $\gamma_{mij} = 0$ otherwise.
$\theta_m(m \in M)$: Linearized binary variables, where $\theta_m = 1$ if surgeries l and k in the department d perform in operating room m, $\theta_m = 0$ otherwise.
$\alpha, \beta_j(j \in J)$: Dual variables.
Formulation
Minimize

$$\sum_{m \in M} t_m \tag{1}$$

subject to

$$\sum_{i \in J} p_j x_{mj} \leq d_m + t_m, \quad \forall m \in M, \tag{2}$$

$$\sum_{m \in M} x_{mj} = 1, \quad \forall j \in J, \tag{3}$$

$$z_{ij} + z_{ji} = 1, \quad i \neq j, \forall i, j \in J, \tag{4}$$

$$z_{ij} + z_{jk} + z_{ki} \leq 2, \quad i \neq j, j \neq k, k \neq i, \forall i, j, k \in J, \tag{5}$$

$$\sum_{j \in J} j x_{(m-1)j} \geq \sum_{j \in J} j x_{mj}, \quad m = 2, ..., |M|, \tag{6}$$

$$\sum_{m \in M} m x_{mi} - \sum_{m \in M} m x_{mj} = (|M| - 1) z_{ji}, \quad i \neq j, \forall i, j \in J, \tag{7}$$

$$\sum_{i \in E_d} \sum_{j \in J} z_{ij} = \frac{n_d(n_d + 1)}{2}, \quad \forall d \in D, \tag{8}$$

$$\sum_{m \in M} \theta_m \geq 1, \tag{9}$$

$$\sum_{j \in J}(p_j - \underline{p_j}) \geq \alpha\tau + \sum_{j \in J}(\overline{p_j} - \underline{p_j})\beta_j, \tag{10}$$

$$\frac{1}{\overline{p_j} - \underline{p_j}}\alpha + \beta_j \geq 1, \quad \forall j \in J, \tag{11}$$

$$\underline{p_j} \leq p_j \leq \overline{p_j}, \quad \forall j \in J, \tag{12}$$

$$\gamma_{mij} + 1 \geq z_{ji} + x_{mi}, \quad \forall m \in M, i \neq j, \forall i, j \in J, \tag{13}$$

$$1 - x_{ml} - x_{mk} - \theta_m \geq 0, \quad \forall l, k \in E_d, \forall m \in M, \forall d \in D, \tag{14}$$

$$x_{ml} - \theta_m \geq 0, \quad \forall l \in E_d, \forall m \in M, \forall d \in D, \tag{15}$$

$$t_m \geq 0, \quad \forall m \in M, \tag{16}$$

$$\alpha \geq 0, \tag{17}$$

$$\beta_j \geq 0, \quad \forall j \in J, \tag{18}$$

$$z_{ij} \in \{0, 1\}, \quad i \neq j, \forall i, j \in J, \tag{19}$$

$$x_{mj} \in \{0, 1\}, \quad \forall m \in M, \forall j \in J, \tag{20}$$

$$\gamma_{mij} \in \{0, 1\}, \quad \forall m \in M, i \neq j, \forall i, j \in J, \tag{21}$$

$$\theta_m \in \{0, 1\}, \quad \forall m \in M. \tag{22}$$

In the above formulation, the objective function (1) minimizes the total delay from the regular closing time of the operating room. Constraint (2) determines delay based upon the surgical duration and regular closing time of operating room m. Constraint (3) ensures that only one surgery is performed at a time. Constraints (4) and (5) are partial circuit constraints for surgery assignments. Constraints (6) and (7) prevent symmetry in surgery assignments. Constraint (8) ensures that surgeries in the same department are performed in succession. Note that the constraint has a hidden constraint; surgeries in the same department are assigned last of the operating room. Constraint (9) ensures that surgeries in the same department are performed in the same operating room. The right-hand side of constraint (10) is the objective function value of the dual problem, and constraint (11) is the constraint for the dual problem. Here, the dual problem is the complement problem of maximizing the total duration of the surgery in the next section. Constraint (12) limits the upper and lower bounds of surgical duration. Constraint (13) is a constraint on the linearization variable γ_{mij}. Constraint (14) and (15) are constraints on the linearization variable θ_m. Constraints (16)–(18) are nonnegative constraints on variables t_m, α, β_j. Constraints (19)–(22) are binary variable constraints on variables z_{ij}, x_{mj}, γ_{mij} and θ_m.

3.2   Surgical Duration Uncertainty

Real-world surgical durations are often subject to uncertainties. A robust optimization model with uncertainty may be more suitable and reasonable for decision-making. We assume that the uncertain surgical duration $\tilde{q}_j$ for surgery j is with respect to the uncertainty set, without assumptions on distribution. This assumption eliminates the need for accurate distribution information and enables the scheduling using only limited information, such as the average, minimum, and maximum values of data. Variable $\tilde{q}_j$ is defined $\tilde{q}_j = p_j - \overline{p_j}$, $\forall j \in J$. The formulations are as follows:

Maximize

$$\sum_{j \in J} \tilde{q}_j \tag{23}$$

subject to

$$\sum_{j \in J} \left(\frac{\tilde{q}_j}{\overline{p_j} - \underline{p_j}} \right) \leq \tau, \tag{24}$$

$$0 \leq \tilde{q}_j \leq \overline{p_j} - \underline{p_j}, \qquad \forall j \in J. \tag{25}$$

The objective function (23) maximizes the total surgical duration. Constraint (24) limits worst-case scenarios by conservative τ. Worst-case scenarios represent the number of surgeries for which the upper bound of the surgical duration is reached. Constraint (25) defines the possible range of uncertain surgical duration $\tilde{q}_j$.

4 Numerical Analysis

4.1 Data

In the following analysis, we used a dataset based on Ito et al. [13]. We solved the scheduling problem of five operating rooms and eleven surgeries using Gurobi 9.5.1. We compared the schedule created using the robust optimization model with that created using the stochastic programming model. The computational equipment is an Intel(R) Core (TM) i7-7500U CPU 2.90 GHz 8.00 GB.

The time from the operating room opening to the time when surgery j should be completed, d_m is 8 h or 480 min. It is desirable that all surgeries be completed within the regular opening time d_m. There are ten departments, i.e., $|D| = 10$. Surgeries 1 and 2 were in the same department. Table 1 shows the expected value $\mathbb{E}_j[p_j]$ and standard deviation σ_j of the duration of surgery j. The upper $\overline{p_j}$ and lower $\underline{p_j}$ bounds on the surgical duration used in robust optimization were the 10th and 90th percentiles of the duration in surgical scenarios. The conservative τ varied from 0 to 10 with 1. We assumed that the occurrence probability of the 1000 scenarios used in the stochastic programming model followed a uniform distribution, and the surgical duration in each scenario followed a log-normal distribution. We used the stochastic programming model proposed by Ito et al. [12].

4.2 Results

Figure 1 shows the results of total delay under different τ of robust optimization. The total delay of the stochastic programming is shorter than that of the robust optimization for all control conservatives because the surgical scenarios in the stochastic programming are used. The surgical scenario refers to a combined set of the duration of surgery. From Fig. 1, when the control conservative τ is 6, the total delay of the robust optimization shows a value equivalent to that of the

Table 1. Expected value and standard deviation of surgical duration (min).

Surgery j	$\mathbb{E}_j[p_j]$	σ_j
1	230	141
2	235	146
3	236	70
4	225	70
5	271	66
6	230	117
7	242	131
8	242	131
9	245	134
10	242	138
11	241	142

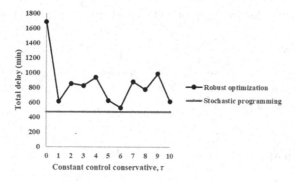

Fig. 1. Results of total delay.

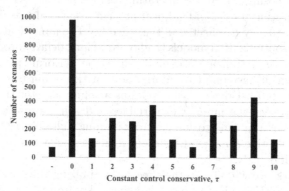

Fig. 2. Number of schedule scenarios with an excessive delay of 1000 min or more, '-': result obtained by stochastic programming.

stochastic programming. Figure 2 shows the number of schedule scenarios with an excessive delay of 1000 min or more. From Fig. 2, when the control conservative τ is 6, the number of schedule scenarios with an excessive delay also shows an equivalent value in robust optimization and stochastic programming. In the above results, the robust optimization, which has only information on the 10th and 90th percentile duration of the scenario, exhibits the same performance as the stochastic programming, which has complete information on the scenario in specific control conservative. The robust optimization does not require the estimation of accurate distribution.

Table 2 provides the CPU times required to obtain robust optimization and stochastic programming solutions for a single instance as the constant control conservative varies. From Table 2, the CPU time of the robust optimization is shorter than that of the stochastic programming in all cases. From the above results, robust optimization can obtain a solution faster than stochastic programming because the number of variables and constraints in robust optimization is smaller than in stochastic programming. The robust optimization involves 1,311 constraints and 759 variables, while the stochastic programming comprises 11,271 constraints and 5,742 variables.

Table 2. CPU times for robust optimization and stochastic programming solutions (seconds).

Model	τ										
	0	1	2	3	4	5	6	7	8	9	10
Robust optimization		0.10	0.11	0.10	0.20	0.10	0.09	0.09	0.10	0.15	0.12	0.12
Stochastic programming	380.38											

5 Conclusion

In this study, we proposed a robust optimization model that minimizes the delay from the regular closing time of the operating room. The proposed model considers multiple operating rooms and surgical procedures. We also verified whether the risk-averse tendency is reflected in the schedule. The numerical analysis suggests that robust optimization models tend to avoid long delays. From the numerical analysis, robust optimization exhibits the same performance as stochastic programming, which has complete information on the scenario in specific control conservative. Robust optimization obtains a solution faster than stochastic programming because the number of variables and constraints in robust optimization is smaller than in stochastic programming. The robust optimization model is more effective for operating room managers who desire to obtain an accurate solution quickly.

In future work, we will clarify the effect of optimizing the surgical sequence on delay reduction. For this purpose, we define delay from the planned surgery end time and modify a part of the proposed model.

References

1. Addis, B., Carello, G., Grosso, A., Tánfani, E.: Operating room scheduling and rescheduling: a rolling horizon approach. Flex. Serv. Manuf. J. **28**, 206–232 (2016)
2. Akiyama, R., Ito, M., Takashima, R., Hoshino, K.: Stochastic programming model for elective surgery planning: an effect of emergency surgery. In: Proceedings of 11th International Conference on Operations Research and Enterprise Systems, pp. 231–235 (2022)
3. Bandi, C., Gupta, D.: Operating room staffing and scheduling. Manuf. Serv. Oper. Manag. **22**(5), 869–1106 (2020)
4. Batun, S., Denton, B.T., Huschka, T.R., Schaefer, A.J.: Operating room pooling and parallel surgery processing under uncertainty. INFORMS J. Comput. **23**(2), 220–237 (2011)
5. Cardoen, B., Demeulemeester, E., Beliën, J.: Operating room planning and scheduling: a literature review. Eur. J. Oper. Res. **201**(3), 921–932 (2010)
6. Denton, B.J., Miller, A.J., Balasubramanian, H.J., Huschka, T.R.: Optimal allocation of surgery blocks to operating rooms under uncertainty. Oper. Res. **58**, 802–816 (2010)
7. Denton, B.J., Viapiano, A.V.: Optimization of surgery sequencing and scheduling decisions under uncertainty. Health Care Manag. Sci. **10**(1), 13–24 (2007)

8. Dexter, F., Macario, A.: Applications of information systems to operating room scheduling. Anesthesiology **85**, 1232–1234 (1996)
9. Gerchak, Y., Gupta, D., Henig, M.: Reservation planning for elective surgery under uncertain demand for emergency surgery. Manag. Sci. **42**(3), 321–334 (1996)
10. Guerriero, F., Guido, R.: Operational research in the management of the operating theatre: a survey. Health Care Manag. Sci. **14**, 89–114 (2011)
11. Ito, M., Kobayashi, F., Takashima, R.: Minimizing conditional-value-at-risk for a single operating room scheduling problems. In: Proceedings of International Multi-Conference of Engineers and Computer Scientists 2018, vol. 2, pp. 968–973 (2018)
12. Ito, M., Kobayashi, F., Takashima, R.: Risk averse scheduling for a single operating room with uncertain durations. In: Ao, S.-I., Kim, H.K., Castillo, O., Chan, A.H., Katagiri, H. (eds.) IMECS 2018, pp. 291–306. Springer, Singapore (2020). https://doi.org/10.1007/978-981-32-9808-8_23
13. Ito, M., Hoshino, K., Takashima, R., Suzuki, M., Hashimoto, M., Fujii, H.: Does case-mix classification affect predictions?: a machine learning algorithm for surgical duration estimation. Healthc. Anal. **2**, 100119 (2022)
14. Namba, Y., Ito, M., Takashima, R.: A robust optimization for a single operating room scheduling problem with uncertain durations. In: Proceedings of the 12th International Conference on Operations Research and Enterprise Systems, pp. 180–184 (2023)
15. Jackson, R.: The bushiness of surgery. Health Manag. Technol. **23**(7), 20–22 (2002)
16. Kamran, M.A., Karimi, B., Dellaert, N.: Uncertainty in advance scheduling problem in operating room planning. Comput. Ind. Eng. **126**, 252–268 (2018)
17. Lamiri, M., Xie, X., Dolgui, A., Grimaud, F.: A stochastic model for operating room planning with elective and emergency demand for surgery. Eur. J. Oper. Res. **185**, 1026–1037 (2008)
18. Macario, A., Vitez, T.S., Dunn, B., McDonald, T.: Where are the costs in perioperative care? Analysis of hospital costs and charges for inpatient surgical care. Anesthesiology **83**, 1138–1144 (1995)
19. Suzuki, A.: Analytics approach to the improvement of the management of hospitals. In: Sinha, B.K., Bagchi, S.B. (eds.) Strategic Management, Decision Theory, and Decision Science, pp. 247–256. Springer, Singapore (2021). https://doi.org/10.1007/978-981-16-1368-5_15
20. Zhu, S., Fan, W., Yang, S., Pei, J., Pardalos, P.M.: Operating room planning and surgical case scheduling: a review of literature. J. Comb. Optim. **37**, 757–805 (2019)

Matheuristic Local Search for the Placement of Analog Integrated Circuits

Josef Grus[1,2(✉)] and Zdeněk Hanzálek[2]

[1] DCE, FEE, Czech Technical University in Prague, Prague, Czech Republic
grusjose@fel.cvut.cz
[2] IID, CIIRC, Czech Technical University in Prague, Prague, Czech Republic
zdenek.hanzalek@cvut.cz

Abstract. The suboptimal physical design of the integrated circuits may not only increase the manufacturing costs due to the larger size of the chip but can also impact its performance by placing interconnected rectangular devices too far from each other. In the domain of Analog and Mixed-Signal Integrated Circuits (AMS ICs), placement automation is lacking behind its digital counterpart, mainly due to the variety of components and complex constraints the placement needs to satisfy. Integer Linear Programming (ILP) is a suitable approach to modeling the placement problem for AMS ICs. However, not even state-of-the-art solvers can create high-quality placements for large problem instances. In this paper, we study how to improve the results of our previous ILP model, first by introducing additional constraints and second by using matheuristics. Given the initial solution we obtain using our original ILP model, we use the solver to perform a local search. We try to improve the criterion by considering only a few spatially close rectangles while keeping the rest of the placement fixed. This local search approach enables us to significantly improve the quality of instances whose solution space we could not sufficiently explore before, even when the computation time reserved for the matheuristic is limited. Finally, we evaluate our revised approach on synthetically generated instances containing more than 200 independent rectangles and on real-life problems.

Keywords: Matheuristics · Placement optimization · Analog circuits

1 Introduction

The importance of ICs for modern civilization is apparent. Advanced computing, Internet-of-Things devices, automotive, and consumer electronics rely on high-performance ICs. Such market pressure further motivates the companies to shorten the design time and lower the development costs to increase their profitability and strengthen their market position. One of the crucial steps in the design of the ICs is the physical design. During this step, the circuit diagram is converted into the geometrical representation of the final product - positions and orientations of the rectangular devices (transistors, resistors, etc.) are determined during the placement phase, and the interconnections between them are planned during the routing phase. While these two steps are commonly solved one after another, the placement phase needs to consider the approximated interconnections to make the final product competitive and high-performant.

F. Liberatore et al. (Eds.): ICORES 2022/2023, CCIS 1985, pp. 178–200, 2024.
https://doi.org/10.1007/978-3-031-49662-2_10

AMS components remain crucial nowadays, as operational amplifiers and analog-to-digital converters are required to convert signals from many sensors surrounding us. The placement phase for the digital ICs has already been successfully automated. Digital devices are in the form of standardized cells, each sharing the same height, and they are placed in rows rather than freely. These properties make the digital ICs' placement similar to the 1D bin packing problem and enable the automation tools to handle thousands of devices.

On the other hand, AMS ICs usually contain tens or hundreds of devices. However, the devices may appear in different sizes and aspect ratios and can be placed freely. They also have different voltage levels, which does not happen in a digital domain. Furthermore, the presence of noise and other negative effects inherent to the analog domain significantly influence the overall performance of the circuits. This is mitigated by additional constraints and rules the engineers must adhere to. Due to these complications, the placement of the AMS ICs has not been largely automated and still remains a time-consuming and error-prone manual process; its automation is pursued by projects funded both by DARPA [8] and EU [7]. It is further complicated by constraints specific to different technologies of the ICs. This paper specifically discusses BCD technology (technology combining analog, digital, and high-voltage components), which means the placer has to consider various minimum distances between devices and isolated pockets, among other features.

ILP offers a formalism to successfully model the placement problem of AMS ICs. Most constraints regarding the sizes of the devices and their mutual proximity or connectivity can be described using linear inequalities, while the non-linear criterion of the circuit's area might be approximated with its half-perimeter. Nevertheless, even the state-of-the-art ILP solvers, which improve every year, cannot sufficiently well explore the space of feasible placements of larger ICs.

In this paper, we build upon our previous work [13], where we used warm-started ILP to place devices of the AMS ICs. We discuss the effect of additional symmetry-breaking and redundant constraints on the model's performance. Finally, we develop a Matheuristic (MH) local search technique, which iteratively optimizes the initial solution obtained by solving the entire model, and which offers significant improvement, especially on the large synthetically generated instances with more than 200 devices to be placed. This paper is structured as follows. In Sect. 2, we mention the previous work done in domains of both placement and matheuristics. In Sect. 3, we formulate the placement problem for BCD technology. Section 4 describes our original ILP model, as well as additional redundant constraints we experimented with. Section 5 describes our MH approach. In Sect. 6, we describe the problem instances and present the experimental results, which show how well the MH approach performs. Also, real-life instances are evaluated and compared with manual benchmarks. Finally, conclusions are drawn in Sect. 7.

2   Related Work

Even though the placement of the AMS ICs is not as automated as in the case of digital ICs, the problem has already been tackled in the past. Many methods use so-called topological representation - the solution is encoded using relative positions between the devices. Then, a packing procedure is used to convert the representation into the actual placement. Sequence pairs are one such representation. Proposed in [25], two permutations of the devices encode the relative positions between devices. Importantly, as was demonstrated in [21], this formulation can be extended to successfully model symmetry groups and other crucial features. Another example of the topological representation is B*-trees, which use binary trees to determine the relative positions between the parent and child nodes. Used in [19,31], this representation offers a low level of redundancy in its search space.

Other methods consider the absolute coordinates of the devices. This makes encoding constraints such as symmetry groups easier; however, it also introduces infeasible solutions to search space. In the early work of [6], the simulated annealing was used to optimize the coordinates of the devices. The criterion contained both the area and wire length of the IC, as well as penalty terms for constraint violations. In [23], a similar approach, using a multi-objective constrained variant of simulated annealing, was also considered. Alternatively, methods described in papers [4, 20] firstly use the global placement phase, where the approximate positions of the devices are determined using non-linear programming, and then the feasible placement without the overlaps is created using Linear Programming (LP). The mentioned core was extended to accommodate the different manufacturing layers of the ICs in [34]. In [17], the neural network was used to estimate the circuit's performance, and it was added to the differentiable criterion.

The force-directed approach was successfully applied to placement in [30], where the attractive and repulsive forces between the devices were derived from the connectivity of the IC and the devices' overlaps, respectively. Machine learning found its applications as well. An end-to-end pipeline of [24] was utilized as a placer of macros, while the learned model performed fine-optimization of the already-placed IC in [22].

While the methods outlined in the previous paragraphs successfully solved their associated placement problems, we cannot directly apply them to BCD technology ICs; these ICs rely on various minimum allowed distances between devices, isolated pockets, and other features that were rather omitted in the previous works. This was also a reason why we used the ILP, which allowed us to model these crucial features easily.

The ILP was applied to placement problems in the past. In [35], the authors used hierarchical decomposition to improve the solver's performance and created high-quality placements. In our previous work [13], we employed Force-Directed Graph Drawing-based (FDGD) method to warm start the solver instead of relying on decomposition. Our proposed MH offers to improve the results produced by other methods even when the warm starting the solver or decomposing the problem is not sufficient or leads to low-quality solutions. Furthermore, ILP is often used to solve subproblems that arise within the problem of placement, such as the determination of the number of fingers of transistors [27].

The placement of AMS ICs much resembles other problems encountered within the domain of operations research. Rectangle packing can be viewed as a simplification of this paper's topic due to the rectangular shape of the devices. Papers [2, 16] used constraint programming to solve the rectangle packing problem. In [14], a genetic algorithm was used together with a Bottom-Left first packing heuristic. Later, the GRASP metaheuristic was applied to strip packing [1]. Even more closely related to our problem is Facility Layout Problem (FLP), where the task is to determine the positions of the facilities while minimizing the travel distances between them. This can be perceived as an analogy to the interconnectivity of the devices. ILP formulations of the FLP were investigated in [15,33]. The latter work optimized the paths between the departments simultaneously with the layout, which resembles the simultaneous optimization of placement and routing in the case of ICs.

MHs, heuristics based on mathematical programming, have been recently successfully applied to many combinatorial problems [12], especially with the ever-increasing performance of the black-box ILP solvers. While the solvers often cannot solve the industrial-size instances, their search capabilities when the model is smaller cannot be ignored. The MHs were used successfully in the domains of scheduling or routing, but the literature regarding their use for packing and cutting is rather sparse [28]. There are many ways how to build the heuristic around the ILP solver. The constructive MHs iteratively solve a series of simpler subproblems and construct the final solution by combining the intermediate results. This was used both for rostering problems [29], as well as for FLP [32]. In the latter, authors fix the relative positions between the already placed facilities and iteratively add the remaining ones until the layout is completed. Evolutionary MHs use mathematical programming to tackle the efficiently solvable subproblems encountered while using metaheuristics. In [26], parallel batch processing scheduling is tackled using a genetic algorithm, and LP is used to improve the solution by solving the minimum cost flow problem.

Finally, the MHs are often used to perform the local search. Given a starting solution to a problem, we try to improve it by solving the restricted variant of the original ILP model. There are several ways how to achieve such restriction. The first way, called local branching, limits how many variables can change its value. Assuming the ILP model only contains binary variables, then the following constraint can be introduced [12]:

$$\sum_{i \in B_0} x_i + \sum_{i \in B_1} (1 - x_i) \leq k \tag{1}$$

where variable x_i was originally assigned to 0, if $i \in B_0$ and vice versa. The restrictiveness depends on the value of k. Local branching was successfully used in the improvement phase of [29]. In [36], the flow-shop problem with time windows was tackled, and local branching was even used to construct the feasible solution from the initial infeasible one.

Another way to restrict the search space is to explicitly fix a subset of variables of the model. This application is very close to Large Neighborhood Search [11] or Ruin and Recreate heuristics [5]; the damaged solution (i.e., the free variables in the restricted ILP model) is repaired using the exact solver. Variable-fixing local search MHs were successfully applied to the scheduling domain, such as in the case of university timetabling [18], flow-shop scheduling [10], and evacuations scheduling [9]. In these works, the choice of free and fixed variables is crucial for the successful application of MHs and often depends on domain-specific information. In this paper, we decided to apply such variable-fixing MH to our placement problem.

3   Problem Formulation

During the placement phase of the physical design of the AMS ICs, the positions and orientations of the devices are determined. The input of the problem, the netlist, contains information about the sizes of the devices, their voltage level, and interconnectivity. The devices have a rectangular shape of fixed size and can be rotated. Furthermore, we need to consider topological structures. These are higher-level building blocks, such as differential pairs or current mirrors, and they consist of several devices that have to be placed in a regular pattern (see two columns of darker rectangles in Fig. 1). Thus, we enumerate all possible variants (with a varying number of rows and columns into which the devices are organized) of such topological structures beforehand, using algorithms based on list scheduling [13]. Afterward, we treat both the single devices and the topological structures as rectangles with multiple variants (in the case of single devices, the only alternative variant is rotation). Further in the text, we refer to both types of these building blocks as rectangles. Given a task to place n rectangles, we describe each one of them with the coordinates of its bottom-left corner (x_i, y_i) and its size (w_i, h_i), which corresponds to one of its m_i variants.

Since we want to create as small a placement as possible, we would like to minimize its area $W \cdot H$. However, due to our use of ILP, we minimize the half perimeter of the placement's bounding box $W + H$ instead.

The overall connectivity is modeled as Half Perimeter Wire Length (HPWL). The core concept of connectivity is a set of nets E - each net $e \in E$ consists of a set of connected rectangles L_e. Each rectangle can be a member of multiple nets. The overall connectivity is formulated as follows:

$$\text{HPWL} = \sum_{\forall e \in E} c_e \cdot \left(\max_{i \in L_e} x_i^c - \min_{i \in L_e} x_i^c + \max_{i \in L_e} y_i^c - \min_{i \in L_e} y_i^c \right) \tag{2}$$

where the centroid coordinates are given by:

$$x_i^c = x_i + w_i/2 \tag{3}$$

$$y_i^c = y_i + h_i/2 \tag{4}$$

Multiplied by its cost c_e, each net contributes to the overall HPWL metric the half of the perimeter of the smallest bounding box that contains all of the net's rectangles' centroids [34]. Altogether, our task is to find a feasible placement that not only minimizes the area of its bounding box but minimizes the HPWL metric as well.

The physical devices (darker rectangles surrounded by lighter shells in Fig. 1), such as transistors, cannot overlap when they are manufactured in the same layer. Furthermore, an increased minimum distance can be imposed between some devices, e.g., to mitigate the effect of the noise on sensitive components. We also need to model additional empty space, or pocket, around the placed structures and devices (the lighter shells around packed devices in Fig. 1). Pockets are needed to isolate devices with different voltage levels, which is common for BCD technology. When the devices do not share their input voltage (BULK) net, and thus their voltage level may differ, we need to place them so their pockets do not overlap. Otherwise, their pockets can be merged

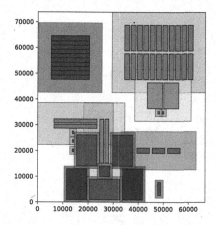

Fig. 1. Example placement with critical constraints of the BCD technology [13].

as long as their internal devices do not overlap, as the yellow and orange rectangles in Fig. 1 demonstrate.

Additional constraints include the control of the aspect ratio of the final placement. Also, the engineer can restrict a subset of rectangles from a part of a canvas; we call this type of constraint a blockage area. An example is shown in the bottom-left corner of Fig. 1, which remained unoccupied due to the use of the blockage area. Furthermore, a group of rectangles may belong to a symmetry group, which shares a common axis of symmetry. An example is a group of darker rectangles with the vertical axis of symmetry located in the bottom part of Fig. 1.

4 ILP Model and Extensions

4.1 Baseline Model

We use our model proposed in [13], which was extended from rectangle packing formulation in [2]. Let $\mathcal{I} = \{1, \ldots, n\}$ be set of rectangles' indices. Four real variables represent each rectangle; coordinates of its bottom-left corner (x_i, y_i) and width and height (w_i, h_i), which has to correspond to one of the m_i pre-defined variants (w_i^k, h_i^k), $k \in \{1, \ldots, m_i\}$. Note that the sizes of rectangles' variants are increased to model the use of the pockets. The selection of variants is made using binary variables s_i^k for each rectangle i and variant k, as is shown in Eqs. (6), (7). k-th variant is selected if $s_i^k = 1$. Placement's width W and height H are variables constrained by the positions of the placed rectangles.

$$x_i + w_i \leq W, \quad y_i + h_i \leq H \qquad\qquad \forall i \in \mathcal{I} \qquad (5)$$

$$\sum_{k=1}^{m_i} s_i^k = 1 \qquad\qquad\qquad\qquad \forall i \in \mathcal{I} \qquad (6)$$

$$w_i = \sum_{k=1}^{m_i} w_i^k \cdot s_i^k, \quad h_i = \sum_{k=1}^{m_i} h_i^k \cdot s_i^k \qquad\qquad \forall i \in \mathcal{I} \qquad (7)$$

$$\sum_{k=1}^{4} r_{i,j}^k \geq 1 \qquad\qquad\qquad \forall i,j \in \mathcal{I} : i < j \qquad (8)$$

$$x_i + w_i + a_{i,j} \leq x_j + M(1 - r_{i,j}^1) \qquad \forall i,j \in \mathcal{I} : i < j \qquad (9)$$

$$y_i + h_i + a_{i,j} \leq y_j + M(1 - r_{i,j}^2) \qquad \forall i,j \in \mathcal{I} : i < j \qquad (10)$$

$$x_j + w_j + a_{i,j} \leq x_i + M(1 - r_{i,j}^3) \qquad \forall i,j \in \mathcal{I} : i < j \qquad (11)$$

$$y_j + h_j + a_{i,j} \leq y_i + M(1 - r_{i,j}^4) \qquad \forall i,j \in \mathcal{I} : i < j \qquad (12)$$

$$x_i, y_i, w_i, h_i \geq 0 \qquad\qquad\qquad \forall i \in \mathcal{I} \qquad (13)$$

$$W, H \geq 0 \qquad\qquad\qquad\qquad\qquad (14)$$

$$s_i^k \in \{0, 1\} \qquad\qquad\qquad \forall i \in \mathcal{I} \; \forall k \leq m_i \qquad (15)$$

$$r_{i,j}^k \in \{0, 1\} \qquad\qquad\qquad \forall i,j \in \mathcal{I} : i < j$$

$$\qquad\qquad\qquad\qquad\qquad\qquad\qquad \forall k \in \{1, 2, 3, 4\} \qquad (16)$$

Non-overlapping of the devices is ensured by binary variables $r_{i,j}^k$ and inequalities (8)–(12), which utilize the big-M approach [3]. At least one of the inequalities, which corresponds to the relationship (left/right/over/under) between rectangles, must be valid ($r_{i,j}^k = 1$). Parameter $a_{i,j}$ defines the minimum allowed distance between rectangles. By setting the parameter $a_{i,j}$ to the negative value, the solver can place associated rectangles with their pockets merged, similarly to device layer-aware placements [34]. Ultimately, the ILP model for feasible placement of n rectangles uses $\sum_{i=1}^{n} m_i$ binary variables to encode variant selection, and $4 \cdot \binom{n}{2} = 2 \cdot n \cdot (n-1)$ binary variables to encode the relative positions between rectangles.

Blockage areas are modeled as additional dummy rectangles. We fix their positions and sizes and define the minimum allowed distance parameters. $a_{i,b} = 0$ if the rectangle i is blocked by the blockage area b; if the rectangle is unaffected by the blockage area b, we simply omit the associated relative position constraints from the model.

We define the final aspect ratio as $\text{AR} = \min\{W, H\} / \max\{W, H\}$, and we want to ensure that $l_R \leq \text{AR} \leq u_R$ holds for chosen aspect ratio parameters $0 \leq l_R \leq u_R \leq 1$. Then, the following additional constraints are needed. The binary variable r_R is used to handle the non-convex solution space that is induced when $u_R \neq 1$. When $u_R = 1$, we omit the associated inequalities entirely.

$$l_R \cdot W \leq H \leq u_R \cdot W + M \cdot (1 - r_R) \qquad (17)$$

$$l_R \cdot H \leq W \leq u_R \cdot H + M \cdot r_R \qquad (18)$$

$$r_R \in \{0; 1\} \qquad (19)$$

To model the symmetry groups, we require another continuous variable per group to represent the axis of symmetry. Assume that G is the symmetry group with the vertical axis of symmetry, whose horizontal position is determined by the real variable x_G. The symmetry group consists of self-symmetric rectangles $(i, -)$ and symmetric pairs (i, j). Then the following equations constrain the symmetry group's rectangles to share the same axis of symmetry:

$$w_i = w_j \qquad \forall (i,j) \in G \qquad (20)$$

$$h_i = h_j \qquad \forall (i,j) \in G \qquad (21)$$

$$y_i = y_j \qquad \forall (i,j) \in G \qquad (22)$$

$$x_i + x_j + w_i = 2 \cdot x_G \qquad \forall (i,j) \in G \qquad (23)$$

$$2 \cdot x_i + w_i = 2 \cdot x_G \qquad \forall (i,-) \in G \qquad (24)$$

HPWL connectivity elements are formulated per net. Thanks to the minimization of the connectivity in the final criterion, no integer variables are needed. For each net e, we create four continuous variables $X_e^M, X_e^m, Y_e^M, Y_e^m \in \mathbb{R}$, which describe the net's bounding box. Then, we formulate the connectivity criterion $\mathcal{L}_C$ using the following constraints for each net $e \in E$, given the set of the net's connected rectangles L_e and net cost c_e:

$$X_e^M \geq x_i + w_i/2 \qquad \forall i \in L_e \qquad (25)$$

$$X_e^m \leq x_i + w_i/2 \qquad \forall i \in L_e \qquad (26)$$

$$Y_e^M \geq y_i + h_i/2 \qquad \forall i \in L_e \qquad (27)$$

$$Y_e^m \leq y_i + h_i/2 \qquad \forall i \in L_e \qquad (28)$$

$$\mathcal{L}_C = \sum_{\forall e \in E} c_e \cdot \left(X_e^M - X_e^m + Y_e^M - Y_e^m \right) \qquad (29)$$

To minimize the area of the placement, which is a non-linear expression $W \cdot H$, we approximate it using the half perimeter of the placement's bounding box:

$$\mathcal{L}_A = W + H \qquad (30)$$

We expect that thanks to the correlation between the perimeter and the area of the bounding rectangle, a solution minimizing $\mathcal{L}_A$ will have a small area as well. Ultimately, the final criterion function is defined as:

$$\mathcal{L} = c_A \cdot \mathcal{L}_A + \frac{c_C}{\sum_{\forall e \in E} c_e} \cdot \mathcal{L}_C \qquad (31)$$

where the c_A, c_C are tunable costs; by tuning them, we can achieve a suitable trade-off between both $\mathcal{L}_A$ and $\mathcal{L}_C$. However, since there are only two criterion elements, we fix $c_A = 1$ and tune only the connectivity cost. Furthermore, we divide $\mathcal{L}_C$ by $\sum_{\forall e \in E} c_e$, so the effect of using a specific value of c_C is less sensitive to a number of nets present in the IC.

4.2 Improving the Performance of the Solver

As we have shown in [13], the presented formulation leads to feasible high-quality placements, but the performance of even the state-of-the-art ILP solvers is insufficient when the number of rectangles grows. We were able to mitigate this problem by providing a solver with an FDGD-based solution as a warm start. In this paper, we want to go even further, and we try to introduce redundant constraints to the original model that do not affect the optimal solutions but could potentially improve the performance of the solver.

Symmetry Breaking. Firstly, we tried to remove the symmetric solutions from the search space. Since all of our constraints are rotation invariant (with the only exception being symmetry groups), we can prune the search space by fixing the orientations or positions of specific rectangles. Firstly, we select a suitable rectangle (the largest one as in [16]); let its index be K. Then, to remove the solutions symmetrical with respect to the $y = x$ axis, we set all variant variables of rectangle K, which correspond to a rotated variant with index r, to zero.

The second approach is concerned with the solution symmetry achieved by swapping the quadrants of the bounding box. For example, from the current solution, another one can be created by simply mirroring it with respect to either $x = \frac{W}{2}$ or $y = \frac{H}{2}$ axes, or by reflecting it with respect to point $(\frac{W}{2}; \frac{H}{2})$ point. To prune these parts of the search tree, we constrain the coordinates of rectangle K so its centroid lies within the first quadrant, closest to the origin:

$$2 \cdot x_K + w_K \leq W \tag{32}$$

$$2 \cdot y_K + h_K \leq H \tag{33}$$

W+H Constraint. If we could predict how large the bounding box of the optimal solution would be, we could prune the search space using constraint:

$$W + H \leq P \tag{34}$$

where P is the upper bound on the half perimeter of the solution obtained from the prediction. There are two reasons why this could improve the performance of the model. Firstly, such a hard constraint prunes some branches of the search tree that would otherwise be investigated, especially when the connectivity metric of the objective function is more emphasized and the LP relaxation does not offer a tight enough lower bound. Such restriction can also be beneficial by allowing the model to employ a much smaller big-M constant than previously possible, which can improve the LP relaxation and mitigate issues with numerical stability.

When the $W + H$ constraint is introduced with a bound P, the big-M value can be set to $M = P + a_M$ without making otherwise feasible solutions infeasible. We set a_M to the maximum of the minimum allowed distances between pairs of rectangles, $a_M = \max_{(i,j)} a_{i,j}$. This way, the constraints (9)–(12) hold even in the most extreme cases. In experiments regarding the W+H constraint, we set the P to half the perimeter of the previously found solution with additional slack to not restrict the solver too much. We discuss the obtained results in Sect. 6.2.

5 Matheuristic as a Local Search

Given the initial solution, which can be provided either by the ILP solver with limited computation time or a suitable heuristic, we try to improve it using the variable-fixing MH. We refer to this improvement phase as intensification.

5.1 Intensification

Rectangle Selection. The choice of which variables should be fixed and which should remain flexible during intensification is crucial. Inspired by the job-window approach of [10], we select a local group of rectangles $\mathcal{G}$. Given a position (x, y) within the placement and size of the group g, the set $\mathcal{G}$ consists of g rectangles closest to the point (x, y). We define the 'proximity' metric of rectangle i to point (x, y) as:

$$\text{proximity}(x, y, i) = \max\left\{|x_i - x|, |x_i + w_i - x|, |y_i - y|, |y_i + h_i - y|\right\} \quad (35)$$

This way, selected rectangles should be located spatially close to each other, and when removed from the placement, mostly unfragmented empty space should appear. This should enable the solver to locally improve the connectivity by modifying the spatially local part of the placement. However, the positions and variants of the selected rectangles are not constrained, giving the solver the freedom to move them significantly if necessary.

ILP Intensification. After the rectangle selection, the solver tries to improve the solution. The used ILP model corresponds to the one shown in Sect. 4.1, so the feasibility of the solution is ensured. We fix the positions and variants of each rectangle $i \notin \mathcal{G}$; thus, their respective relative position variables $r_{i,j}^k$ or variant variables s_i^k are not necessary. The selected rectangles belonging to $\mathcal{G}$ still have all their associated variables free. Therefore, the number of binary variables associated with n rectangles decreases from:

$$\sum_{i=1}^{n} m_i + 2 \cdot n \cdot (n - 1) \quad (36)$$

to significantly smaller:

$$\sum_{i \in \mathcal{G}} m_i + 2 \cdot g \cdot (g - 1) + 4 \cdot g \cdot (n - g) \quad (37)$$

Before the optimization, the solver is warm-started with the current placement. For a sufficiently small value of g, the solver is able to solve the restricted model optimally or at least find an improvement in a short time. Since the growing number of rectangles n may slower intensification significantly, we impose a time limit on optimization.

LP Fine Optimization. To account for gaps between rectangles that can emerge by the variable fixing approach, we follow the previous step with LP optimization, which can lead to a lower value of HPWL and make the placement more compact. For each pair of rectangles, we find the least violated relative position constraint (9)–(12), and its associated variable $r^k_{i,j}$. Then, we optimize the original model of Sect. 4.1, fixing the chosen relative position variables to 1 and the variant variables to select the variants present in the current solution. Thus, the model does not contain binary variables, and the optimization is done quickly, even for large instances.

Overall Intensification. After each successful intensification iteration, the improved solution replaces the previous one. Then, a new selection point (x, y) is sampled, and the process repeats until the computation budget is exhausted. In this paper, we generate the selection points by sampling uniformly from interval $\langle 0; W \rangle$, $\langle 0; H \rangle$ respectively. While such a simplistic strategy performed well, a more informed approach could yield better results.

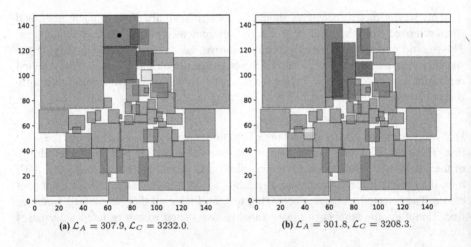

(a) $\mathcal{L}_A = 307.9, \mathcal{L}_C = 3232.0$.   (b) $\mathcal{L}_A = 301.8, \mathcal{L}_C = 3208.3$.

Fig. 2. Placement before and after ILP intensification. The black dot shows where the selection position (x, y) was sampled. Rectangles modified during the process are highlighted. See the decrease in height after intensification. (Color figure online)

The process of ILP intensification is demonstrated in Figs. 2. The current placement is shown in Fig. 2a. The sampled position (x, y), shown as a black dot, is located near the top side of the bounding box, and 5 rectangles were selected (red, purple, green, blue, and yellow). After the ILP intensification step, the new, improved placement is shown in Fig. 2b. We can see that the selected rectangles both moved and changed their variants. Both the half perimeter of the bounding box and the HPWL were decreased by this step, as is reported in the captions.

5.2 Diversification

While the ILP solver guarantees us that the local neighborhood of the current solution is thoroughly searched, the algorithm can get stuck in the local minimum. In that case, it is beneficial to divert from the current solution significantly and try to reach another potentially better local minimum.

To perform a diversification step, we try to swap the positions of the rectangles so the overall placement changes, but we still try to keep the placement competitive. To do this, we create a swapping ILP model. In this model, each rectangle i is associated with its centroid coordinates (x_i^c, y_i^c), as well as its area $A_i = w_i \cdot h_i$. Note that chosen variant and coordinates of the rectangles are retrieved from the current solution. Then, the ILP model is formed as follows:

$$\min \quad c_A \cdot \mathcal{L}_A + \frac{c_C}{\sum_{\forall e} c_e} \cdot \mathcal{L}_C + c_\xi \cdot \xi \tag{38}$$

$$\sum_{j=1}^{n} p_i^j = 1, \quad \sum_{j \in T_i} p_i^j = 1 \qquad \forall i \in \mathcal{I} \tag{39}$$

$$\sum_{i=1}^{n} p_i^j = 1 \qquad \forall j \in \mathcal{I} \tag{40}$$

$$x_i^s = \sum_{j=1}^{n} x_j^c \cdot p_i^j, \quad y_i^s = \sum_{j=1}^{n} y_j^c \cdot p_i^j \qquad \forall i \in \mathcal{I} \tag{41}$$

$$\xi \geq N - (n - \sum_{i=1}^{n} p_i^i) \tag{42}$$

$$\xi \geq 0 \tag{43}$$

$$x_i^s, y_i^s \geq 0 \qquad \forall i \in \mathcal{I} \tag{44}$$

$$p_i^j \in \{0, 1\} \qquad \forall i, j \in \mathcal{I} \tag{45}$$

Binary variable p_i^j is equal to one if the rectangle i should be placed to the position of the rectangle j (thus, $p_i^i = 1$ means the rectangle i does not move). To disallow the situation when a large rectangle would be placed in a position of the small one, we create a set of allowed swapping indices T_i for each rectangle i. Note that $i \in T_i$ for each rectangle i.

$$T_i = \left\{ j \in \mathcal{I} \ \middle| \ \frac{|A_i - A_j|}{\min \{A_i, A_j\}} \leq A_{\text{diff}} \right\} \tag{46}$$

Maximum relative difference ($A_{\text{diff}} = 0.25$) limits the search space of the model significantly. Variables x_i^s, y_i^s track the new centroid positions of the swapped rectangles that are used to calculate the half perimeter and connectivity criteria, using additional constraints shown in Sect. 4.1. Finally, ξ is used to penalize the insufficient number of swaps performed, i.e., when $p_i^i = 1$ for too many rectangles. If the less than the expected minimum number of swaps N is performed (we use $N = n/3$), additional penalty $c_\xi \cdot \xi$ is applied; we set the cost c_ξ to quite a large value $\max \{W, H\}$, so the solver is motivated to perform the swaps.

After determining which swaps should be performed, we modify the current solution so the centroids of the swapped rectangles are moved to their associated positions.

However, this can make the solution infeasible due to possible overlaps. To make the solution feasible, we use the original ILP model of Sect. 4.1 again. As in LP fine optimization of the intensification phase, we find the least violated relative position constraint for each pair of rectangles, and we warm start the solver with the corresponding variables set to 1. The values of variant variables are also obtained from the previous solution. The feasible result of the diversification phase is obtained by solving the model for a limited time. Afterward, we continue with intensification.

Since our intensification implementation does not exhaustively search all possible neighborhoods, we need a mechanism to decide when to perform the diversification and when to keep searching locally. Whenever the local search does not improve the solution's quality, we increment the counter. When the counter reaches 10, we perform the diversification and reset the counter. The counter is also reset when the improvement is achieved during the intensification.

6 Experiments

6.1 Methodology and Data

We utilized the Gurobi ILP solver v9.5.1, using four threads in each experiment. The project was implemented using Python 3.7. Experiments were performed on an Intel Xeon E5-2690.

We generated several sets of instances inspired by the structure of real-life ones. Sets S_{50} and S_{100} were already discussed in our previous work [13]. Additional sets S_{200} and S_{200}^{sym} contain a larger number of rectangles, and the latter also contains several symmetry groups as a part of each instance. Each instance contains both the smaller rectangles, which only allow rotation, and larger ones with multiple variants. In total, 120 instances were evaluated. The computation time was fixed for each instance, depending on its set (shown in Table 1). When the MH was used, the initial solution was obtained by optimizing the original ILP model for a third of the computation time, and the rest was reserved for MH. The time required for warm starting the original model with the FDGD method was included in the total computation time. The costs in the criterion function were set to $c_A = 1.0$, and $c_C \in \{0.1, 1.0, 8.0\}$ respectively.

As baseline results, the methods proposed in [13] were used. The baseline model without any improvement, denoted as **ILP**, was run only on the instance set S_{50} and S_{100}, as it could not recover any solution for larger instances within the given runtime. FDGD warm-started variant **FDGD-ILP** solved all the instances.

Table 1. Description of synthetically generated instances.

instance set	# instances	# rectangles	symmetry	comp. time
S_{50}	60	20, 30, 50	No	10 min
S_{100}	20	100	No	20 min
S_{200}	20	200	No	40 min
S_{200}^{sym}	20	200+	Yes	40 min

To compare the results obtained on the synthetically generated instances, we use the average relative difference (aRD) of the criterion, calculated for method m and instance set S as:

$$\text{aRD}_S^m = \frac{1}{|S|} \cdot \sum_{i \in S} \frac{\mathcal{L}^{i,m} - \mathcal{L}^{i,best}}{\mathcal{L}^{i,best}} \cdot 100 \ [\%] \tag{47}$$

where $\mathcal{L}^{i,m}$ is the value of criterion achieved on instance i by method m, and $\mathcal{L}^{i,best}$ is the lowest value of criterion of among studied methods. Therefore, aRD refers to the ratio of the method's and best-known solution's criterion values averaged over the entire instance set. The best hits metric (BH) tells us how many times a specific method achieved the best-known value of the criterion.

6.2   Performance with Redundant Constraints

To study how the additional constraints affect the performance of the ILP solver, we performed experiments on instance sets S_{50} and S_{100}. In the case of set S_{100}, only results for $c_C \in \{0.1, 1.0\}$ are reported, as for $c_C = 8.0$, not all methods found a feasible solution for each instance. The baseline **ILP** model is compared with symmetry-breaking one **SB-ILP** from Sect. 4.2, and the model **WH-ILP** using the W+H constraint from Sect. 4.2. Note that parameter P used to define the W+H constraint was derived from the half perimeter of the feasible solution obtained using **FDGD-ILP**, which we increased by 20 %. Furthermore, the studied instances did not contain symmetry groups; thus, utilizing symmetry breaking did not cause any problems.

Table 2. Comparison of solutions obtained using baseline **ILP** model and the models with additional constraints, with reported values of aRD (BH) for each instance set and connectivity cost c_C.

method	S_{50}			S_{100}	
	$c_C = 0.1$	$c_C = 1.0$	$c_C = 8.0$	$c_C = 0.1$	$c_C = 1.0$
ILP	2.01 (19)	1.69 (20)	4.41 (22)	**1.81 (13)**	**2.29 (11)**
SB-ILP	1.90 (21)	3.16 (21)	5.55 (21)	4.56 (12)	6.07 (12)
WH-ILP	**0.92 (31)**	**1.56 (33)**	**2.81 (29)**	–	–

As shown in Table 2, the results are rather inconclusive. The symmetry-breaking constraints help a little for $c_C = 0.1$ on S_{50} scenario, but lead to worse solutions on average. The W+H constraint leads to better solutions, but we were not able to find a feasible solution consistently for S_{100} instances. In the case of the $c_C = 0.1$ experiment on S_{100} instance set, the feasible solution was found only for 10 of 20 instances. Furthermore, the average time needed to find the first feasible solution was 356 s. We concluded that imposing the upper bound on the half perimeter of the bounding box, and thus also on the big-M value, can improve the results. However, without passing the initial solution to a solver, the solver has a problem finding any feasible solution.

6.3   Matheuristics on Synthetic Data

Our MH approaches rely on several parameters which may significantly influence the outcome of the local search. We fixed several parameters beforehand. When the diversification is used, we apply it after 10 non-improving intensification attempts. The maximum time reserved for each intensification and diversification optimization step was set to 10 s.

We performed experiments with four different MH settings. The settings **MH-5**, **MH-10**, and **MH-10D** used **FDGD-ILP** to find the initial solution for local search. Settings **MH-5** and **MH-10** relied only on intensification and differed in the number of rectangles g selected to be optimized in each iteration (see Sect. 5.1). The first setting **MH-5** used $g = 5$, and the second setting **MH-10** used $g = 10$. The larger value of g was not used, as the complexity of the larger model decreased the performance of the ILP solver significantly. The third setting **MH-10D** also used $g = 10$ and employed diversification.

Finally, the remaining setting **MH-10B** used the baseline **ILP** method instead of the warm-started one to generate the initial solution. Then, it only uses intensification with $g = 10$, thus being comparable to **MH-10**.

Choice of Suitable Setting. Firstly, we tried to determine how the value of g and the use of diversification affects the results. We ran the experiments on all instance sets for all three values of the c_C. The experiments were performed with **MH-5**, **MH-10**, and **MH-10D** settings, and with **FDGD-ILP** serving as a baseline. The results are reported in Table 3.

We can see the baseline **FDGD-ILP** was outperformed on all instance sets. Furthermore, the improvement provided by MHs seems to be much more significant when the connectivity cost is high. This corresponds to the expected behavior of the intensification phase. Since we only free up to 10 rectangles in each iteration, and they are selected locally close to each other, there often remains a fixed rectangle that keeps the half perimeter of the bounding box unchanged. On the other hand, the connectivity of a single net can be significantly changed by moving even a single rectangle.

The improvements provided by the MHs are especially important in the case of instance set S_{200}^{sym}, where the differences between the baseline results and the proposed methods are the largest - 30 % on average. We believe that the main reason is the rigid handling of the symmetry groups our FDGD warm start uses. To create a feasible initial solution, each symmetry group is handled as a single entity, which, however, may lead to low-quality placement shown in Fig. 3a (note, that we do not show internal devices inside the rectangles). Then, the solver cannot sufficiently improve the solution within the provided computation time due to the complexity of the model. On the other hand, the MH approach is able to decrease the value of the criterion significantly, and the overall placement looks more compact, see Fig. 3b. We also demonstrate this in Fig. 4, which shows how the criterion value changes as the computation progresses. We can see that both shown MH settings, after their initialization phase, lower the criterion rapidly, while the solver optimizing the entire model struggles. This holds true even from the area-wise point of view; MH transforms the FDGD-produced circular placement to a more compact rectangular one.

Table 3. Comparison of different MH settings and **FDGD-ILP** baseline, with reported values of aRD (BH) for each instance set and connectivity cost c_C.

method	S_{50}			S_{100}		
	$c_C = 0.1$	$c_C = 1.0$	$c_C = 8.0$	$c_C = 0.1$	$c_C = 1.0$	$c_C = 8.0$
FDGD-ILP	2.06 (5)	3.51 (0)	10.35 (1)	2.08 (0)	5.50 (0)	12.51 (0)
MH-5	1.71 (6)	2.34 (4)	5.77 (2)	0.85 (9)	1.17 (7)	1.81 (6)
MH-10	**0.14 (44)**	**0.26 (43)**	2.48 (15)	**0.48 (11)**	**0.20 (13)**	**0.26 (14)**
MH-10D	2.12 (5)	1.15 (13)	**0.70 (42)**	3.80 (0)	3.81 (0)	3.82 (0)
method	S_{200}			S_{200}^{sym}		
	$c_C = 0.1$	$c_C = 1.0$	$c_C = 8.0$	$c_C = 0.1$	$c_C = 1.0$	$c_C = 8.0$
FDGD-ILP	3.79 (3)	8.02 (1)	15.50 (0)	27.10 (0)	28.72 (0)	31.61 (0)
MH-10	**0.33 (16)**	**0.03 (19)**	**0.68 (19)**	**0.68 (15)**	**0.75 (11)**	2.53 (1)
MH-10D	5.38 (1)	5.83 (0)	5.38 (1)	1.51 (5)	0.85 (9)	**0.04 (19)**

From the experiments on sets with less complex instances S_{50}, S_{100}, we found out that while freeing only 5 rectangles leads to significant improvements and shorter optimization time per iteration, using $g = 10$ yields better results on average. Therefore, we omitted the **MH-5** from the experiments on larger instances. Then we studied the effect of diversification. **MH-10** without diversification worked well in all cases, while the **MH-10D** was less predictable. However, for two instance sets with $c_C = 8.0$, the **MH-10D** offered the best results, as is shown in Table 3. Also, the larger diversification step can lead to significant improvements, as is illustrated in Fig. 4, where the first time the diversification step is used (computation time 1100), the criterion drops significantly. We concluded that diversification offers advantages that could be more thoroughly exploited. However, due to the consistency of its results, we used the **MH-10** setting in the rest of the paper instead.

Importance of the FDGD Warm Start. After the previous experiments, we wanted to study whether it is still necessary to use the FDGD warm start to find the initial solution for MH. To do so, evaluated the original ILP model without warm start **ILP**, as well as its MH variant **MH-10B** on instances from S_{50} and S_{100}. The results are reported in Table 4.

Table 4. Comparison of FDGD-warm started and non-warm started MHs and ILP baselines, with reported values of aRD (BH) for each instance set and connectivity cost c_C.

method	S_{50}			S_{100}		
	$c_C = 0.1$	$c_C = 1.0$	$c_C = 8.0$	$c_C = 0.1$	$c_C = 1.0$	$c_C = 8.0$
ILP	5.62 (2)	10.29 (1)	19.49 (1)	19.76 (0)	45.16 (0)	46.98 (0)
FDGD-ILP	2.74 (1)	3.78 (0)	9.41 (1)	2.21 (2)	5.36 (0)	12.33 (0)
MH-10B	1.93 (26)	2.43 (15)	2.95 (24)	7.46 (4)	5.60 (4)	7.08 (2)
MH-10	**0.82 (31)**	**0.52 (44)**	**1.62 (34)**	**0.62 (14)**	**0.07 (16)**	**0.10 (18)**

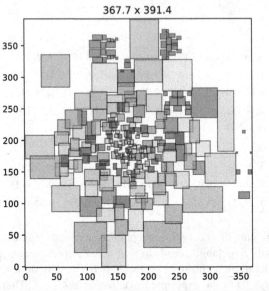

(a) **FDGD-ILP** result, $\mathcal{L} = 1190.09$, area $= 143894$, HPWL $= 128885$.

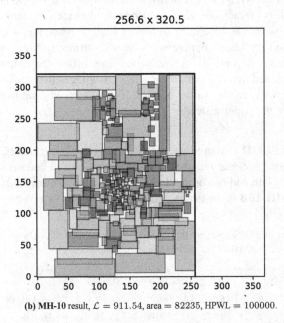

(b) **MH-10** result, $\mathcal{L} = 911.54$, area $= 82235$, HPWL $= 100000$.

Fig. 3. Comparison of final placements obtained by **FDGD-ILP** and **MH-10** respectively, on instance from set S_{200}^{sym} with $c_C = 1.0$. Both experiments' computation time was set to 2400 s.

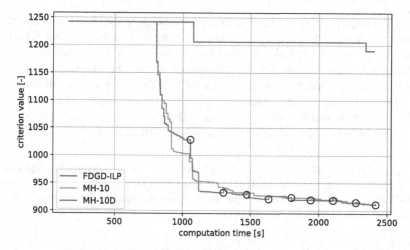

Fig. 4. Value of criterion during optimization, for instance shown in Fig. 3. Black circles show when **MH-10D** performed diversification.

From the provided table, we can see that the MH local search significantly improves the ILP baseline; it is even able to outperform the **FDGD-ILP** setting. When we compare the **MH-10B** with our main setting **MH-10**, we see that the FDGD warm start still provides some benefits. The warm-started variant of MH outperforms its non-warm-started counterpart, and this becomes more prevalent for more complex instances (where the **ILP** may not even find any solution).

6.4 Improvement on Real Life Instances

Afterward, we studied how well MH works on real-life instances that were provided by industry partner STMicroelectronics and which we used previously in [13]. 17 instances were provided, each consisting of up to 60 independent rectangles, and we ran two different experimental settings for each instance, either allowing or forbidding the use of pocket merging. Thus, the total number of experiments was 34. As in our previous work, the optimization was limited to 8 min. Three connectivity costs $c_C \in \{0.1, 1.0, 8.0\}$ were considered for each experiment.

In Table 5, we report the metrics of manual designs and our solutions (the use of pocket merging depended on the manual design). The shown metrics are the half perimeter of the bounding box W+H, the placement area, and the connectivity metric HPWL. We found a solution dominating the metrics of the manual one for 12 out of 17 instances, matching our previous results. However, when we focus on the average ratios between automated and manual designs and compare them to results generated by **FDGD-ILP** in [13], we can see that we were able to quite significantly lower the connectivity while keeping the area and half-perimeter competitive.

To highlight the differences between solutions found by **FDGD-ILP** and **MH-10**, we show Table 6. We can see that with the exception of the $c_C = 0.1$ scenario, the MH approach, on average, reduced the criterion of the final solution and found the better solution in a majority of the cases. This again corresponds to the observations we made in Sect. 6.3. The real-life instances also contain only up to 60 rectangles, and as we have shown, the effect of the MH shows off when the instances are more complex.

Furthermore, note the values reported in columns DOM. These correspond to a number of occurrences when the method found a solution that had both a smaller area and HPWL than the solution found by the other method; such a solution is objectively better given our two main metrics. We can see that **MH-10** was able to do so in more cases, which again highlights the power of local search performed by the ILP solver.

Table 5. Values of $W + H$ in μm, area in μm^2 and HPWL in μm for each instance, and average ratios of automated and manual metrics, obtained using **MH-10**. The average ratios obtained by FDGD-ILP in [13] are shown in the last row for comparison. Solutions dominating the manual one, given all three metrics, are highlighted.

| instance | manual | | | MH-10 | | | | | | | | |
| | | | | $c_{conn} = 0.1$ | | | $c_{conn} = 1.0$ | | | $c_{conn} = 8.0$ | | |
	W+H	area	HPWL	W+H	area	HPWL	W+H	area	HPWL	W+H	area	HPWL
1	158	6118	1850	157	6172	1636	157	6183	1562	166	6889	1478
2	116	2710	1784	**88**	**1936**	**1024**	91	2070	928	106	2757	797
3	106	2650	906	**85**	**1779**	**660**	89	1968	654	92	2119	547
4	129	4096	812	**112**	**3117**	**782**	114	3256	717	131	4064	662
5	207	8972	13797	**159**	**6351**	**9955**	165	6789	8141	169	7120	7863
6	178	7698	4039	**169**	**7167**	**3666**	167	7009	3647	174	7224	3615
7	168	6580	2908	164	6756	2633	168	7093	2314	173	7466	2307
8	173	7294	1501	**160**	**6399**	**1224**	169	6973	1068	173	7139	1093
9	243	14129	4705	**225**	**12647**	**4205**	234	13664	4003	241	14487	3882
10	205	10214	28386	191	9093	38626	194	9446	32363	236	13714	24930
11	225	9922	28527	197	9356	29074	205	10313	17864	241	13717	13210
12	155	5953	3824	**123**	**3803**	**2315**	126	3937	2162	159	6298	1597
13	162	6511	2061	**153**	**5855**	**1822**	155	6002	1665	155	6008	1693
14	247	15235	2399	**193**	**9212**	**1720**	193	9263	1557	211	10657	1363
15	123	3758	1619	115	3309	1817	113	3178	1852	116	3385	1712
16	232	12397	2676	**215**	**11551**	**1973**	223	12318	1792	221	12143	1944
17	247	12525	4586	**225**	**12172**	**3313**	235	13708	3008	252	15790	2964
avg ratio **MH-10**	1.00	1.00	1.00	0.89	0.84	0.86	0.91	0.89	0.77	0.98	1.02	0.71
avg ratio **FDGD-ILP** [13]	1.00	1.00	1.00	0.88	0.84	0.93	0.91	0.89	0.82	0.99	1.04	0.74

Table 6. Comparison of **FDGD-ILP** and **MH-10** for all 34 experiments performed on real-life instances. DOM shows in how many cases the method dominated the other one with respect to both the area and HPWL.

| | $c_C = 0.1$ | | $c_C = 1.0$ | | $c_C = 8.0$ | |
	aRD (BH)	DOM	aRD (BH)	DOM	aRD (BH)	DOM
FDGD-ILP	0.46 (22)	1	1.82 (11)	1	3.84 (14)	0
MH-10	0.60 (12)	8	**0.11 (23)**	10	**0.49 (20)**	11

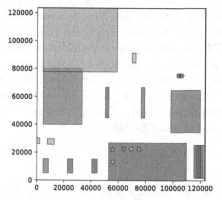

(a) Manual design, area $= 15235\mu m^2$, HPWL $= 2399\mu m$.

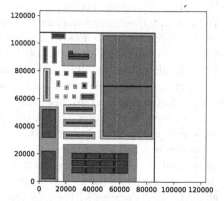

(b) FDGD-ILP, area $= 9212\mu m^2$, HPWL $= 1898\mu m$.

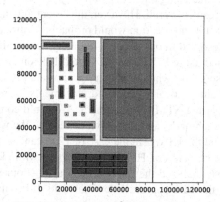

(c) MH-10, area $= 9212\mu m^2$, HPWL $= 1720\mu m$.

Fig. 5. Comparison of manual and automated placements, obtained for $c_C = 0.1$. Shown instance corresponds to the 14th row in Table 5.

We illustrate the mentioned results with Figs. 5, which show instance number 14. Three figures correspond to the manual solution, the solution obtained using **FDGD-ILP**, and finally using **MH-10**. Note that the manual design does not show the positions of the physical devices within the rectangles. We can see that both automatically generated solutions dominated the manual design. The differences between both automated solutions are subtle; their areas are actually equal. However, even these subtle changes in the positions of smaller rectangles are enough to dramatically decrease the HPWL in the case of the solution generated by **MH-10**.

7 Conclusion

In this paper, we extended our previous work on the automation of the placement of AMS ICs. We studied the effect of additional redundant constraints on the performance of the state-of-the-art ILP solver. While the symmetry-breaking constraints did not enhance the solver's performance, imposing an additional constraint on the maximum value of half perimeter of the placement led to improvement on the smaller instances. However, for larger instances, such constraint made the solver unable to find any feasible solution in a given computation time, even though the bound was derived from a known feasible solution. Therefore, we need to provide a solver with an initial solution if we would like to exploit the half-perimeter constraint in the future.

Our experiments with MHs were more successful. We proposed applying the ILP solver to perform a local search in the created placement. The intensification phase of the MH relied on freeing variables associated with a few spatially close rectangles while fixing the other. We showed an additional ILP model that we used to perform the diversification step, to diverge further from the current solution when the local minimum is reached. We evaluated several different MH settings on synthetically generated instances. We concluded that using intensification only and freeing 10 rectangles in each iteration led to the best results overall. However, the potential benefits of diversification cannot be overlooked, but its application would probably require a more advanced control mechanism than presented in our paper. Ultimately, we significantly improved our previous results, obtained using FDGD-warm started ILP, on large instances with 200 and more rectangles, especially when symmetry groups are present in the instance.

Finally, we created automatically generated placements for real-life instances provided by our industry partner STMicroelectronics. We could compare our results with the manually created benchmarks and our previous results. We were again able to outperform both the area and the HPWL in the case of 12 instances; furthermore, we were able to reduce the average value of HPWL even further while keeping the area metric unaffected. When we analyzed the improvement against our previous results more closely, we found that the MH approach dominated its ILP-only counterpart regarding both the HPWL and area in one-third of the experiments performed on real-life instances. This again suggests that the use of MH could be beneficial not only in the specific domain of AMS IC placement but in the domains of packing and cutting as well, where only the area is minimized.

Acknowledgements. This work was supported by the Grant Agency of the Czech Republic under the Project GACR 22-31670S. This work was co-funded by the European Union under the project ROBOPROX - Robotics and advanced industrial production (reg. no. CZ.02.01.01/00/22_008/0004590). We would like to thank the STMicroelectronics company, namely Dalibor Barri and Patrik Vacula, for providing real-life instances and helpful discussions about a problem.

References

1. Alvarez-Valdes, R., Parreño, F., Tamarit, J.: Reactive GRASP for the strip-packing problem. Comput. Oper. Res. **35**(4), 1065–1083 (2008)
2. Berger, M., Schröder, M., Küfer, K.H.: A constraint-based approach for the two-dimensional rectangular packing problem with orthogonal orientations. In: Fleischmann, B., Borgwardt, K.H., Klein, R., Tuma, A. (eds.) Operations Research Proceedings 2008, vol. 2008, pp. 427–432. Springer, Heidelberg (2009). https://doi.org/10.1007/978-3-642-00142-0_69
3. Camm, J.D., Raturi, A.S., Tsubakitani, S.: Cutting big M down to size. Interfaces **20**(5), 61–66 (1990)
4. Chen, T.C., Jiang, Z.W., Hsu, T.C., Chen, H.C., Chang, Y.W.: NTUplace3: an analytical placer for large-scale mixed-size designs with preplaced blocks and density constraints. IEEE Trans. Comput. Aided Des. Integr. Circ. Syst. **27**(7), 1228–1240 (2008)
5. Christiaens, J., Vanden Berghe, G.: Slack induction by string removals for vehicle routing problems. Transp. Sci. **54**, 417–433 (2020)
6. Cohn, J., Garrod, D., Rutenbar, R., Carley, L.: KOAN/ANAGRAM II: new tools for device-level analog placement and routing. IEEE J. Solid-State Circ. **26**(3), 330–342 (1991)
7. CORDIS: Analog/mixed signal back end design automation based on machine learning and artificial intelligence techniques (AMBEATion). https://cordis.europa.eu/project/id/101007730. Accessed 16 May 2023
8. DARPA: Intelligent Design of Electronic Assets (IDEA). https://www.darpa.mil/program/intelligent-design-of-electronic-assets. Accessed 16 May 2023
9. Deghdak, K., T'kindt, V., Bouquard, J.L.: Scheduling evacuation operations. J. Sched. **19**(4), 467–478 (2016)
10. Della Croce, F., Grosso, A., Salassa, F.: A matheuristic approach for the two-machine total completion time flow shop problem. Ann. Oper. Res. **213**(1), 67–78 (2014)
11. Dumez, D., Lehuédé, F., Péton, O.: A large neighborhood search approach to the vehicle routing problem with delivery options. Transp. Res. Part B: Methodological **144**, 103–132 (2021)
12. Fischetti, M., Fischetti, M.: Matheuristics. In: Martí, R., Pardalos, P.M., Resende, M.G.C. (eds.) Handbook of Heuristics, pp. 121–153. Springer, Cham (2018). https://doi.org/10.1007/978-3-319-07124-4_14
13. Grus., J., Hanzálek., Z., Barri., D., Vacula., P.: Automatic placer for analog circuits using integer linear programming warm started by graph drawing. In: Proceedings of the 12th International Conference on Operations Research and Enterprise Systems, pp. 106–116 (2023)
14. Hopper, E., Turton, B.: A genetic algorithm for a 2D industrial packing problem. Comput. Ind. Eng. **37**(1), 375–378 (1999)
15. Klausnitzer, A., Lasch, R.: Optimal facility layout and material handling network design. Comput. Oper. Res. **103**, 237–251 (2019)
16. Korf, R., Moffitt, M., Pollack, M.: Optimal rectangle packing. Ann. Oper. Res. **179**(1), 261–295 (2010)
17. Lin, Y., et al.: Are analytical techniques worthwhile for analog IC placement? In: 2022 Design, Automation & Test in Europe Conference & Exhibition (DATE), pp. 154–159 (2022)

18. Lindahl, M., Sørensen, M., Stidsen, T.: A fix-and-optimize matheuristic for university timetabling. J. Heuristics 24(4), 645–665 (2018)
19. Lourenco, N., Vianello, M., Guilherme, J., Horta, N.: LAYGEN - automatic layout generation of analog ICs from hierarchical template descriptions. In: 2006 Ph.D. Research in Microelectronics and Electronics, pp. 213–216 (2006)
20. Lu, J., et al.: ePlace: electrostatics-based placement using fast Fourier transform and Nesterov's method. ACM Trans. Des. Autom. Electron. Syst. 20(2), 1–34 (2015)
21. Ma, Q., Xiao, L., Tam, Y.C., Young, E.F.Y.: Simultaneous handling of symmetry, common centroid, and general placement constraints. IEEE Trans. Comput. Aided Des. Integr. Circuits Syst. 30(1), 85–95 (2011)
22. Mallappa, U., Pratty, S., Brown, D.: RLPlace: deep RL guided heuristics for detailed placement optimization. In: Proceedings of the 2022 Conference & Exhibition on Design, Automation & Test in Europe, pp. 120–123 (2022)
23. Martins, R., Lourenço, N., Horta, N.: Multi-objective optimization of analog integrated circuit placement hierarchy in absolute coordinates. Expert Syst. Appl. 42(23), 9137–9151 (2015)
24. Mirhoseini, A., Goldie, A., Yazgan, M., et al.: A graph placement methodology for fast chip design. Nature 594(7862), 207–212 (2021)
25. Murata, H., Fujiyoshi, K., Nakatake, S., Kajitani, Y.: VLSI module placement based on rectangle-packing by the sequence-pair. IEEE Trans. Comput. Aided Des. Integr. Circuits Syst. 15, 1518–1524 (1996)
26. Mönch, L., Roob, S.: A matheuristic framework for batch machine scheduling problems with incompatible job families and regular sum objective. Appl. Soft Comput. 68, 835–846 (2018)
27. Ou, H.C., Tseng, K.H., Liu, J.Y., Wu, I.P., Chang, Y.W.: Layout-dependent effects-aware analytical analog placement. IEEE Trans. Comput. Aided Des. Integr. Circuits Syst. 35(8), 1243–1254 (2016)
28. Polyakovskiy, S., M'Hallah, R.: A lookahead matheuristic for the unweighed variable-sized two-dimensional bin packing problem. Eur. J. Oper. Res. 299(1), 104–117 (2022)
29. Smet, P., Wauters, T., Mihaylov, M., Vanden Berghe, G.: The shift minimisation personnel task scheduling problem: a new hybrid approach and computational insights. Omega 46, 64–73 (2014)
30. Spindler, P., Schlichtmann, U., Johannes, F.M.: Kraftwerk2-a fast force-directed quadratic placement approach using an accurate net model. IEEE Trans. Comput. Aided Des. Integr. Circ. Syst. 27(8), 1398–1411 (2008)
31. Strasser, M., Eick, M., Grab, H., Schlichtmann, U., Johannes, F.M.: Deterministic analog circuit placement using hierarchically bounded enumeration and enhanced shape functions. In: 2008 IEEE/ACM International Conference on Computer-Aided Design, pp. 306–313 (2008)
32. Xiao, Y., Xie, Y., Kulturel-Konak, S., Konak, A.: A problem evolution algorithm with linear programming for the dynamic facility layout problem-a general layout formulation. Comput. Oper. Res. 88, 187–207 (2017)
33. Xie, W., Sahinidis, N.V.: A branch-and-bound algorithm for the continuous facility layout problem. Comput. Chem. Eng. 32(4), 1016–1028 (2008)
34. Xu, B., et al.: Device layer-aware analytical placement for analog circuits. In: Proceedings of the 2019 International Symposium on Physical Design, pp. 19–26. ISPD 2019 (2019)
35. Xu, B., Li, S., Xu, X., Sun, N., Pan, D.Z.: Hierarchical and analytical placement techniques for high-performance analog circuits. In: Proceedings of the 2017 ACM on International Symposium on Physical Design, pp. 55–62. ISPD 2017 (2017)
36. Yang, F., Leus, R.: Scheduling hybrid flow shops with time windows. J. Heuristics 27(1), 133–158 (2021)

Applications

An Urban-Scale Application of the Problem of Designing Green Tourist Trips with Time Windows

Annarita De Maio[1]([⊠])(iD), Roberto Musmanno[2](iD), and Aurora Skrame[2](iD)

[1] Department of Economy, Statistics and Finance "G. Anania",
University of Calabria, Rende, Italy
annarita.demaio@unical.it

[2] Department of Mechanical, Energy and Management Engineering,
University of Calabria, Rende, Italy
{roberto.musmanno,aurora.skrame}@unical.it

Abstract. In this paper, we deal with the problem of designing tourist trips with time windows. The proposed optimization model is characterized by the presence of three different objectives: 1) the minimization of transportation costs and expense incurred to access the tourist sites to be visited; 2) the minimization of CO_2 emissions generated by the vehicles used for travel; and 3) the maximization of the attractiveness of the tourist sites to be visited, where the attractiveness of each site is measured qualitatively through a predefined index on a homogeneous scale of values. The constraints of the problem concern, among others, the need to comply with predefined time windows for visiting tourist sites and the maximum total time available for the tourist trips. For each tourist trip, it is also possible to differentiate the tourist sites to be visited belonging to different categories and also provide for the possibility of a lunch break.

The proposed model has been applied to a real case concerning the city of Florence, Italy. The results obtained demonstrate the correctness of the model and allow, in particular, to assess the impact of the objective of attractiveness of POIs to visit compared to the other two proposed.

Keywords: Green tourist trip · Time windows · Multiobjective

1 Introduction

The performance of the Tourism industry in the world has been heavily impacted by the COVID-19 pandemic, with many countries imposing travel restrictions and tourist attractions being closed for long periods. As a result, the number of international tourist arrivals has decreased significantly in 2020 and early 2021, compared to previous years. According to the World Tourism Organization (UNWTO), international tourist arrivals decreased by 72% in 2020 compared to 2019, representing a loss of 1.3 billion international arrivals. This has resulted in an estimated loss of $ 1.3 trillion in tourism export revenues, with many tourism-dependent economies feeling the effects.

However, the pandemic period was also an opportunity for the tourism industry to evolve more rapidly toward Tourism 2.0. The shift from tourism to Tourism 2.0 is a result of technological advancements and changes in consumer behavior. The following are some key steps that have led to this paradigm shift:

© The Author(s), under exclusive license to Springer Nature Switzerland AG 2024
F. Liberatore et al. (Eds.): ICORES 2022/2023, CCIS 1985, pp. 203–219, 2024.
https://doi.org/10.1007/978-3-031-49662-2_11

- emergence of mobile technology: with the proliferation of smartphones and tablets, travelers can now access travel information and services anywhere, at any time. Tourism 2.0 has leveraged mobile technology to create apps that provide personalized experiences and real-time information;
- introduction of big data and artificial intelligence: Tourism 2.0 relies heavily on data-driven insights and machine learning algorithms to provide personalized recommendations and insights into traveler behavior. With the use of big data, tourism companies can anticipate traveler needs and provide customized services;
- shift towards eco-tourism and sustainable tourism: sustainability and eco-tourism have become increasingly important in recent years, and Tourism 2.0 has adapted to this trend. Many tourism companies are now adopting sustainable practices and promoting eco-friendly travel options to appeal to conscious travelers.

Overall, the steps from tourism to Tourism 2.0 have seen the industry undergo significant transformations. Tourism 2.0 has disrupted traditional tourism models, providing tourists with more personalized experiences and empowering them to play an active role in the tourism ecosystem.

As a consequence, in the last years, it is possible to register the growth of different Tourist Recommender Systems ([23] and [31]), that are technology-based tools using data analytics and machine learning algorithms to provide personalized recommendations to travelers. These systems are designed to help tourists make informed decisions about travel destinations, accommodation, and activities based on their preferences and behavior ([24]).

Finally, the concept of sustainability in tourism is closely linked to the United Nations Agenda 2030 (https://www.unwto.org/tourism-in-2030-agenda), which aims to promote sustainable economic growth, social development, and environmental protection. Tourism 2.0 can contribute to achieving the Sustainable Development Goals outlined in Agenda 2030 (environmental, social, economic, cultural sustainability, education and awareness).

Numerical optimization techniques have significant potential for improving the Tourism 2.0 industry, in particular, for building appealing and sustainable tours. The problem of helping tourists in the definition of a customized tour is defined Tourist Trip Design Problem (TTDP) ([25]). In this problem a tourist usually wants to visit a city, rich of attractions, in a limited amount of time. The scope is to build a personalized tour, according to the tourist's preferences, selecting a sequence of POIs to visit. Each POI is characterized by a score related to its attractiveness, a cost and a time window for the visit. Usually, the tourist can also impose limits about the total budget available for the visits. Moreover, the tourist is able to move from a POI to another using a variety of transportation modes, each of them with different costs (also in terms of CO_2 emissions) and travel times. The scope of this paper is to introduce a variant of the Green Tourist Trip Design Problem (GTTDP) in which the sustainability of the tour is considered, introduced for the first time by [26]. Under this respect, there are different aspects to be taken into account: the potential attractiveness of the tour, the number of POIs to visit, the time windows of the POIs, the cost and the time duration of the tour, and the quantity of CO_2 released during the movements between POIs. From another perspective, smart and inclusive cities should take care to the needs of the so-called vulnerable road users (VRUs). Traditionally, pedestrians, bicyclists, and motorcyclists are considered VRUs, as well as people with disability, elder people and children ([4]). Under this respect, the research interest is more concentrated to reduce crashes involving VRUs ([1]), to build personalized trips for people with mobility impairments ([3]) or to define information systems and mobile services sharing real-time information and integrating a tour design for VRUs ([2]). As argued by [3], it is necessary in the next future to propose fully accessible and personalized tours in a broader multimodal mobility context for VRUs, as well as to include useful functionalities into the smart services for the different users' needs.

In order to consider all these aspects, a multimodal multiobjective GTTDP with time windows is presented, which is an extended version of the problem described in [27]. The rest of the paper is organized as follows: Sect. 2 reports the literature review in the field and the main contributions of this work. Section 3 describe the methodology used for the POIs evaluation, Sect. 4 describes the mathematical formulation, whereas Sect. 5 introduces the case study settings and the computational results. Finally, Sect. 6 reports the conclusions and the future developments.

2 Literature Review

Under the tourism perspective, the TTDP is one of the most investigated class of optimization problems, deriving from the more general class of the Vehicle Routing Problems (VRP) with profits ([22]), in which two different decisions have to be taken: 1) the more convenient customers to be served (not defined a-priori like it happens in the classical VRP) and 2) the routes to be used. If only one route is built, the problem becomes a variant of the Travelling Salesman Problem (TSP), described in the scientific literature by using different alternative denominations. The most diffused denomination is the Orienteering Problem (OP), introduced by [21] and deriving from the well known orienteering sport, in which each participant has to maximize the total collected prizes associated with the visited points, returning to the starting point within a certain time. Alternative definitions and variants of the same problem are: the Maximum Collection Problem ([20]), the Selective Travelling Salesperson Problem ([19]) and the Bank Robber Problem ([18]). Note that, when multiple vehicles are involved, the most investigated variant is the so-called Team Orienteering Problem (TOP). An overview of the OPs that is worth mentioning is reported in the survey of [25], in which several variants are classified and described: classical OP, TOP ([17]), TOP with time windows ([15]), time dependent TOP ([16]), stochastic OP ([14]) and the generalized OP ([13]). The authors, in their conclusions, also identified the tourism trip design as one of the major practical applications of this problem. The TTDP is illustrated in the surveys of [12] and [11]. The authors underlined that the TTDP can be mainly classified in single-objective and multiobjective problems. In the single-objective case, the scope is the maximization of the benefits associated with visiting the POIs. Under this respect, several variants are investigated: presence of time windows denoting the opening and closing times of each POI ([10]); score of POIs deriving from personal preferences or sensitive analysis ([9]); hotel selection option ([8]); lunch scheduling ([28]); restaurant selection ([29]); and time and budget constraints. In the multiobjective case, different objective functions are considered in the TTDP formulations: minimizing transportation and visiting costs, minimizing waiting times at POIs, maximizing attractiveness of the tour, minimizing travel time between POIs and maximizing the diversity of the selected POIs (see [6,7] and [5]). Finally, the authors introduced some future perspective in the tourist trip design, indicating the *green* variant as one of the most promising research areas in the next future. Green and sustainable tourism can be declined in different forms: for example, taking into account CO_2 emission during movements between POIs, suggesting alternative mobility options to the users, protecting natural areas or potentiating the development of nature tourism and ecotourism. This discussion underlines the timeliness of this work, and a GTTDP with time windows is introduced in the following sections. From the best of our knowledge, the GTTDP was previously investigated only by [26]. The authors introduced a multiobjective and multimodal GTTDP where a mixed-integer linear model is formulated, considering three objective functions balancing the total score of the trip, the total cost and the total CO_2 emission produced by the trip. Under this respect, various transportation modes are considered as possible choice for the tourists in order to move between POIs. Constraints related to travel time are also considered. A first extension of the cited work was introduced by [27]. The authors propose three different objective functions: minimizing the total CO_2 emissions and maximizing both the number of visited POIs and the total scores associated with visited POIs. Moreover, additional constraints are incorporated in the model, in order to customize the tour considering only the transportation modes chosen by the tourists, or considering their preferences related to the own physical possibilities (for example, families with children would not like moving by using bicycles or elder people would like avoiding long walking paths). In this paper, a further extension of the problem already introduced by [27] is presented, focusing mainly on the need to personalize the digital user journey. In details:

- three different objective functions, related to different perspectives from [27]: minimizing both the total CO_2 emissions and the total costs (consisting in the cost for travelling between the POIs and the ticked cost for visiting the POIs); maximizing the quality associated with the attractiveness of the tour;
- the scores associated to the POIs are determined using a multi-criteria method (for more details see Sect. 3);
- the tour is designed considering a further rich variety of operational constraints: a minimum number of POIs to be included in the tour, the time windows constraints related to the opening and closing time of the POIs, the tourists preferences for the starting and ending times of the tour; the possibility to schedule a lunch break at a food spot, in a specific time window of the day;
- POIs are classified into different categories (e.g., museums, religious attractions, seasides, etc.). It is possible to impose a minimum and maximum number of POIs to visit for each category (further details on the classification procedure are described in the sequel);
- a larger variety of transportation modes is taken into account (car, public transport, bike, feet and push scooter);
- the model is tested on a large real-setting case study in an urban context (city of Florence).

3   A Multi-criteria Method for the POI Evaluation

To evaluate the POIs, a multi-criteria method can be used, corresponding to the well-known weighted scoring method ([30]). It is considered accurate and fast enough for the definition of the attractiveness parameter for each POI, which is only one of the input data in a larger decision-making process where the proposed optimization model comes into play (see Sect. 4). Note that, multi-criteria evaluation is successfully applied in different context, see for example [33, 34] and [35].

The weighted scoring method adopted for the evaluation of POIs is outlined below.

1. Establishing criteria. The first step is to identify the criteria that will be used to assess the POI. The criteria can be economic, social, environmental or cultural and can be different for each category of POI.
   In the sequel, we refer to the following notation: N^+ is the set of POIs; H is the number of categories adopted to classify the POIs. In this way, N^+ is assumed to be partitioned into the following subsets: $N_1^+, \ldots, N_H^+; R_t, t = 1, \ldots, H$, is the set of criteria used to evaluate the attractiveness of the POIs belonging to N_t^+.
2. Weighting the criteria. The second step is to assign a weight w_r to each criterion $r \in R_t, t = 1, \ldots, H$, based on its relative importance. Criteria that are more critical to stakeholders will have more significant weights. The sum of the weights of the criteria for each POI category should be equal to one, that is, $\sum_{r \in R_t} w_r = 1$, for each $t = 1, \ldots, H$.
3. Assigning scores. The third step is to assign a score $s_{ir}, i \in N_t^+, r \in R_t, t = 1, \ldots, H$. This can involve collecting data from various sources, such as surveys, field trips, and expert evaluations. Scores can be assigned on a numerical scale, such as $1-10$.
4. Aggregating the scores. The fourth step is to aggregate the weighted scores to provide an overall assessment of the attractiveness of each POI:

$$p_i = \sum_{r \in R_t} w_r s_{ir}, \forall i \in N_t^+, t = 1, \ldots, H. \tag{1}$$

4   Mathematical Formulation

In order to formalize the model under investigation we adopt the following further notation with respect to that introduced in Sect. 3. Let N_{H+1}^+ be the set of food spots where tourists can take a lunch break. Let $N^{++} = N^+ \cup N_{H+1}^+$ and let $|N^{++}| = n$. In this way, $N = N^{++} \cup \{0, n+1\}$, where 0 is the starting point of the tour, duplicated as $n + 1$ also to denote the ending point of the tour. A is the arc set of the complete directed graph induced by N. K is the set of possible transportation modes to realize the visit (car, bus/metro, bike, on foot). For each POI $i \in N^+$ the following parameters are assumed to be known: a score p_i (defined as introduced in the Sect. 3), a cost f_i, i.e., the price of the entrance ticket, a time visit v_i, e_i and l_i, that is, respectively, the opening and closing times. The parameters f_i, v_i, e_i and l_i are also defined for each food spot $i \in N_{H+1}^+$ (in particular, f_i is the cost of lunch consumed at food spot i).

Other parameters are: e_0 and l_0, respectively, the earliest and latest times at which the visit can be planned (the time interval $[l_0 - e_0]$ corresponds to the time window at node 0); T is the maximum travel time determined by all movements between nodes, whereas T_k is the maximum travel time determined by all movements between nodes considering the transportation mode k (clearly, $T_k \leq T, \forall\, k \in K$); c_{ijk}, t_{ijk} and s_{ijk} are, respectively, the cost, the time and the level of CO_2 emission associated with arc $(i, j) \in A$ when travelled by using transportation mode k; M is an arbitrarily large constant.

Moreover, α_h and δ_h are the minimum and the maximum number of POIs for each category $h = 1, \ldots, H$ to be visited, and β a binary constant equal to one if the tourist wishes a lunch break, and zero otherwise. The decision variables are the following: z_i, $i \in N^{++}$, are binary, each of them equal to one if the node i (POI or lunch spot) is visited, zero otherwise; x_{ijk}, $(i, j) \in A$, $k \in K$, are binary, each of them equal to one if (i, j) is travelled by using transportation mode k, zero otherwise; u_i, $i \in N$, are continuous, each of them is the arrival time at node i and, if i is a POI (or a lunch spot), corresponds to the starting time of the visit at POI (or lunch spot) i.

The three-objective problem is formulated as follows:

$$\text{Minimize} \sum_{(i,j) \in A} \sum_{k \in K} c_{ijk} x_{ijk} + \sum_{i \in N^{++}} f_i z_i \tag{2}$$

$$\text{Minimize} \sum_{(i,j) \in A} \sum_{k \in K} s_{ijk} x_{ijk} \tag{3}$$

$$\text{Maximize} \sum_{i \in N^+} p_i z_i \tag{4}$$

subject to

$$\sum_{i \in N \setminus \{0\}} \sum_{k \in K} x_{0ik} = \sum_{i \in N \setminus \{n+1\}} \sum_{k \in K} x_{i,n+1,k} = 1 \tag{5}$$

$$\sum_{i \in N \setminus \{j\}} \sum_{k \in K} x_{ijk} = \sum_{i \in N \setminus \{j\}} \sum_{k \in K} x_{jik} = z_j, \forall\, j \in N^{++} \tag{6}$$

$$\sum_{k \in K} x_{ijk} \leq 1, \quad \forall\, (i, j) \in A \tag{7}$$

$$\sum_{(i,j) \in A} \sum_{k \in K} t_{ijk} x_{ijk} \leq T \tag{8}$$

$$\sum_{(i,j) \in A} t_{ijk} x_{ijk} \leq T_k, \quad \forall\, k \in K \tag{9}$$

$$u_i + v_i + \sum_{k \in K} t_{ijk} x_{ijk} \leq u_j + M\left(1 - \sum_{k \in K} x_{ijk}\right),$$
$$\forall\, i \in N, j \in N \setminus \{i\} \tag{10}$$

$$u_i + v_i \leq l_i z_i, \quad \forall\, i \in N^{++} \tag{11}$$

$$e_i z_i \leq u_i, \quad \forall\, i \in N^{++} \tag{12}$$

$$u_0 \geq e_0 \tag{13}$$

$$u_{n+1} \leq l_0 \tag{14}$$

$$\alpha_h \leq \sum_{i \in N_h^+} z_i \leq \delta_h, \quad \forall\, h = 1, \ldots, H \tag{15}$$

$$\sum_{i \in N_r^+} z_i = \beta \tag{16}$$

$$x_{ijk} \in \{0, 1\}, \quad \forall\, (i, j) \in A, k \in K \tag{17}$$

$$z_i \in \{0, 1\}, \quad \forall\, i \in N^{++} \tag{18}$$

$$u_i \geq 0, \quad \forall\, i \in N \tag{19}$$

The objective function (2) is the total cost of the tour, whereas the objective function (3) corresponds to the total level of CO_2 emission associated with the visit. Both (2) and (3) should be minimized. The objective function (4) corresponds to the total attractiveness of the tour that needs to be maximized. The two constraints (5) state that the visit starts from node 0 and ends to node $n + 1$, respectively. Constraints (6) ensure the connectivity of the tour. They also provide the logic condition that an arc incident in any node j can be travelled using any transportation mode only if POI (or lunch spot) j is selected for the visit. Constraints (7) establish that each arc (i, j) can be travelled by using one transportation mode at most. Constraint (8) ensures that the total travel time of the tour does not exceed the value of T (the constraint is assumed because the visitor does not wish to waste a lot of travel time). Constraints (9) impose a maximum time budget to be spent for any transportation mode. These constraints allow to limit (or completely forbid) the use of a particular transportation mode, in accordance with the specific needs of the tourist who can be a VRU.

Constraints (10) are the subtour elimination constraints. They state that if an arc (i, j) is visited (i.e., travelled by one transportation mode), then the arrival time at node i plus the duration of the visit at node i and the travel time from node i to node j cannot be greater than the arrival time at node j. Of course, if the arc (i, j) is not visited, then, thanks to the arbitrarily large constant M, the corresponding constraint is redundant. Constraints (11) and (12) are time windows constraints at each visited POI (or lunch spot). In particular, constraints (11) impose that if POI (or lunch spot) i is visited, then the starting time of the visit at node i plus the time visit should be within the closing time of node i and, thanks to constraints (12), the starting time of the visit should be after the opening time of node i. Constraints (13) and (14) ensure that the visit starts and ends within the time window $[l_0 - e_0]$ established for the visit. Constraints (15) impose that a minimum and a maximum number of POIs for each category has to be visited. Since β is a binary constant, the constraint (16) imposes the possibility of whether or not a lunch spot can be included in the tour. Constraints (17) and (18) impose that some decision variables are binary, whereas constraints (19) define the non-negative conditions on the remaining ones. In the following section, some computational experiments related to the real case study are presented.

5 Case-Study Setting and Computational Results

The mathematical model illustrated in Sect. 4 has been applied to the city of Florence in Italy. A set N^+ of 30 POIs has been considered, whose features, in terms of geographical coordinates, categories, price, average duration of the visit, time windows and score, are reported in Table 1. Note that all data are extracted by using the support of available databases: for the positions, the travel times and distances between POIs (not reported in details for the sake of brevity), a `Python` routine linked with `Google Maps` has been used. The spatial distribution of the selected POIs is represented in Fig. 1.

Fig. 1. Locations of some POIs and food outlets chosen in the city of Florence.

The score of each POI $i \in N_t^+$, $t = 1, \ldots, H$, has been computed considering the application of the weighted scoring method illustrated in Sect. 3. In the sequel, we refer to the steps of this method. In particular, we have considered $H = 3$ categories:

1. squares and parks ($|N_1^+| = 6$);
2. museums ($|N_2^+| = 9$);
3. religious attractions ($|N_3^+| = 5$).

In addition, the set N_4^+ of food outlets is composed of 10 items. The set R_t (see Step 1) consists of the same three criteria for each POI category $t = 1, 2, 3$ (i.e., $R_t = R$, $t = 1, 2, 3$):

1. a performance criterion defined by the Italian Ministry of Tourism (https://catalogo. beniculturali.it/);

Table 1. Features of the 20 POIs chosen in the city of Florence.

ID	POI i	Latitude	Longitude	Category	Cost f_i, (€)	Time window $[e_i, l_i]$	Visiting time v_i, (minutes)	Score p_i
1	Piazza Duomo	43.772904	11.257733	squares and parks	20	[09:30; 17:30]	150	9.84
2	Piazza della Signoria	43.769306	11.255592	squares and parks	22	[08:15; 18:30]	150	9.71
3	Ponte Vecchio	43.768459	11.253549	squares and parks	0	[00:00; 24:00]	30	9.67
4	Galleria degli Uffizi	43.768665	11.25568	museums	20	[08:15; 18:50]	150	9.96
5	Palazzo Strozzi	43.771527	11.251981	museums	16	[09:30; 17:30]	60	9.10
6	Museo Galileo	43.767776	11.256297	museums	10	[09:30; 18:00]	60	8.87
7	Dante's House	43.771016	11.257115	museums	8	[10:00; 18:00]	60	7.98
8	Basilica of Holy Spirit	43.766924	11.247824	religious attractions	0	[08:30; 18:00]	90	8.88
9	Synagogue and Jewish Museum	43.773316	11.265698	religious attractions	10	[10:00; 17:30]	120	9.65
10	English Cemetery	43.77685	11.268549	religious attractions	3	[14:00; 18:00]	60	8.28
11	Marini's Museum	43.771968	11.249687	museums	6	[10:00; 19:00]	120	7.36
12	Piazza della S. Annunziata	43.776338	11.260921	religious attractions	4	[07:30; 19:00]	60	7.54
13	Horne Museum	43.767522	11.259347	museums	6	[10:00; 18:00]	30	7.42
14	Clet's Workshop	43.764474	11.260829	museums	0	[13:30; 19:30]	60	7.03
15	Basilica di Santa Croce	43.768737	11.26203	religious attractions	9	[09:30; 17:30]	90	9.41
16	Parco delle Cascine	43.782497	11.219324	squares and parks	5	[11:00; 19:00]	150	8.34
17	Complesso di Palazzo Pitti	43.765343	11.249898	museums	22	[08:10; 19:10]	150	9.44
18	Enzo Pazzagli Art Park	43.768647	11.324149	squares and parks	6	[11:00; 21:00]	150	7.93
19	Piazzale Michelangelo	43.766762	11.264762	squares and parks	0	[00:00; 00:00]	30	9.28
20	Gardens and Villa of Castello	43.819098	11.229186	museums	8	[08:30; 18:30]	150	9.25
21	La Bucchetta Food & Wine	43.767364	11.25929	food outlets	20	[12:00; 14:30]	40	–
22	Braceria All'11	43.766903	11.246419	food outlets	20	[12:00; 14:30]	40	–
23	Il Bufalo Trippone	43.769761	11.2585	food outlets	20	[12:00; 14:30]	40	–
24	All'Antico Vinaio	43.768434	11.257409	food outlets	20	[12:00; 14:30]	40	–
24	All'Antico Vinaio	43.768434	11.257409	food outlets	20	[12:00; 14:30]	40	–
25	Ristorante il Ricettario	43.768885	11.254372	food outlets	20	[12:00; 14:30]	40	–
26	Antica Trattoria dell'Oca	43.790338	11.222449	food outlets	20	[12:00; 14:30]	40	–
27	Giotto Pizzeria-Bistrot	43.786691	11.231321	food outlets	20	[12:00; 14:30]	40	–
28	Il Farro	43.800181	11.221528	food outlets	20	[12:00; 14:30]	40	–
29	Trattoria il Bargello	43.769308	11.258122	food outlets	20	[12:00; 14:30]	40	–
30	Braceria Ristorante Fumo e Fiamme	43.773795	11.248032	food outlets	20	[12:00; 14:30]	40	–

2. a criterion based on the opinion of a group of experts in the tourism industry;
3. a criterion based on the European system of indicators for sustainable tourist destinations (https://single-market-economy.ec.europa.eu/sectors/tourism/eu-funding-and-businesses/funded-projects/sustainable/indicators_en).

The weights chosen for each criterion (see Step 2) are the following: $w_1 = 0.5$, $w_2 = 0.3$ and $w_3 = 0.2$. The score assigned to each POI for each criterion (see Step 3) is reported in Tables 2, 3 and 4.

Finally (see Step 4), the score of each POI $i \in N_t^+$, $t = 1, 2, 3$, obtained applying (1), is reported in the last column of Tables 2, 3 and 4.

Table 2. Multi-criteria analysis on square and parks category.

POI	Criterion 1	Criterion 2	Criterion 3	Weighted score
i	s_{i1}	s_{i2}	s_{i3}	p_i
Piazza Duomo	9.80	10.00	9.50	9.84
Piazza della Signoria	10.00	9.70	9.30	9.71
Ponte Vecchio	9.60	9.70	9.70	9.67
Parco delle Cascine	9.10	8.10	7.80	8.34
Enzo Pazzagli Art Park	8.20	7.90	7.60	7.93
Piazzale Michelangelo	9.00	9.60	8.90	9.28
Weight w_r	0.3	0.5	0.2	

Table 3. Multi-criteria analysis on museums category.

POI	Criterion 1	Criterion 2	Criterion 3	Weighted score
i	s_{i1}	s_{i2}	s_{i3}	p_i
Galleria degli Uffizi	10.00	10.00	9.80	9.96
Palazzo Strozzi	9.30	9.10	8.80	9.10
Museo Galileo	8.10	9.40	8.70	8.87
Dante's House	8.60	7.80	7.50	7.98
Marini's Museum	7.90	7.30	6.70	7.36
Horne Museum	7.60	7.40	7.20	7.42
Clet's Workshop	7.20	7.10	6.60	7.03
Complesso di Palazzo Pitti	9.70	9.30	9.40	9.44
Gardens and Villa of Castello	8.80	9.50	9.30	9.25
Weight w_r	0.3	0.5	0.2	

Table 4. Multi-criteria analysis on religious attractions category.

POI	Criterion 1	Criterion 2	Criterion 3	Weighted score
i	s_{i1}	s_{i2}	s_{i3}	p_i
Basilica of Holy Spirit	8.90	9.00	8.70	8.88
Synagogue and Jewish Museum	9.60	9.50	9.90	9.65
English Cemetery	8.00	8.70	8.00	8.28
Piazza della S. Annunziata	8.30	7.30	7.10	7.54
Basilica di Santa Croce	9.80	9.20	9.30	9.41
Weight w_r	0.3	0.5	0.2	

For a more detailed description of how the values shown in the Table have been chosen, see Sect. 3.

Five different transportation modes are considered: car, walking, traditional bike (not electric version), public transportation (tram and bus) and push scooter. The amount of produced CO_2 per transportation mode is extracted from the CO_2 Connect website (www.co2nnect.org): 0.183 kg/km for car, 0 kg/km for walking and bike, 0.065 kg/km for public transportation and 0.126 kg/km for push scooter. Finally, the kilometric cost per transportation mode is equal to € 0.25 for car, € 0.00 for walking, € 0.20 for bike, € 0.35 for public transportation and € 0.50 for push scooter. Note that the costs for bike and push scooter are referred to the most common services of bike sharing and push scooter sharing. It is assumed that the tourist has indicated the maximum duration of the daily itinerary (in our test case 10.5 hours, with a feasible time window set to [9:00;19:30]), the maximum time to be spent for travelling between POIs (90 minutes), the maximum time to be spent with each transportation mode (30 minutes), the minimum number of POIs to be visited for each category (equal to one), the necessity to have a lunch break at a food spot and the departure/arrival point of the tour, represented by Santa Maria Novella railway station.

In order to determine a set of Pareto optimal solutions of the problem (2)–(19), the ϵ-constraint method has been applied, in the way as introduced in [32]. This is a posteriori method, in which a set of efficient solutions of the problem is generated and then the decision maker selects, from these, the preferred one.

A preliminary step is to transform the objective function (4) into a constraint, that is,

$$\sum_{i \in N^+} p_i z_i \geq P. \tag{20}$$

In this way, a minimum level P of total attractiveness is imposed on the tour.

The payoff table (see [32] for more details) defines the ranges of the objective function (2) and (3) values.

We then divide the range of each objective function in four equal intervals and we use the five grid points as the value of ϵ in the ϵ-constraint method.

In this way, for a fixed value of P, we solve ten different problems, in which we optimize, alternatively, one of the two objective functions (2) and (3) using the other as a constraint.

The implementation of any optimization problem has been carried out by using GAMS 24.7.4 (GAMS Development Corporation) as the algebraic modelling system, with CPLEX 12.6 (IBM Corporation) as the solver. The code has been executed on a PC Intel Core i7 (2.3 GHz) with 16 GB of RAM.

Several tests have been carried out, varying the parameter P, to construct several efficient solutions. Specifically, P has been set equal to 45, 50, 55, 60, 65 respectively. Below, only the most relevant results obtained by setting the minimum level of total tour attractiveness to the highest value ($P = 65$) are reported. In this case, the average execution time of each test has been on the order of tens of seconds, which is entirely acceptable when considering the implementation of the algorithm within a mobile app available to a tourist.

The payoff table for the objective functions (2) and (3) is given in Table 5.

Table 6 reports the corresponding five grid points for each objective function (2) and (3) used as the value of ϵ in the ϵ-constraint method.

Table 5. Payoff table considering the minimum level of total tour attractiveness $P = 65$.

	f_1	f_2
	Costs [€]	CO_2 Emissions [kg]
Min f_1	40.614	0.069
Min f_2	80.625	0.000

Table 6. Grid points in the *epsilon*-constraint method.

	ϵ_1	ϵ_2	ϵ_3	ϵ_4	ϵ_5
f_1	40.614	50.616	60.619	70.622	80.625
f_2	0.000	0.017	0.035	0.052	0.069

A total of five efficient solutions have been obtained, shown in Table 7. The table has the following structure: the column **ID** identifies the solutions, the column **Cost** corresponds to the total cost of the tours (value of the objective function (2)), and the column CO_2 **Emission** refers to the calculation of the total CO_2 emissions of the tours (value of the objective function (3)). The results are also shown in Fig. 2. Finally, Tables 8 and

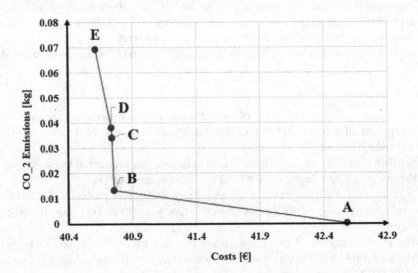

Fig. 2. Efficient solutions obtained through the ϵ-constraint method.

Table 7. Efficient solutions.

Solution ID	f_1	f_2
	Cost [€]	CO_2 Emission [kg]
A	42.585	0.00
B	40.757	0.013
C	40.743	0.034
D	40.736	0.038
E	40.614	0.069

Table 8. Transportation modes activated within the solutions.

Solution ID	Car	Walk	Bicycle	Public transport	Push scooter
A		x	x		
B		x	x	x	
C		x	x	x	
D	x	x	x		
E		x	x	x	

Table 9. Number and category of visited POIs within the solutions.

Solution ID	Squares and parks	Museums	Religious attractions	Food outlets
A	2	3	3	1
B	2	3	3	1
C	2	3	3	1
D	2	3	3	1
E	2	3	3	1

9 report the types of transportation modes used and the number of POIs per category visited within the different solutions, respectively.

Examining the results obtained, it can be observed that, despite an increase in cost of 4.8% when moving from solution E to solution A, the latter involves entirely green transportation modes, thus maximizing the sustainability of the tour (see Fig. 2). However, even in cases where partially green transportation modes are used (solutions B, C, D and E), the total CO_2 emission level remains below 0.1 kg for the entire tour, as car and public transportation are only used for limited distances.

Note that the push scooter mode, which has the worst cost-CO_2 emission trade-off, is never activated (see Table 8).

Furthermore, it should be highlighted that, although the minimum number of POIs to visit has been set to three (one for each category) plus a food outlet, every generated tour includes the visit of eight POIs, thus meeting the needs of the most eclectic tourists (see Table 9). Figures 3 and 4 show the geography of the two tours corresponding to solutions A and E, representing the two extreme points of the Pareto front. This allows for a better appreciation of the differences between the two tours in terms of multi-modality and categories of visited POIs.

Finally, it is worth highlighting that, from the tests carried out on the city of Florence and as already mentioned in [27], of which the present work is an extension, the described model allows for the construction of ad-hoc tourist tours for users, by including/excluding particular transportation modes to satisfy the needs of even the most VRUs, as well as including/excluding specific categories of POIs to visit, thus guaranteeing a win-win strategy for defining the tour in terms of cost-score-CO_2 emissions.

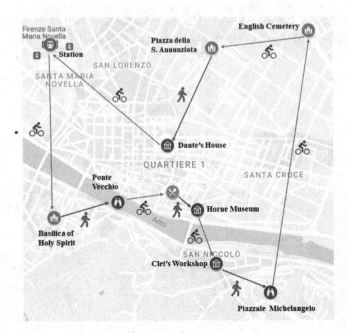

Fig. 3. Tour corresponding to solution A.

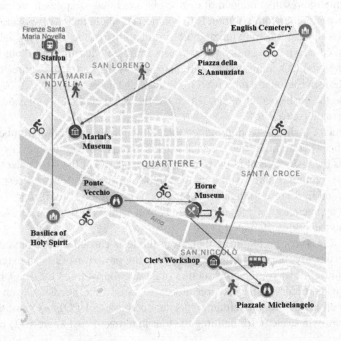

Fig. 4. Tour corresponding to solution E.

6 Conclusion

This work addresses the problem of determining a multimodal green tourist trip in an urban setting that takes into account the preferences and needs of different types of tourists. The proposed multi-objective optimization model and the approach used for its solution provide a valid decision support tool for tourists who aim for environmental sustainability in the planning of tourist itineraries, as well as cost reduction.

Further studies are underway to extend the application of the model to cases where the number of potentially visitable POIs is higher, and to consider the possibility of planning tours over multiple days. In this case, although the mathematical model can be easily extended, its increasing complexity requires the use of heuristic approaches for its solution.

Finally, a further future development concerns the possibility of considering the time dependency and stochasticity of some of the parameters used, in particular, travel times and the duration of visits to POIs.

Acknowledgements. The work of Annarita De Maio is partially supported by MUR (Italian Minister of University and Research) under the grant H25F21001230004. This support is gratefully acknowledged.

References

1. Oxley, J., Langford, J., Charlton, J.: The safe mobility of older drivers: a challenge for urban road designers. J. Transp. Geogr. Spec. Issue Mobil. Older People **18**(5), 642–648 (2010)
2. Scholliers, J., van Sambeek, M., Moerman, K.: Integration of vulnerable road users in cooperative ITS systems. Eur. Transp. Res. Rev. **9**(15) (2017)
3. Darko, J., et al.: Adaptive personalized routing for vulnerable road users. IET Intell. Transp. Syst. WILEY. **16**, 1011–1025 (2022)
4. Methorst, R.: Vulnerable road users: new approaches needed. In: 15th ICTCT Workshop, Transport Research Centre, Rotterdam (2002)
5. Huang, T., Gong, Y.J., Zhang, Y.H., Zhan, Z.H., Zhang, J.: Automatic planning of multiple itineraries: a niching genetic evolution approach. IEEE Trans. Intell. Transp. Syst. **21**(10), 4225–4240 (2019)
6. Castillo, L., et al.: SAMAP: an user-oriented adaptive system for planning tourist visits. Expert Syst. Appl. **34**(2), 1318–1322 (2008)
7. Lim, K.H., Chan, J., Leckie, C., Karunasekera, S.: Personalized trip recommendation for tourists based on user interests, points of interest visit durations and visit recency. Knowl. Inf. Syst. **54**(2), 375–406 (2017)
8. Zheng, W., Ji, H., Lin, C., Wang, W., Yu, B.: Using a heuristic approach to design personalized urban tourism itineraries with hotel selection. Tour. Manag. **76**(422), 103956 (2020)
9. Zheng, W., Liao, Z.: Using a heuristic approach to design personalized tour routes for heterogeneous tourist groups. Tour. Manag. **72**(555), 313–325 (2019)
10. Abbaspour, R.A., Samadzadegan, F.: Itinerary planning in multimodal urban transportation network. J. Appl. Sci. **9**(10), 1898–1906 (2009)
11. Ruiz-Meza, J., Montoya-Torres, J.R.: A systematic literature review for the tourist trip design problem: extensions, solution techniques and future research lines. Oper. Res. Perspect. **9**, 100228 (2022)

12. Gavalas, D., Konstantopoulos, C., Mastakas, K., Pantziou, G.: A survey on algorithmic approaches for solving tourist trip design problems. J. Heuristics **20**(3), 291–328 (2014)

13. Geem, Z.W., Tseng, C.-L., Park, Y.: Harmony search for generalized orienteering problem: best touring in China. In: Wang, L., Chen, K., Ong, Y.S. (eds.) ICNC 2005. LNCS, vol. 3612, pp. 741–750. Springer, Heidelberg (2005). https://doi.org/10.1007/11539902_91

14. Ilhan, T., Iravani, S.M.R., Daskin, M.S.: The orienteering problem with stochastic profits. IIE Trans. **40**(4), 406–421 (2008)

15. Labadie, N., Mansini, R., Melechovsky, J., Wolfler Calvo, R.: The team orienteering problem with time windows: an LP-based granular variable neighborhood search. Eur. J. Oper. Res. **220**(1), 15–17 (2012)

16. Verbeeck, C., Sörensen, K., Aghezzaf, E.H., Vansteenwegen, P.: A fast solution method for the time-dependent orienteering problem. Eur. J. Oper. Res. **236**(2), 419–432 (2014)

17. Pessoa, A., Poggi, M., Uchoa, E.: A robust branch-cut-and-price algorithm for the heterogeneous fleet vehicle routing problem. Networks **54**(4), 167–177 (2009)

18. Awerbuch, B., Azar, Y., Blum, A., Vempala, S.: New approximation guarantees for minimum-weight k-trees and prize-collecting salesmen. SIAM J. Comput. **28**, 254–262 (1998)

19. Laporte, G., Martello, S.: The selective travelling salesman problem. Discret. Appl. Math. **26**, 193–207 (1990)

20. Butt, S.E., Cavalier, T.M.: A heuristic for the multiple tour maximum collection problem. Comput. Oper. Res. **21**, 101–111 (1994)

21. Golden, B.L., Levy, L., Vohra, R.: The orienteering problem. Nav. Res. Logist. **34**, 307–318 (1987)

22. Archetti, C., Speranza, M.G., Vigo, D.: Vehicle Routing Problems with Profits. Vehicle Routing: Problems, Methods and Applications. SIAM, pp. 273–296 (2014)

23. Ardito, L., Cerchione, R., Del Vecchio, P., Raguseo, E.: Big data in smart tourism: challenges, issues and opportunities. Curr. Issues Tour. **22**, 1805–1917 (2019)

24. Kontogianni, A., Kabassi, K., Virvou, M., Alepis, E.: Smart tourism through social network user modeling: a literature review. In: 2018 9th International Conference on Information, Intelligence, Systems and Applications (IISA), pp. 1–4. IEEE (2018)

25. Gunawan, A., Lau, H.C., Vansteenwegen, P.: Orienteering problem: a survey of recent variants, solution approaches and applications. Eur. J. Oper. Res. **255**, 315–332 (2016)

26. Divsalar, G., Ali Divsalar, A., Jabbarzadeh, A., Sahebi, H.: An optimization approach for green tourist trip design. Soft Comput. **26**, 4303–4332 (2022)

27. De Maio, A., Musmanno, R., Skrame, A.: The green tourist trip design problem with time windows: a model application on a urban scale. In: Proceedings of the 12th International Conference on Operations Research and Enterprise Systems (ICORES 2023), pp. 62–70 (2023). https://doi.org/10.5220/0011669500003396

28. Ciancio, C., De Maio, A., Laganà, D., Santoro, F., Violi, A.: A genetic algorithm framework for the orienteering problem with time windows. In: Daniele, P., Scrimali, L. (eds.) New Trends in Emerging Complex Real Life Problems. ASS, vol. 1, pp. 179–188. Springer, Cham (2018). https://doi.org/10.1007/978-3-030-00473-6_20

29. Choachaicharoenkul, S., Coit, D., Wattanapongsakorn, N.: Multi-objective trip planning with solution ranking based on user preference and restaurant selection. IEEE Access **10**, 10688–10705 (2022). https://doi.org/10.1109/ACCESS.2022.3144855

30. Dean, M.: Chapter Six - multi-criteria analysis. In: Mouter, N. (ed.), Advances in Transport Policy and Planning, Academic Press, vol. 6, pp. 165–224 (2020). ISSN 2543–0009, https://doi.org/10.1016/bs.atpp.2020.07.001

31. De Maio, A., Fersini, E., Messina, E., Santoro, F., Violi, A.: Exploiting social data for tourism management: the SMARTCAL project. Quality and Quantitythis link is disabled (2020). https://doi.org/10.1007/s11135-020-01049-8

32. Mavrotas, G.: Effective implementation of the ϵ-constraint method in multi-objective mathematical programming problems. Appl. Math. Comput. **213**, 455–465 (2009). https://doi.org/10.1016/j.amc.2009.03.037
33. Skrame, A., Ciancio, C., Corvello, V., Musmanno, R.: A quantitative model supporting socially responsible public investment decisions for sustainable tourism. Int. J. Financ. Stud. **8**, 33 (2020). https://doi.org/10.3390/ijfs8020033
34. Gebre, S.L., Cattrysse, D., Van Orshoven, J.: Multi-criteria decision-making methods to address water allocation problems: a systematic review. Water **13**, 125 (2021). https://doi.org/10.3390/w13020125
35. Ho, W., Xiaowei, X., Dey, P.K.: Multi-criteria decision making approaches for supplier evaluation and selection: a literature review. Eur. J. Oper. Res. **202**(1), 16–24 (2010). https://doi.org/10.1016/j.ejor.2009.05.009

Optimization of the Storage Location Assignment Problem Using Nested Annealing

Johan Oxenstierna[1,4(✉)] ⓘ, Louis Janse van Rensburg[3], Peter J. Stuckey[2] ⓘ, and Volker Krueger[1] ⓘ

[1] Department of Computer Science, Lund University, Lund, Sweden
{johan.oxenstierna,volker.krueger}@cs.lth.se
[2] Faculty of Information Technology, Monash University, Melbourne, Australia
peter.stuckey@monash.edu
[3] Flux Robotics, Caltowie , Australia
louis@fluxrobotics.ai
[4] Kairos Logic AB, Lund, Sweden

Abstract. The Storage Location Assignment Problem (SLAP) has a significant impact on the efficiency of warehouse operations. We propose a multi-phase optimizer for the SLAP, where the quality of an assignment is based on distance estimates of future-forecasted order-picking. Candidate assignments are first sampled using a Markov Chain accept/reject method. Order-picking Traveling Salesman Problems (TSPs) are then modified according to the assignments and solved. The model is graph-based and generalizes to any obstacle layout in two dimensions. We investigate whether optimization speed-ups are possible using methods such as cost approximation, rejection of samples with low approximate cost and restarts from local minima. Results demonstrate that these methods improve performance, with total travel-cost reductions of up to 30% within 8 h of CPU-time. We share a public repository with SLAP instances and corresponding benchmark results on the generalizable TSPLIB format.

Keywords: Storage location assignment problem · Nested annealing · Hamming distances

1 Introduction

The Storage Location Assignment Problem (SLAP) concerns the search for suitable locations for products in a warehouse. There exist dozens of proposed versions and optimization methods for the SLAP [5]. We work with a standard picker-to-parts scenario where racks and other obstacles can be laid out freely on a two-dimensional plane and where vehicles may start and end their paths at any location. In order to evaluate the quality of a location assignment, we combine two costs. The first cost consists of the travel distance needed to complete a given *picking-log*, i.e., a set of pick-rounds (sequences of product visits). A pick-round is equivalent to a *Steiner* Traveling Salesman Problem (TSP) [36], where the origin and destination locations may be different

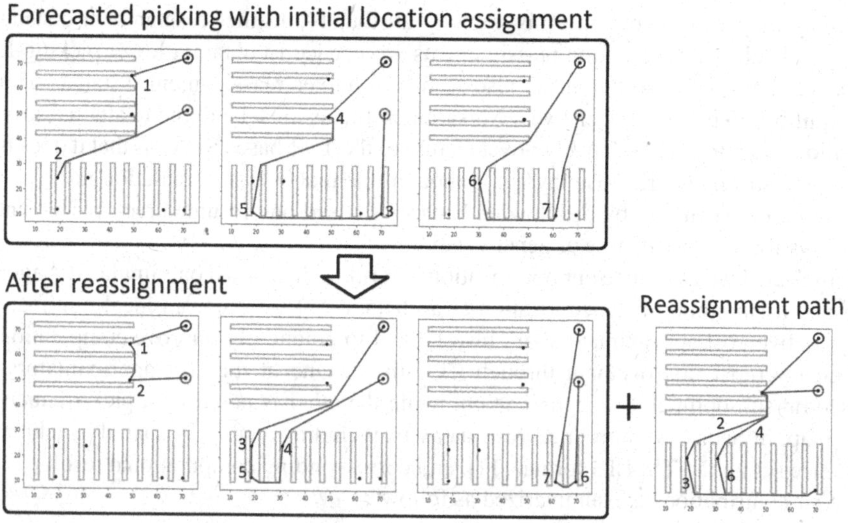

Forecasted picking with initial location assignment

After reassignment

Reassignment path

Fig. 1. Example of a SLAP with products enumerated 1–7 and an unconventional obstacle-layout [28]. The picking-log consists of three pick-rounds (TSPs) and their optimal solutions give the picking-log distance. The initial baseline assignment (top) has a longer picking-log distance compared to a candidate (sample) assignment (bottom left). In this example, the *reassignment path* needed to move the products according to the sample (bottom right), is longer than any possible savings concerning the picking-log (more pick-rounds are needed for savings).

and where the same location may be revisited by one or several vehicles. We obtain the picking-log distance by solving all TSPs given a location assignment of products. The second cost is the travel distance needed to move the products such that the assignment is obtained, in a single *reassignment path*. We refer to this model as the TSP-based SLAP. A visualization of the TSP-based SLAP is provided in Fig. 1.

In Sect. 2 we discuss strengths and weaknesses of proposed SLAP models in the literature. The TSP-based SLAP can be compared to the closely related Order Batching Problem (OBP)-based SLAP [26], where the picking-log is replaced by a set of orders (where an order is a set of products). The OBP-based SLAP requires the batching of orders into pick-rounds, as well as the subsequent TSP optimization of these pick-rounds, before quality of a proposed location assignment can be estimated. While the theoretical optimization gains may be higher in the OBP-based SLAP, its larger search space also adds significant challenges [17,33].

Choice of SLAP-model is inevitably a trade-off between simplicity, on the one hand, and complexity, on the other. Regarding the former, there is a need in research to discuss what a relatively simple and standardized version of the SLAP should entail, since there is little consensus on the matter [5]. Apart from order batching, examples of other optional features include various forms of dynamicity, warehouse layout, vehicle types, cost functions and reassignment scenarios. The TSP-based SLAP excludes order-batching and dynamicity and uses distance instead of more realistic but complex cost alternatives, such as time-based costs. Nevertheless, the TSP-based SLAP still poses a highly intractable problem. This is partly attributable to the reassignment distance. Hypothetically, more location reassignments are needed to obtain a lower picking-log distance, but more reassignments also lead to a longer reassignment distance. Thus, an equilibrium point between two adversarial problems must be found to attain a strong solution. One final and relatively novel feature of the TSP-based SLAP is that it does not assume a specific warehouse layout. Although this makes cost calculation more computationally expensive, by disallowing heuristics based on presumed rack-placements, it allows for a higher degree of generalization.

In Sect. 5 we introduce our optimization algorithm. It is based on Simulated Annealing and a Hamming-distance location-swap heuristic. Restarts from local minima, as well as two cost approximators, are investigated to potentially improve computational efficiency (cost improvement through CPU-time). One of the cost approximators is based on sub-optimal TSP optimization, while the other is based on a pick-frequency heatmap. In Sect. 6 we introduce three datasets, including a publicly shared benchmark instance set on the TSPLIB format [11], and corresponding computational results.

Our contributions are summarized as follows:

1. A SLAP optimizer using a novel version of the Simulated Annealing algorithm and experiments to test its computational efficiency.
2. Performance comparison of two cost-approximators utilized within the optimizer.
3. A publicly shared SLAP instance set on the TSPLIB format.

This paper is an extension of a ICORES-2023 paper [28]. Apart from a thorough revision of the text, the extension includes new data (dataset 3 in Sect. 6.3), a new cost approximator (Sect. 5.3), re-runs of previous experiments, as well as new experiments and results (Sect. 6 and Sect. 8).

2   Literature Review

In this section we discuss how the SLAP has been described and optimized in previous work. We particularly refer to the extensive literature review by Charris et al. [5]. There are several strategies for conducting a storage location assignment. These include *Dedicated*, *Random* and *Class-based*.

– *Dedicated*: The locations of products are assumed to never change. This strategy is suitable if the collection of products does not change much through time. If human picking is used, this approach has the advantage that pickers can learn to associate products with locations, allowing for speed-ups in picking [43].
– *Random*: Products can be assigned any location in the warehouse. This is particularly suitable if the collection of products changes frequently.
– *Class-based (zoning)*: Each product is assigned a class and the warehouse is divided into zones. Each zone contains one or several classes of products. Class-based storage can incorporate dedicated and random strategies for certain zones and/or classes [23]

The quality of a location assignment can be modeled in several ways. In a human picking scenario, Larco et al. [18] show that there exists a relationship between the height at which products are placed and worker welfare. Worker welfare can be quantified by estimating parameters such as "ergonomic loading", "discomfort" or "expenditure of human energy" [5]. On a similar note and for autonomous vehicle or shuttle based storage and retrieval systems (AVS/R), there exists a model which has the objective to minimize "energy consumption" [2].

Another way to judge solution quality is through datamining, using computations such as support, confidence and lift [25]. These can also be used to propose concrete location assignments [14,43]. Datamining is primarily focused on the statistical analysis of products and their relationships, but it is often combined with order-picking in a SLAP.

A third proposal studies the effect of traffic congestion. Bottlenecks can be caused if, for example, too many products with high pick-frequency are placed close to the depot. Lee et al. [19], propose Correlated and Traffic Balanced Storage Assignment (C&TBSA), a multi-objective SLAP model which aims to minimize traffic congestion, while also minimizing aggregate order-picking cost.

Order-picking has many variations, depending on obstacle layout, picking strategy and travel conventions [5,23,31,41]. Concerning obstacle layout, we distinguish between two types: *Conventional* and *Unconventional*. In the conventional layout, warehouse racks are assumed to be organized in Manhattan style blocks with parallel aisles and cross-aisles. Conventional layouts are used in the majority of research on both order-picking and the SLAP [5,15]. The unconventional layout includes the "fishbone" and "cascade" layouts [4,5], as well as all other layouts that are not conventional. Regardless of layout, the picking path of a vehicle can be formulated as a Traveling Salesman Problem (TSP) where paths cannot intersect obstacles [12,31]. For conventional layouts, the TSP is often optimized using S-shape or Largest-Gap algorithms [32]. For unconventional layouts, Google OR-tools or Concorde have been proposed [27,31].

As mentioned in Sect. 1, the SLAP can be optimized as a joint problem with an Order Batching Problem (OBP). Proposals include Kübler et al. [17], Xiang et al. [40] and Maruyama et al. [24]. While these authors argue for this approach, arguments also exist against it, at least for certain settings [23]. One issue with the OBP-based SLAP is that the OBP is highly intractable in its own right, thus adding to the difficulties involved in optimizing an already challenging problem.

If order-batching is not included in the SLAP, heuristics such as Cube per Order Index (COI) [13] and Order Oriented Slotting (OOS) [23] have been proposed. COI assumes that products with relatively high pick-frequency and low volume should be placed close to depot. COI does not include associations between products and is therefore mainly suitable for pick-rounds with few picks, such as pallet-picking or certain AVS/R systems [2]. OOS, on the other hand, is specifically designed for scenarios where orders may contain more than one product. Mantel et al. [23] introduce a Quadratic Assignment Problem (QAP) heuristic which computes distances between products and the number of times products appear in the same order. The quality of a candidate location assignment can then be estimated using QAP. Similar methods to OOS are used by Žulj et al. [44], Fontana and Nepomuceno [8] and Lee et al. [19].

The SLAP usecase can be divided into two categories depending on the number of products that are to be moved. "Re-warehousing" is the case when a large proportion of products are moved, whereas a smaller proportion is moved in "healing" [14]. Movements can be conducted in many ways, each accompanied by a (re)assignment "effort". Kübler et al. [17] propose the following (re)assignment effort scenarios:

i Product A is moved to an unoccupied location.
ii Product A swaps location with product B.
iii Product A is moved to a location occupied by product B. Product B is moved to a new location. If there is a product C occupying the new location, the procedure continues until a final product is placed at an empty location.

Scenario i comes with the least (re)assignment effort and the effort grows through scenarios ii and iii. Apart from travel distance, time used for product removal/placement on shelves as well as administrative time, can be added to the effort computation [17].

When it comes to optimization algorithms for the SLAP, both exact and non-exact methods have been proposed. The exact algorithms include dynamic programming, branch and bound algorithms and Mixed Integer Linear Programming (MILP) [5]. The SLAP search space is often reduced in scope when exact solutions are sought. These include restricting the number of locations [38], number of products [9,21] or by only working with conventional warehouse layouts [3].

More commonly, non-exact heuristic or meta-heuristic algorithms are used. Proposals include Particle Swarm Optimization (PSO) [17], Genetic and Evolutionary Algorithms [7,19,20] and Simulated Annealing [14,43]. The SLAP is often optimized in multiple phases using these methods. One example is to first generate candidate products for location assignments using datamining, and to then evaluate various candidate assignments using order-picking optimization [14,39].

It is challenging to judge optimization results in previous work due to the multitude of variations in SLAP models [5]. For results including reassignment costs, conventional warehouse layouts, dynamic picking patterns and meta-heuristic optimization, Kofler et al. [14] report best savings around 21%. In a similar scenario, Kübler et al. [17] report best savings around 22%. Excluding reassignment costs, Zhang et al. [43] report best savings around 18% on simulated data with thousands of product locations. In similar settings, Trindade et al. [35] report best savings around 33%, using a multiphase optimizer, and Chiang et al. [25] report best savings around 13% using datamining heuristics and integer programming.

3   Simulated Annealing

Simulated Annealing, which draws inspiration from the annealing process in metallurgy [14], has a useful analogue with SLAP optimization: A poor storage assignment can be viewed as more energetic as it leads to more travel for picking in the warehouse. As the SLAP is optimized, products are reassigned to new locations using a decreasing temperature. As temperature cools, products become fixed in a lower energy state where picking travel costs are reduced. There are many complicating factors in the SLAP which can prevent a smooth decent toward an improved storage assignment, however. In the remainder of this section, we describe the Simulated Annealing algorithm and how it may be modified to help attain stronger results in the SLAP.

A key component in Simulated Annealing (Algorithm 1) is the *sample* function. In each iteration i, sample x_{i+1} is drawn based on a desired distance to sample x_i. This distance is computed using the probability distribution $q(x_{i+1}|x_i)$, without involving the cost of the samples (henceforth we refer to this as the *feature-distance*). The q distribution is often chosen to be Normal, so that the distance between x_i and x_{i+1} is low with high probability [22]. The *cost** function computes/retrieves the cost (f^*) of the new/previous sample (the first sample is retrieved from memory after the first iteration). The accept probability α^* is based on a *cost-distance* function Δ (which outputs a negative value if the new cost is lower than the previous) and a decreasing temperature function T [29]. Functions for q, T and Δ are further discussed in Sect. 5.

Algorithm 1. Simulated Annealing.

1: x_i: Sample (an assignment).
2: $f^*(x_i)$: Ground truth cost of sample x_i.
3: q: Feature-distance function.
4: Δ: Cost-distance function.
5: N: Number of iterations.
6: T: Temperature function.
7: x_1: Initial sample (baseline).
8: **for** $i = 1, ..., N$ **do**
9:     $t \leftarrow T(i)$
10:    $x_{i+1} \leftarrow sample(q(x_{i+1}|x_i))$
11:    $f^*(x_i), f^*(x_{i+1}) \leftarrow cost^*(x_i, x_{i+1})$
12:    $\alpha^* \leftarrow exp(-c_1\Delta(f^*(x_{i+1}), f^*(x_i))/t)$
13:    $u \leftarrow \mathcal{U}(0, 1)$ // random uniform
14:    **if** $u < \alpha^*$ **then** // sample *accepted*
15:        $x_i \leftarrow x_{i+1}$
16:    **end if**
17: **end for**

Simulated Annealing is a type of Markov Chain Monte Carlo (MCMC) method and one advantage of this type of method is that its bias-variance tradeoff can be tuned using relatively few parameters [10]. A known disadvantage is that only two samples are stored in memory at any given time, which risks leading the Markov Chain to convergence on weak local minima [22]. Several methods have been proposed to reduce this risk, including mode-jumping [34], Nested Annealing [29] and Basin Hopping [37]. These methods split the search space into regions which are then subjected to local search. Another method is the Restart Strategy (SARS), which restarts the search from a random new sample whenever a "non-improving" local minimum is found [42].

Simulated Annealing can be modified to include a cost approximator, f, which provides fast cost estimates of f^*, to potentially increase computational efficiency. Christen and Fox [6] propose to use f to reject new samples that are unlikely to yield an improvement in f^* over the previous sample. The common MCMC *accept* method is accordingly split into two parts: *Promote* (f^* cost evaluation for a sample with a strong f) and *accept* (update x_i for the next iteration to be a sample with a strong f^*). In our optimization algorithm (Sect. 5), we utilize this concept and split Simulated Annealing into *promote* based on a fast and less accurate costs computed in f, and *accept* based on slow and more accurate costs computed in f^*.

4 Problem Formulation

4.1 Objective Function

The objective in the TSP-based SLAP is to minimize the aggregate travel distance to:

1. Complete a given picking-log (a set of pick-rounds) $\mathcal{B}$.
2. Carry out any proposed location reassignments in a single reassignment path $\mathcal{R}$.

Each pick-round $b \in \mathcal{B}$ is a list of products. The set of all locations (including pick-locations, origin and destinations and obstacle corners in 2D Cartesian space) is denoted $\mathcal{L}$ and the set of all pick-locations is denoted $\mathcal{L}(\mathcal{P})$. The set of all products in $\mathcal{B}$ is denoted $\mathcal{P}$. Each product $p \in \mathcal{P}$ is defined as a tuple including a unique key (Stock Keeping Unit), a pick location $l(p) \in \mathcal{L}(\mathcal{P})$ and a positive pick frequency count $F(p)$. Each pick location is a tuple consisting of a unique key, a capacity and a location (represented as a node in a graph). A product is located at strictly one location and a location stores strictly one product. A product is allowed to move from its initial location to a new one as long as the new location's capacity is not exceeded.

A SLAP solution candidate (also referred to as sample or assignment) is represented as permutation vector $x \in X$, where the elements are enumerated product keys and the indices are enumerated locations. For an example warehouse with 3 locations, sample $x = (2, 1, 3)$ means that product 2 is assigned location 1, 1 assigned 2 and 3 assigned 3. Each sample x contains positive permutation integers in range 1 to m, $2 \leq m \leq |\mathcal{P}|$ and each permutation x has ground truth cost $f^*(x)$. m denotes the number of products that are subject to location change, and it does not necessarily have to be equal to the number of products in the warehouse, but could instead be manually set to limit the search space. Sample x_1 represents the baseline product location assignment (the initial locations of the products). In order to evaluate performance in optimization experiments, costs $f^*(x_2), f^*(x_3), ..., f^*(x_N)$ are compared against $f^*(x_1)$.

The objective in the TSP-based SLAP, is to find a sample assignment x such that picking-log cost $\sum_{b \in \mathcal{B}} D(b)$ and reassignment cost $D(\mathcal{R})$ are minimized:

$$\underset{x}{\operatorname{argmin}}((\sum_{b \in \mathcal{B}} D(b)) + \lambda D(\mathcal{R})) \qquad (1)$$

Constant λ is used to weigh the two cost terms. Below we show how the picking-log and reassignment costs are computed using Euclidean distances.

4.2 Picking-Log Distance

The cost of all pick-rounds in picking-log $\mathcal{B}$ is computed using distance $\sum_{b \in \mathcal{B}} D(b)$. $D(b)$ is the distance of the solution to the Traveling Salesman Problem (TSP) represented by product locations in b:

$$D(b) = d_{l(origin_b),l(p_1)} + d_{l(p_{|b|}),l(destination_b)} + \sum d_{l(p_i),l(p_j)}, j = i+1, 0 < i < |b| \tag{2}$$

where $d_{l(p_i),l(p_j)}$ denotes the distance between the locations of $p_i, p_j \in b$, and where $d_{l(origin_b),l(p_1)}$ connects an origin location and $d_{l(p_{|b|}),l(destination_b)}$ a destination location to the path. The location of a product $l(p_i)$ is obtained from an index in the location assignment sample x. This index is stored for each product and updated whenever it changes location. We assume shortest distances and corresponding shortest paths (needed if path visualization is sought) between pairs of locations are queryable from Random Access Memory (RAM). All shortest paths and distances are pre-computed using the Floyd-Warshall graph algorithm, using a warehouse digitization process beyond the scope of this paper [31]. This process includes capability for uni-directed and mixed graphs, but in this paper we only work with bi-directed graphs (meaning that the distance between two locations is equal in both directions). We allow the origin and destination locations in the pick-rounds to be any locations in $\mathcal{L}$ (concerning TSP optimization, this is sometimes referred to as a Multi-Depot TSP or Dial-a-ride Problem). In Sect. 5 we describe how TSP optimization works for the multi-depot requirement.

4.3   Reassignment Distance

Reassignment path $\mathcal{R}$ and its distance $D(\mathcal{R})$ is based on direct and indirect exchange scenarios (scenarios ii and iii in Sect. 2) with the following assumptions: Since there are an equal amount of products and locations in the SLAP, scenarios ii and iii represent a bijective relationship between products and locations. When products change locations, the bijection can take three forms: Direct exchange, e.g. $x_1 = (1, 2)$ to $x_2 = (2, 1)$ (product 2 goes to location 1 and 1 goes to 2), indirect exchange, e.g. $x_1 = (1, 2, 3)$ to $x_2 = (3, 1, 2)$ (1 goes to 2, 2 goes to 3 and 3 goes to 1), or a combination of both. We also assume that the operation to change locations of products, using direct and indirect exchanges, can be carried out by a single vehicle traveling along a single path through the warehouse, without intermediate stops at the depot. Algorithm 2 shows how this single reassignment path can be constructed, just from information in the initial assignment x_1 and a subsequent sample x_{1+i}, generated during optimization iteration $i < N$.

r denotes a sub-cycle of locations (a sequence that starts and ends at the same location). The $add\_to\_subcycle$ function has two cases:

1. If the r sequence is empty, a random new element is removed from x_m and its initial location (the index for that product in x_1) is added to r.
2. If r is not empty, the new location of the last added product in r is first found in x and added to r. The product located at that "next" location is found in x_1, matched in and then removed from x_m.

If the added location to r is equivalent to the first one in r, the sub-cycle is completed and r is added to $\mathcal{R}$. After x_m is emptied, $\mathcal{R}$ is first randomly shuffled and then flattened (the inner lists of sub-cycles are converted into a single list). The distance $D(\mathcal{R})$ is then

---

Algorithm 2. Reassignment Path and Distance.

1: x_1: Initial assignment sample (baseline solution).
2: x: Sample obtained during SLAP optimization.
3: $x_m \leftarrow copy(x)$
4: $D(\mathcal{R}_{best}) \leftarrow \infty$
5: **for** $j = 1, ..., K$ **do** // iterations.
6:     $\mathcal{R} \leftarrow list()$
7:     **while** x_m *not_empty* **do**
8:         $r \leftarrow list()$
9:         **while** *not_completed*(r) **do**
10:            $add\_to\_subcycle(r, x, x_m, x_1)$
11:        **end while**
12:        $\mathcal{R}+ = r$
13:    **end while**
14:    *shuffle_and_flatten*$(\mathcal{R})$
15:    $D(\mathcal{R}_{best}) \leftarrow update\_best(\mathcal{R}, \mathcal{R}_{best})$
16: **end for**

---

computed as the sum of all location to location distances in $\mathcal{R}$, plus the distance from an origin depot location to the first location in $\mathcal{R}$ and the last location in $\mathcal{R}$ to a destination depot location. At each iteration, the *update_best*$(\mathcal{R}, \mathcal{R}_{best})$ function updates the lowest minimum found by comparing distance $D(\mathcal{R})$ and distance $D(\mathcal{R}_{best})$. For Algorithm 1 and our modifications to it in Algorithm 3, $D(\mathcal{R})$ is included in the $cost^*$ and $cost$ functions.

In summary, reassignment path $\mathcal{R}$ is a solution to a constrained, linked-list TSP where a product is dropped off and another product picked up at each location. The vehicle conducting the reassignment path is assumed to be able to carry the whole quantity (frequency $F(p)$ in our case) of any single product located at any single location. A model of the reassignment path involving vehicle-capacities, enforcing return trips to depot when a product quantity exceeds vehicle capacity, is left for future work.

5   Optimization Algorithm

5.1   Assignment Sampling Using Markov Chain Monte Carlo (MCMC) and Hamming Distances

As described in Sect. 3, the Simulated Annealing algorithm includes two distributions to describe the amount of distance between samples x_i and x_{i+1}: Feature-distance q and cost-distance Δ. For sampling to be effective, there should exist some degree of proportionality between these two distributions. If the feature-distance between x_i and x_{i+1} is relatively low, the distance between costs $f^*(x_i)$ and $f^*(x_{i+1})$ should also be relatively low. The cost-distance in a SLAP is in the domain $\mathbb{R}^+$, as it represents Euclidean travel distances in the warehouse. The feature-distance between two samples is represented by the difference between two assignments. We hypothesize that the feature-distance can be computed using a Hamming distance heuristic. Hamming distance is a count of the number of non-identical elements between two permutation vectors (which are equivalent to assignments) [30]. The following sampling distribution is then proposed to utilize this Hamming distance (based on bounds proposed by Christen and Fox [6]):

$$q(x_{i+1}|x_i) = e^{-CH_d(x_i, x_{i+1})^P} \tag{3}$$

where C and P are hyperparameters in $\mathbb{R}^+$, and H_d denotes Hamming distance. We propose to use this sampling function within Algorithm 1. Below we propose methods which may improve computational efficiency (cost reduction through CPU-time) of Algorithm 1.

5.2   TSP Optimization and Cost Caching

We utilize two TSP optimizers to compute the picking-log distances of assignment samples. For optimal TSP solutions we use the Concorde TSP solver[1] [1]. For approximate TSP solutions we use the OR-tools TSP optimization suite[2] [16]. In order to limit the CPU-time of OR-tools, we use the *solution_limit* parameter. For both these TSP optimizers, multi-depot scenarios are handled by modifying the input distance matrix with a dummy location whose distance is zero to the origin and destination, and whose other distances are set to infinity.

Before we apply TSP optimization to compute picking-log distance of an assignment sample, we reduce CPU-time through a filtering technique. Given the usage of sampling distribution q (Eq. 3), we note that many pick-rounds will often not contain products that had their location changed. For example, assume we start with assignment $x_1 = (2, 1, 3)$ and two pick-rounds in the picking-log, one containing products 1 and 2 and the other containing product 3. Picking-log distance is then computed by TSP-optimizing the two pick-rounds (to keep the example small, we disregard the fact that TSP optimization only yields savings for longer pick-rounds). Assume we then swap locations of products 1 and 2: $x_2 = (1, 2, 3)$. Since product 3 remains at its initial location, there is no need to re-optimize the pick-round which contains that product. To enable this reduction of redundant TSP-optimization, we cache the TSP costs (both optimal and approximate) of any pick-round once computed. These costs are then queried for the pick-round until one or several product locations are changed, at which point the TSP gets re-optimized and the costs updated (only after promotion in the case of f^*).

5.3   Heatmap-Based Approximation

In order to motivate SLAP optimization, results need to be as interpretable and visually representable as possible. One problem with TSP optimization within a SLAP is that results cannot be easily visualized. Visualizing TSPs entails showing them before and after SLAP optimization. Figure 1 and Fig. 8 (Appendix) are examples. Interpretation of these types of figures becomes very challenging when the picking-log contains hundreds of pick-rounds.

One possible way with which to visualize SLAP optimization in a single figure, is a *heatmap*. Figure 2 is an example which shows number of picks at 2700 locations

---

[1] https://math.uwaterloo.ca/tsp/concorde/downloads/downloads.htm, collected 27-05-2022.
[2] https://developers.google.com/optimization/routing/tsp, collected 12-06-2022.

(several locations share a single cell in the heatmap). The lower picture shows the result after SLAP optimization. To achieve this movement of the "hotter" products closer to depot, a dot product is first computed between the pick frequency count of each product $F(p)$ and the distance of their locations from an origin location and to a destination location:

$$\sum_{i=1}^{|\mathcal{L}|} F(p_i)(d_{l(origin),l(p_i)} + d_{l(p_i),l(destination)}) \tag{4}$$

Location swaps are then conducted based on this dot product. For the heatmap example in Fig. 2, 200 swaps were conducted to achieve a reduction of cost, according to Eq. 4, of around 35%. Apart from the visual interpretability, an additional advantage of using Eq. 4 is that it is very fast to compute. In Sect. 6 we conduct an experiment to investigate whether there is any correlation between this approximation and optimal TSP cost. The predictive quality of Eq. 4 is likely weak, but if CPU-time is low enough it could still outmatch the alternative f approximation achieved by the OR-tools TSP optimizer. Note that this approach only works for cases when all pick-rounds in the picking-log use the same origin and destination location.

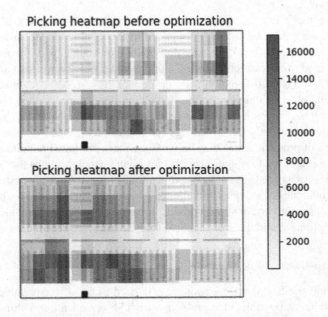

Fig. 2. Heatmap of picking in a warehouse with a single depot location (the black square). The colorbar shows how many picks occur within a given cell.

5.4   Nested Annealing

In Sect. 3 we suggested that the computational efficiency of Simulated Annealing (Algorithm 1) can be increased if there exists a function f which can quickly estimate f^*. We then proceeded to propose two suggestions for such an f: One using sub-optimal TSP óptimization (OR-tools), and one using a heatmap based approxima-tion. In Algorithm 3, we show how either of these can be utilized within a modified Simulated Annealing algorithm:

Algorithm 3. Nested Annealing (based on computational efficiency in cost estimation.

1: x_i: Sample (candidate solution).
2: $f(x_i)$: Less accurate fast cost estimate.
3: $f^*(x_i)$: More accurate slow cost estimate.
4: q: Feature-distance function.
5: Δ: Cost-distance function.
6: α: Probability that sample x_{i+1} is *promoted*.
7: α^*: Probability that sample x_{i+1} is *accepted*.
8: N: Number of iterations.
9: T: Temperature function.
10: x_1: Initial assignment sample (baseline).
11: **for** $i = 1, ..., N$ **do**
12:　　$t \leftarrow T(i)$
13:　　$x_{i+1} \leftarrow sample(q(x_{i+1}|x_i))$
14:　　$f(x_i), f(x_{i+1}) \leftarrow cost(x_i, x_{i+1})$
15:　　$\alpha \leftarrow exp(-c_1 \Delta(f(x_{i+1}), f(x_i))/t)$
16:　　$u \leftarrow \mathcal{U}(0, 1)$ // random uniform
17:　　**if** $u < \alpha$ **then** // sample *promoted*
18:　　　　$f^*(x_i), f^*(x_{i+1}) \leftarrow cost^*(x_i, x_{i+1})$
19:　　　　$\alpha^* \leftarrow exp(-c_2 \Delta(f^*(x_{i+1}), f^*(x_i))/t)$
20:　　　　$u \leftarrow \mathcal{U}(0, 1)$
21:　　　　**if** $u < \alpha^*$ **then** // sample *accepted*
22:　　　　　　$x_i \leftarrow x_{i+1}$
23:　　　　**end if**
24:　　**end if**
25: **end for**

After a sample x_{i+1} is generated, its *cost* is estimated using f. If the sample passes the *promote* filter on Line 17, *cost** is computed using f^*. Note that the *cost* and *cost** functions include reassignment distance $D(\mathcal{R})$ (Algorithm 2). Since Algorithm 2 does not guarantee optimality for $D(\mathcal{R})$, *cost** does not guarantee optimality either, and hence we refer to f^* as "more accurate" rather than optimal. Hyperparameters $c_1, c_2 \in \mathbb{R}^+$ may be set differently. Christen and Fox [6] suggest setting $c_1 > c_2$ so that the promotion of a sample is less likely than the acceptance of a promoted sample. For the temperature function T we use a shifted and scaled reverse sigmoid (decreasing) that gives temperatures in range $[1, 0]$. For the cost-distance function Δ we use a shifted and scaled sigmoid that gives values in range $[0, 1]$. Nested Annealing was first intro-duced by Rajasekaran and Reif [29], but they do not use cost approximation and base the nesting on variable set temperatures in local search regions. Algorithm 3 offers an alternative nesting strategy, based on a trade-off between predictive speed and accuracy.

5.5 Restarts

Due to the large search space of the SLAP, the MCMC sampling function $x_{i+1} \leftarrow sample(q(x_{i+1}|x_i))$, may benefit from occasional *restarts* (Sect. 3). Yu et al. [42], propose restarts from randomly generated samples. Their test-problems do not include reassignment distances, however, and in the SLAP, randomly generated samples can be expected to have a significantly higher cost than x_1 due to reassignment distance $D(\mathcal{R})$. As a solution to this problem, we instead propose restarts from local minima. The best minimum found through optimization is denoted x_{best} and it is used as restart sample with an increasing probability. Forcing restarts from x_{best} is motivated because its local neighbourhood cannot be extensively searched for in any but the smallest SLAP test-instances. A second minimum is denoted x_{lowR} and it is used as a restart sample with a decreasing probability. Forcing restarts from x_{lowR} is designed to target a low reassignment distance $D(\mathcal{R})$. The first such local minimum is $x_{lowR} = x_1$, whose $D(\mathcal{R}) = 0$. $x_{lowR} = x_1$ can be assumed to be a strong local minimum, due to its lack of reassignment distance, but after $f^*(x_1)$ has been beaten by $f^*(x_{1+i})$, x_{lowR} is updated at regular intervals to a previously generated sample which has a relatively low f^* cost and $D(\mathcal{R})$. In Sect. 6 we propose probability distributions for x_{best} and x_{lowR}, as well as optimization results with and without the use of restarts.

6 Experiments

6.1 Overview

We carry out experiments to investigate the following topics with regard to computational efficiency (cost reduction through CPU-time), in chronological order specified below:

1. Utility of *Hamming-distance* based sampling (q).
2. Utility of *restarts*.
3. Comparison of two cost approximators for use within Algorithm 3.
4. Comparison of Algorithm 1 and Algorithm 3 (using best settings from 2 and 3).
5. Other features (such as layout and number of products and pick-rounds).

All experiments are carried out using Intel Core i7-4700MQ, 2.40 GHz, 4 cores and Python3 (with heavy use of Cython) and C.

6.2 Parameters

For all experiments, the number of products open for location reassignment (m) is set to be equivalent to the number of products in the test-instance. The number of reassignment path optimization iterations (K in Algorithm 2) is set to 300. After optimization has completed, the reassignment path is re-optimized with K set to 10000. The accept probability computation is set to be equivalent between Algorithm 1 and 3 ($c_2 = 1$ and equivalent Δ and T functions). The Δ function is set to approach 1 when the ratio of the distance between a new sample and a previous sample exceeds 1.05: This means that if a new sample has a distance 5% higher than the previous sample, it is unlikely to be promoted and/or accepted. c_1 in Algorithm 3 is set to 2, which makes it more difficult for a sample to be promoted than accepted once promoted. The reverse sigmoid probability distribution q, which gives the number of location changes between a new and a previous sample, is set to approach zero when number of location changes exceeds 20. For all experiments where a restart strategy is used, sample x_{i+1} can be built from either x_i, x_{best} or x_{lowR} (Sect. 5). The probability to pick one of the latter two is governed by a sigmoid and reverse sigmoid, respectively, with probabilities in range $[0, 0.2]$ and $[0.2, 0]$, stretched over N iterations. In all iterations where neither x_{best} nor x_{lowR} is picked, x_i is used (no restarts). λ and N are set depending on the dataset.

6.3 Datasets

The following three datasets are used:

1. 266 TSPLIB instances[3] modified for the SLAP and shared in a public repository[4]. These instances include 6 different types of warehouse layouts (including one with no obstacles). The number of products open for location reassignment vary between 5–427 in these instances. The initial locations for all products (baseline assignment x_1) in these instances is selected using a random uniform distribution. Solution proposals are uploaded for each of these instances using Algorithm 3 after a maximum of 20000 iterations (N). Experiments to test utility of Hamming distances and restarts are conducted on this dataset. λ is set to 1 for experiments on this dataset.
2. Data from a real warehouse with a conventional layout. The provided picking-log includes 260 unique products and 260 product locations. There are 200 pick-rounds and most products are picked in several pick-rounds. The experiments where Algorithm 1 and 3 are compared are run on this dataset. Algorithm 1 and 3 are run 10 times each on this dataset, with varying random seeds and a maximum CPU-time set to 8 h. λ is set to 1 for experiments on this dataset.
3. Data from a real warehouse with an unconventional layout. Specific to this dataset is that there is only a single origin/destination and that some products are not located in the warehouse apriori. These products are assigned random initial locations. There are also more locations than products in this dataset. The empty locations are utilized in optimization by placing a mock product at each of them. By flagging these products, they can be excluded from cost computation, while remaining open for product locations swaps. This dataset also contains longer pick-rounds than the other two (with an average of 29 picks per pick-round). The experiments where the two cost approximators are compared are conducted on this dataset, using a maximum CPU-time of 4 h. λ is set to 0 for experiments on this dataset: This removes the reassignment distance and thus ensures that the two approximators can be compared against an optimal f^*.

In all three datasets, the capacity of all locations is assumed to be identical, meaning that any product can be placed at any location. We compare costs of samples against the baseline x_1, where each product is fixed to its initial location, where optimal picking costs are computed in $D(\mathcal{B})$ and where $D(\mathcal{R}) = 0$.

---

[3] https://github.com/johanoxenstierna/OBP/instances, collected 19-10-2022.
[4] https://github.com/johanoxenstierna/L40_266, collected 14-11-2022.

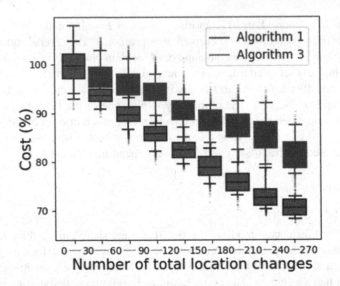

Fig. 3. The total number of product location reassignments needs to be large to achieve the best total travel costs in $f^*(x_{best})$ (dataset 2).

6.4 Experiment Results

Utility of Hamming-Distance Based Sampling. Results show that many location reassignments are needed to reach the best reductions in travel cost (Fig. 3). Also, more reduction in cost is achieved when the Hamming distance (number of location changes) between a previous sample and a new one is relatively low (Fig. 4). On average, the cost of sample $f^*(x_{i+1})$ is more reduced compared to a previous sample $f^*(x_i)$ if fewer location changes are attempted. This result empirically validates the Hamming distance distribution $q(x_{i+1}|x_i)$ and its bias toward conducting fewer location changes at each step in the Markov Chain (Eq. 3).

Utility of *Restarts*. Results with and without *restarts* (Sect. 5) are shown in Fig. 5. Given the same amount of optimization iterations ($N = 30000$) on dataset 2, the best results for both Algorithm 1 and 3 are obtained using restarts. Restarts enforce revisits to local minima with relatively short total travel costs f^* or reassignment costs $D(\mathcal{R})$ (Sect. 5). Since fewer reassignments mean that fewer pick-rounds contain products whose locations change, total TSP optimization CPU-time is significantly lower when restarts are used. This is achieved by the caching of TSP costs (Sect. 5). Furthermore, few reassignments mean that the optimization of the reassignment path requires less CPU-time to reach a relatively strong solution. As can be observed, Algorithm 1 and 3 without restarts (lighter blue and green) quickly jump up in cost. This is mainly attributed to the relatively low cost in assignment x_1, where $D(\mathcal{R}) = 0$, which is never revisited once stepped away from and never improved on (without restarts).

Comparison of the Two Cost Approximators for Algorithm 3. Results on dataset 3 are summarized in Table 2. We first study the coefficient of determination R^2 (goodness of fit) between approximations f against f^*. For OR-tools, $R^2 = 0.97$ and for the

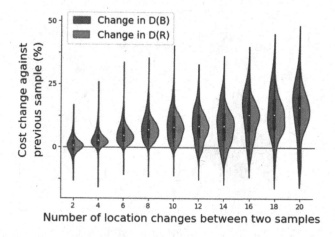

Fig. 4. Distribution (violin) plot showing number of location changes against picking-log distance $D(\mathcal{B})$ (blue) and reassignment distance $D(\mathcal{R})$ (orange) when moving from a previous sample to a new sample in the Markov Chain. The mean cost of both $D(\mathcal{B})$ and $D(\mathcal{R})$ increase when more location changes are attempted in new samples. This plot excludes any x_i and x_{i+1} pairs where either were restarts back to a local minimum. (Color figure online)

heatmap, $R^2 = 0.15$. Even though the heatmap approximation is thousands of times faster to compute compared to TSP-optimizing the picking-log using OR-tools, OR-tools still results in more savings than the heatmap approximation. Due to its high speed, the heatmap approximation allows for more samples to be generated and higher initial savings, but due to its weaker predictive quality it, in the end, loses out to the TSP approximation.

The weakness of the heatmap approximation can be attributed to a combination of two factors. The first is that a swap of two products may result in a frequently picked product being located further from the depot, incurring an increased heatmap cost, while TSP distance, on the contrary, is reduced (this can be observed in Fig. 8). The second factor is its bias to promote samples where high-frequency products are moved closer to depot. This type of bias risks leading the search to a pre-mature convergence on a local minimum. In order to prevent convergence on a local minimum, many samples are needed which temporarily increase TSP costs, but these types of samples are not often promoted in Algorithm 3 when the heatmap approximation is used.

Although OR-tools outperforms the heatmap approximation, one noted issue with the former is its high minimal CPU-time. The CPU-times of OR-tools are averaging 0.1 s to optimize a single TSP, whereas the corresponding CPU-time for Concorde is averaging 0.2 s. We could not achieve a lower value using the *solution_limit* parameter after several tests. On dataset 1 and 2, this CPU-time is potentially advantageous, since OR-tools delivers TSP distances within 1–2% of optimality (Table 2). This high approximation quality is explainable since pick-rounds $b \in B$ rarely exceed 15 locations in length in those datasets. On dataset 3, when the pick-rounds are 29 products on average, OR-tools is within 6% of optimality. We did not attempt to tune the CPU-time and the *solution_limit* parameter in OR-tools to maximize its utility within Algorithm 3.

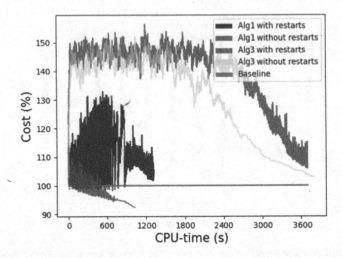

Fig. 5. Algorithm 1 and Algorithm 3 with and without *restarts* for 30000 iterations on dataset 2 [28]. The costs shown are for $f^*(x_{i+1})$.

Finally, we compute goodness of fit between both cost approximators and $R(D)$ for any generated samples (while λ was set to zero for dataset 3, $R(D)$ was still computed and logged). In both cases, R^2 was close to zero. While this may seem disadvantageous, it also means that $\mathcal{R}(\mathcal{D})$ has a high variance and low bias, thus preventing Algorithm 3 from converging on weak local minima. We also note that R^2 increases for promoted samples and even more so for accepted samples (reaching as high as $R^2 = 0.57$ for accepted samples). This provides further validation for Algorithm 3 and its cost function (Eq. 1): The Markov Chain tends to converge on regions where picking-log cost is low and where reassignment costs are low as well.

Algorithm 1 Compared to Algorithm 3. When the best settings found are utilized in Algorithm 3 (Nested Annealing with the OR-tools TSP cost approximation and restarts), it outperforms Algorithm 1 (Simulated Annealing without cost approximation and with restarts) within the given CPU-time (Fig. 6). The Markov chain in Algorithm 3 is more biased compared to the one in Algorithm 1, due to more samples being rejected. Algorithm 1 searches through less attractive search regions, which reduces risk of convergence on local minima, so if given more CPU-time it could reach stronger results.

Other Features. Aggregate averages of results on the generated instances (dataset 1) and Algorithm 3 are shown in Table 1 (Appendix). The elements for columns $f(x_i)$, $f^*(x_i)$, $f(x_{i+1})$, $f^*(x_{i+1})$, $f^*(x_{best})$, $D(R)_1$ $D(R)_{300}$ are all shown as percentages against the distance of the baseline cost $f^*(x_1)$ (100%). $D(R)_1$ and $D(R)_{300}$ denote the distance of the reassignment path after Algorithm 2 has been run for 1 and 300 iterations, respectively. The rows are aggregated averages based on number of products shown in column 1, from a total of 5279885 samples on the instance set (with 3–12 min CPU-time on each instance).

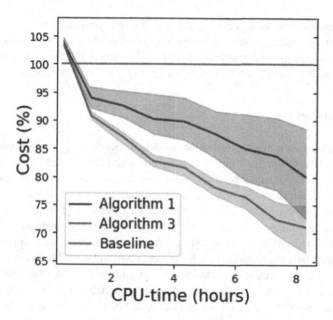

Fig. 6. Aggregate CPU-time against shortest total travel cost ($f^*(x_{best})$) on the real warehouse dataset (20 optimization runs): Blue is Algorithm 1, green is Algorithm 3 and red is the cost of baseline assignment x_1 (100%). The shadowed areas represent 95% confidence intervals [28].

The relationship between number of location changes and $D(\mathcal{R})$ can be seen in Fig. 7. As more location swaps are carried out, the amount of reassignment distance increases, but the rate of increase slows down. One possible misconception is that the gradient should go down to zero as the reassignment path cannot exceed some hypothetical maximum. This is unlikely to occur, however, since the reassignment path may need to go back and forth through the warehouse several times to perform many reassignments.

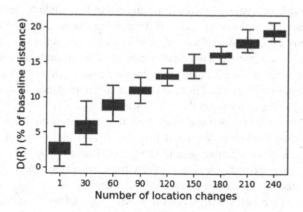

Fig. 7. Number of location changes vs. reassignment distance (as a percentage of baseline costs) (Algorithm 3 and dataset 3).

No correlation was found between the warehouse layout and features such as total cost improvement, reassignment distance and/or number of final proposed location reassignments. This is explainable since both TSP-optimizers (OR-tools and Concorde) and the reassignment path optimizer (Algorithm 2) are *layout-agnostic* (Sect. 1).

7   Conclusion

This paper proposes a new optimization model for the Storage Assignment Location Problem (SLAP). In the Traveling Salesman Problem (TSP)-based SLAP, future forecasted picking is assumed to be static, while the warehouse rack layout can have any shape in two dimensions. In order to optimize the TSP-based SLAP, we propose a Nested Annealing algorithm. The algorithm is an extension of Simulated Annealing and generates assignment samples using a Hamming distance function and two sample filters. The algorithm requires fast and reasonable accurate cost approximations, and we propose two alternatives: One based on sub-optimal TSP optimization, and the other based on a pick-frequency heatmap. In order to reduce risk of convergence on a weak local minimum, we propose a *restart* heuristic, which forces occasional revisits to previously generated and relatively strong samples. Since products cannot be reassigned to new locations for free, a model for the *reassignment path* and *reassignment distance* is proposed. This cost is computed and added to the cost of any generated sample.

To evaluate the proposed optimizer using various SLAP scenarios and optimization settings, experiments were conducted on three datasets: A set of publicly shared test-instances on the generalizable TSBLIB format, as well as two datasets from real warehouses. Results show that Nested Annealing yields cost savings of up to 30% within 8 h of CPU-time. This result is in line with results in prior work, where strong assumptions are made with regard to warehouse layout (but where dynamicity may be included or where number of products is larger) [14,17,35]. Concerning the cost approximators, results show that sub-optimal TSP optimization outperforms the pick-frequency heatmap approach. While the former is thousands of times slower than the latter, it nevertheless achieves a better result due to its higher predictive accuracy.

For future work, heuristics to increase the amount of bias could be investigated. One cause of high variance in the proposed algorithm is that any product is allowed to swap location with any other product. Instead, products could be set up to be allocated to certain areas in the warehouse. This type of *zoning* is not trivially achieved, however, and could, if not carefully handled, lead to premature convergence on local minima. We concluded this after early tests and instead pursued cost approximation and the *promote* filter as another means to constrain the search space.

A topic which we did not explore extensively in this paper is the λ constant (Sect. 4) and its effect on optimization. We set it to either 0 or 1 and only concluded that it significantly slows down effective search when used. Instead of a constant, it could be set to change during optimization to potentially improve performance. For example, λ could be set to start at a low value and then grow linearly.

A final proposal involves analysis of the picking log and how it relates to potential cost savings. Zhang et al. [43] and Kofler et al. [14] use datamining heuristics to show that potential cost savings (the "reassignment potential") are correlated to the way in which products in pick-rounds are distributed. It is challenging to make use of such heuristics to make concrete proposals for reassignments in a Markov Chain, however. The SLAP remains a highly intractable problem.

Acknowledgements. This work was partially supported by the Wallenberg AI, Autonomous Systems and Software Program (WASP) funded by the Knut and Alice Wallenberg Foundation. We also convey thanks to Kairos Logic AB for software.

A   Appendix

Examples of pick-rounds before and after 100 iterations of SLAP optimization (left and right respectively). The SLAP can be challenging even when there are only six pick-

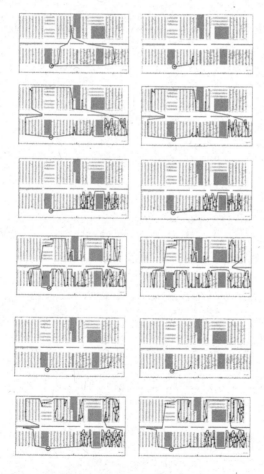

Fig. 8. Pictures of optimally solved pick-rounds (TSP's) before (left) and after SLAP optimization (right). The product which is picked in all pick-rounds is the lower-rightmost one in the upper two pictures (before and after it was moved).

rounds in the picking-log. While it is relatively easy to spot suitable swaps of locations for pick-rounds involving few products, it is more difficult when pick-rounds are long. One of the products is picked in all of the pick-rounds, and as that product is moved, it affects total distance in an unforseeable manner.

Table 1. Aggregate averages of results from 5279885 generated samples for optimization runs on the 266 publicly shared instances. The results are aggregated based on ranges of number of products (the first column).

Num_products	Num_pick-rounds	Num_samples	CPU-time (s)	$f^*(x_i)$	$f(x_i)$	$f^*(x_{i+1})$	$f(x_{i+1})$	$f^*(x_{best})$	$D(R)_1$	$D(R)_{300}$	Number of location changes
5	7	39997	215	89	89	109	98	85.3	27.62	24.83	7.3
14	9	319966	254	89	89	110	99	88.0	26.80	24.31	8.5
23	11	439980	293	91	91	106	101	88.4	24.48	22.35	12.9
32	14	439983	281	94	93	107	102	91.8	21.06	19.27	12.9
41	19	459987	269	96	96	107	103	93.1	17.35	15.85	13.2
50	23	519994	295	96	96	104	102	93.5	15.89	14.58	15.2
60	22	459991	344	97	97	105	103	94.5	15.33	13.94	14.3
69	26	339997	313	97	97	104	103	94.8	13.37	12.17	14.2
78	25	279995	335	97	97	105	103	94.1	13.44	12.23	15.1
87	28	380000	372	98	98	104	103	96.3	12.08	10.95	14.0
96	33	200000	373	98	98	104	102	94.9	12.31	11.20	15.2
106	24	119999	410	97	97	103	102	95.4	12.22	10.99	12.9
115	29	179998	462	99	99	105	103	97.4	10.43	9.40	12.5
124	32	180000	487	99	99	104	103	97.6	9.83	8.78	11.4
133	32	120000	455	99	99	104	103	97.1	9.27	8.40	12.5
142	36	60000	449	98	98	103	102	97.5	9.87	8.95	11.8
151	35	79999	527	98	98	103	102	97.4	11.21	10.11	15.8
161	39	60000	476	99	99	103	102	98.6	7.48	6.74	12.3
170	38	119999	551	99	99	103	102	97.7	9.30	8.40	14.5
179	37	80000	576	100	100	104	103	98.3	7.24	6.50	10.7
188	37	60000	583	100	100	104	103	99.7	7.21	6.49	11.3
197	32	20000	693	100	100	104	103	99.7	8.88	7.93	11.0
216	44	20000	552	101	100	104	103	99.8	5.68	5.03	8.3
225	38	40000	655	100	100	104	103	98.7	7.99	7.17	11.9
234	36	20000	682	99	98	103	102	97.7	8.11	7.31	13.8
252	39	20000	825	100	100	104	103	99.7	6.36	5.70	10.2
262	44	20000	624	100	99	103	102	99.0	6.66	5.95	12.0
289	40	40000	821	100	100	104	103	99.4	6.38	5.68	11.5
317	161	20000	629	100	100	101	101	99.2	4.35	3.95	17.9
326	162	20000	636	99	99	101	101	99.2	3.66	3.32	15.5
381	206	20000	723	100	100	101	101	99.3	2.89	2.62	18.2
390	207	20000	719	100	100	101	101	99.6	2.78	2.50	13.9
399	205	20000	735	100	100	101	101	99.6	3.11	2.81	16.3
418	211	40000	757	100	100	101	101	99.7	2.73	2.45	13.1
427	222	20000	806	100	100	101	101	99.9	2.41	2.15	13.7

Table 2. Aggregate averages of results on dataset 3, where the two cost approximators are compared. The CPU-times are here for predictions of single TSPs.

CPU-time total (s)	Heatmap					OR-tools					CPU-time Concorde (s)
	CPU-time (s)	$f(x_i)$	$f^*(x_i)$	$f(x_{i+1})$	$f^*(x_{i+1})$	CPU-time (s)	$f(x_i)$	$f^*(x_i)$	$f(x_{i+1})$	$f^*(x_{i+1})$	
0	3.7E-05	91.51	106.63	96.27	112.34	0.098	107.33	101.58	110.58	104.33	0.20
270	3.5E-05	103.32	98.90	111.67	106.26	0.096	105.97	98.70	112.11	103.82	0.26
543	3.6E-05	97.47	96.78	104.30	102.54	0.096	103.85	95.65	110.98	102.11	0.24
816	3.8E-05	88.73	96.29	95.32	103.28	0.102	101.93	95.01	107.07	99.13	0.15
1089	3.7E-05	93.56	92.87	100.61	98.94	0.099	102.67	95.13	106.29	98.45	0.28
1362	3.9E-05	100.55	95.32	107.56	100.74	0.101	101.87	93.99	104.47	96.32	0.37
1635	3.7E-05	98.19	89.44	100.68	93.41	0.095	101.53	93.16	105.50	97.02	0.24
1908	3.7E-05	92.69	94.06	90.60	99.75	0.101	99.82	92.16	104.77	96.55	0.12
2181	3.9E-05	97.68	93.58	95.73	97.06	0.098	99.32	91.68	103.81	95.37	0.23
2454	3.9E-05	92.99	93.09	90.73	97.05	0.098	99.44	91.90	102.96	95.09	0.33
2727	4.2E-05	95.97	91.94	93.89	97.27	0.099	98.29	90.27	102.37	93.51	0.30
3000	4.2E-05	99.26	97.54	97.64	103.29	0.104	97.28	87.64	100.91	92.80	0.32
3273	3.9E-05	98.11	99.43	97.15	105.25	0.101	93.52	87.75	99.65	92.12	0.27
3546	3.7E-05	103.19	96.21	102.50	101.50	0.104	93.65	87.56	97.72	92.41	0.19
3819	3.8E-05	101.53	90.65	99.49	96.67	0.103	93.47	87.05	98.73	92.77	0.22
4092	3.7E-05	97.16	82.51	95.38	87.16	0.106	93.21	86.11	97.84	91.10	0.26
4365	3.5E-05	89.30	89.68	87.11	93.44	0.100	92.52	83.55	96.28	89.34	0.16
4638	3.6E-05	96.11	86.36	94.77	91.63	0.105	92.44	83.72	96.29	89.56	0.34
4911	3.5E-05	87.10	90.53	85.77	96.55	0.102	90.74	83.19	94.42	89.14	0.15
5184	3.5E-05	95.06	80.34	93.42	84.97	0.101	89.85	85.54	94.84	88.84	0.26
5457	4.2E-05	92.76	84.32	92.28	88.70	0.105	90.21	85.04	94.44	88.84	0.21
5730	3.9E-05	89.06	84.80	91.24	89.46	0.100	90.47	85.48	94.24	88.71	0.29
6003	3.9E-05	92.65	86.15	93.86	92.10	0.102	89.95	84.55	93.92	87.65	0.22
6276	3.9E-05	99.31	85.65	99.20	91.94	0.100	88.25	83.61	91.95	86.30	0.18
6549	3.8E-05	100.67	78.19	100.20	85.76	0.102	87.33	82.98	91.57	85.45	0.18
6822	3.7E-05	99.41	79.06	100.38	83.74	0.105	84.69	82.24	88.97	84.57	0.10
7095	3.8E-05	89.81	77.54	89.49	83.44	0.105	84.93	81.31	89.00	83.14	0.19
7368	4.1E-05	97.47	83.69	95.78	88.54	0.100	83.83	78.13	88.31	83.35	0.21
7641	3.8E-05	82.96	83.08	78.77	88.70	0.103	83.15	78.36	87.61	83.59	0.22
7914	3.6E-05	94.68	86.11	90.21	90.42	0.098	83.51	78.08	86.89	82.65	0.25
8187	3.7E-05	99.96	85.77	95.34	91.37	0.101	82.51	75.72	86.83	81.83	0.16
8460	3.6E-05	96.36	82.89	91.15	88.68	0.099	81.45	75.62	85.25	80.90	0.11
8733	3.7E-05	90.84	81.15	85.80	88.06	0.099	82.36	74.71	85.72	80.28	0.08
9006	3.7E-05	100.12	79.78	95.60	85.21	0.098	80.37	74.88	83.76	79.62	0.21
9279	3.5E-05	98.74	80.18	93.87	85.31	0.101	80.55	74.82	83.82	78.79	0.21
9552	3.6E-05	89.82	79.70	85.77	84.68	0.105	79.34	74.82	82.63	72.64	0.27
9825	3.5E-05	91.59	77.13	87.20	81.96	0.098	79.89	73.46	82.04	72.33	0.15
10098	3.4E-05	95.74	83.77	93.14	87.98	0.099	80.07	73.34	82.27	72.58	0.15
10371	3.3E-05	93.30	78.10	93.16	83.05	0.097	80.03	75.46	82.61	72.64	0.08
10644	3.3E-05	87.79	77.62	86.76	82.77	0.099	79.11	72.41	81.55	72.11	0.18
10917	3.3E-05	83.07	88.14	82.73	81.81	0.101	78.38	71.94	80.03	71.55	0.12
11190	3.3E-05	91.08	86.66	90.38	81.21	0.105	77.78	71.47	79.69	71.82	0.12
11463	3.2E-05	83.91	86.18	84.08	80.54	0.100	77.02	71.35	79.19	71.97	0.24
11736	3.3E-05	81.51	82.23	80.71	86.52	0.107	77.88	71.12	80.14	71.40	0.18
12009	3.3E-05	80.99	84.06	80.05	88.76	0.104	76.13	73.06	78.06	72.34	0.16
12282	3.3E-05	90.28	85.41	89.69	80.27	0.102	74.89	71.93	77.25	72.07	0.26
12555	3.3E-05	77.18	84.25	76.41	80.76	0.106	74.14	71.24	77.00	71.76	0.21
12828	3.4E-05	71.68	88.43	71.20	85.31	0.103	73.66	71.03	76.55	71.13	0.31
13101	3.4E-05	69.98	83.28	69.62	88.12	0.102	73.12	70.92	76.07	71.07	0.20

References

1. Applegate, D., Cook, W., Dash, S., Rohe, A.: Solution of a min-max vehicle routing problem. INFORMS J. Comput. **14**, 132–143 (2002)
2. Azadeh, K., De Koster, R., Roy, D.: Robotized and automated warehouse systems: review and recent developments. Transp. Sci. **53**, 917–945 (2019)
3. Boysen, N., Stephan, K.: The deterministic product location problem under a pick-by-order policy. Discret. Appl. Math. **161**(18), 2862–2875 (2013)
4. Cardona, L.F., Rivera, L., Martínez, H.J.: Analytical study of the fishbone warehouse layout. Int. J. Log. Res. Appl. **15**(6), 365–388 (2012)
5. Charris, E., et al.: The storage location assignment problem: a literature review. Int. J. Ind. Eng. Comput. **10**, 199–224 (2018)
6. Christen, J.A., Fox, C.: Markov Chain Monte Carlo using an approximation. J. Comput. Graph. Stat. **14**(4), 795–810 (2005). https://www.jstor.org/stable/27594150
7. Ene, S., Öztürk, N.: Storage location assignment and order picking optimization in the automotive industry. Int. J. Adv. Manuf. Technol. **60**, 1–11 (2011). https://doi.org/10.1007/s00170-011-3593-y
8. Fontana, M.E., Nepomuceno, V.S.: Multi-criteria approach for products classification and their storage location assignment. Int. J. Adv. Manuf. Technol. **88**(9), 3205–3216 (2017)
9. Garfinkel, M.: Minimizing multi-zone orders in the correlated storage assingment problem. PhD Thesis, School of Industrial and Systems Engineering, Georgia Institute of Technology (2005)
10. Gidas, B.: Nonstationary Markov chains and convergence of the annealing algorithm. J. Stat. Phys. **39**(1), 73–131 (1985). https://doi.org/10.1007/BF01007975
11. Hahsler, M., Kurt, H.: TSP - infrastructure for the traveling salesperson problem. J. Stat. Softw. **2**, 1–21 (2007)
12. Henn, S., Wäscher, G.: Tabu search heuristics for the order batching problem in manual order picking systems. Eur. J. Oper. Res. **222**(3), 484–494 (2012)
13. Kallina, C., Lynn, J.: Application of the cube-per-order index rule for stock location in a distribution warehouse. Interfaces **7**(1), 37–46 (1976). https://www.jstor.org/stable/25059400
14. Kofler, M., Beham, A., Wagner, S., Affenzeller, M.: Affinity based slotting in warehouses with dynamic order patterns. In: Klempous, R., Nikodem, J., Jacak, W., Chaczko, Z. (eds.) Advanced Methods and Applications in Computational Intelligence. Topics in Intelligent Engineering and Informatics, vol. 6, pp. 123–143. Springer, Heidelberg (2014)
15. Koster, R.D., Le-Duc, T., Roodbergen, K.J.: Design and control of warehouse order picking: a literature review. Eur. J. Oper. Res. **182**(2), 481–501 (2007)
16. Kruk, S.: Practical Python AI Projects: Mathematical Models of Optimization Problems with Google OR-Tools. Apress, New York (2018)
17. Kübler, P., Glock, C., Bauernhansl, T.: A new iterative method for solving the joint dynamic storage location assignment, order batching and picker routing problem in manual picker-to-parts warehouses. Comput. Ind. Eng. **147**, 106645 (2020)
18. Larco, J.A., Koster, R.D., Roodbergen, K.J., Dul, J.: Managing warehouse efficiency and worker discomfort through enhanced storage assignment decisions. Int. J. Prod. Res. **55**(21), 6407–6422 (2017). https://doi.org/10.1080/00207543.2016.1165880
19. Lee, I.G., Chung, S.H., Yoon, S.W.: Two-stage storage assignment to minimize travel time and congestion for warehouse order picking operations. Comput. Ind. Eng. **139**, 106129 (2020) https://doi.org/10.1016/j.cie.2019.106129, https://www.sciencedirect.com/science/article/pii/S0360835219305984
20. Li, J., Moghaddam, M., Nof, S.Y.: Dynamic storage assignment with product affinity and ABC classification-a case study. Int. J. Adv. Manuf. Technol. **84**(9), 2179–2194 (2016). https://doi.org/10.1007/s00170-015-7806-7

21. Liu, C.M.: Clustering techniques for stock location and order-picking in a distribution center. Comput. Oper. Res. **26**(10), 989–1002 (1999). https://doi.org/10.1016/S0305-0548(99)00026-X, https://www.sciencedirect.com/science/article/pii/S030505489900026X

22. Mackay, D.J.C.: Introduction to Monte Carlo methods. In: Jordan, M.I. (ed.) Learning in Graphical Models. NATO ASI Series, vol. 89, pp. 175–204. Springer, Dordrecht (1998). https://doi.org/10.1007/978-94-011-5014-9_7

23. Mantel, R., et al.: Order oriented slotting: a new assignment strategy for warehouses. Eur. J. Ind. Eng. **1**, 301–316 (2007)

24. Maruyama, K., Yamazaki, T.: Improved efficiency of warehouse picking by co-optimization of order batching and storage location assignment. J. Adv. Mech. Des. Syst. Manuf. **16**(5), JAMDSM0052-JAMDSM0052 (2022). https://doi.org/10.1299/jamdsm.2022jamdsm0052

25. Ming-Huang Chiang, D., Lin, C.P., Chen, M.C.: Data mining based storage assignment heuristics for travel distance reduction. Expert Syst. **31**(1), 81–90 (2014)

26. Oxenstierna, J., Krueger, V., Malec, J.: New benchmarks and optimization model for the storage location assignment problem. In: 3rd International Conference on Innovative Intelligent Industrial Production and Logistics, IN4PL 2022. SciTePress (2022)

27. Oxenstierna, J., Malec, J., Krueger, V.: Analysis of computational efficiency in iterative order batching optimization. In: Proceedings of the 11th International Conference on Operations Research and Enterprise Systems - ICORES, pp. 345–353. SciTePress (2022). https://doi.org/10.5220/0010837700003117

28. Oxenstierna, J., Rensburg, L.V., Stuckey, P., Krueger, V.: Storage assignment using nested annealing and hamming distances. In: Proceedings of the 12th International Conference on Operations Research and Enterprise Systems - ICORES, pp. 94–105. SciTePress (2023). https://doi.org/10.5220/0011785100003396. backup Publisher: INSTICC ISSN: 2184-4372

29. Rajasekaran, S., Reif, J.H.: Nested annealing: a provable improvement to simulated annealing. Theoret. Comput. Sci. **99**(1), 157–176 (1992)

30. Rathod, A.B., Gulhane, S.M., Padalwar, S.R.: A comparative study on distance measuring approches for permutation representations. In: 2016 IEEE International Conference on Advances in Electronics, Communication and Computer Technology (ICAECCT), pp. 251–255. IEEE (2016)

31. Janse van Rensburg, L.J.V.: Artificial intelligence for warehouse picking optimization - an NP-hard problem. Master's thesis, Uppsala University (2019)

32. Roodbergen, K.J., Koster, R.: Routing methods for warehouses with multiple cross aisles. Int. J. Prod. Res. **39**(9), 1865–1883 (2001)

33. Schapire, R.: Using Output Codes to Boost Multiclass Learning Problems (2001)

34. Tak, H., Meng, X.L., Dyk, D.A.V.: A repelling-attracting metropolis algorithm for multimodality. J. Comput. Graph. Stat. **27**(3), 479–490 (2018). https://doi.org/10.1080/10618600.2017.1415911

35. Trindade, M.A.M., Sousa, P., Moreira, M.: Ramping up a heuristic procedure for storage location assignment problem with precedence constraints. Flex. Serv. Manuf. J. **34**, 646–669 (2022). https://doi.org/10.1007/s10696-021-09423-w

36. Valle, C., Beasley, J.E., da Cunha, A.S.: Optimally solving the joint order batching and picker routing problem. Eur. J. Oper. Res. **262**(3), 817–834 (2017)

37. Wales, D.J., Doye, J.P.K.: Global optimization by basin-hopping and the lowest energy structures of Lennard-jones clusters containing up to 110 atoms. J. Phys. Chem. A **101**, 5111–5116 (1997)

38. Wu, J., Qin, T., Chen, J., Si, H., Lin, K.: Slotting optimization algorithm of the stereo warehouse. In: Proceedings of the 2012 2nd International Conference on Computer and Information Application (ICCIA 2012), pp. 128–132. Atlantis Press (2014). https://doi.org/10.2991/iccia.2012.31

39. Wutthisirisart, P., Noble, J.S., Chang, C.A.: A two-phased heuristic for relation-based item location. Comput. Ind. Eng. **82**, 94–102 (2015) https://doi.org/10.1016/j.cie.2015.01.020, https://www.sciencedirect.com/science/article/pii/S036083521500039X

40. Xiang, X., Liu, C., Miao, L.: Storage assignment and order batching problem in Kiva mobile fulfilment system. Eng. Optim. **50**(11), 1941–1962 (2018). https://doi.org/10.1080/0305215X.2017.1419346

41. Yu, M., Koster, R.B.M.D.: The impact of order batching and picking area zoning on order picking system performance. Eur. J. Oper. Res. **198**(2), 480–490 (2009)

42. Yu, V.F., Winarno, Maulidin, A., Redi, A.A.N.P., Lin, S.W., Yang, C.L.: Simulated Annealing with Restart Strategy for the Path Cover Problem with Time Windows. Mathematics **9**(14) (2021). https://doi.org/10.3390/math9141625, https://www.mdpi.com/2227-7390/9/14/1625

43. Zhang, R.Q., et al.: New model of the storage location assignment problem considering demand correlation pattern. Comput. Ind. Eng. **129**, 210–219 (2019). https://doi.org/10.1016/j.cie.2019.01.027, https://www.sciencedirect.com/science/article/pii/S0360835219300294

44. Žulj, I., Glock, C.H., Grosse, E.H., Schneider, M.: Picker routing and storage-assignment strategies for precedence-constrained order picking. Comput. Ind. Eng. **123**, 338–347 (2018). https://doi.org/10.1016/j.cie.2018.06.015, https://www.sciencedirect.com/science/article/pii/S0360835218302869

Inventory Management Optimization in Multi-Stage Supply Chains Under Uncertainty

Beatrice Ietto[1,2] and Valentina Orsini[3(✉)]

[1] Department of Management, Università Politecnica delle Marche, Ancona, Italy
b.ietto@univpm.it
[2] Weizenbaum Institute, Berlin, Germany
[3] Department of Information Engineering, Università Politecnica delle Marche, Ancona, Italy
vorsini@univpm.it

Abstract. We deal with the optimal inventory control problem for Multi-Stage Supply Chains (MSSC) with uncertain dynamics. The two sources of uncertainty we consider are about the perishability factor of stored products and on the customer prediction information. The control problem consists in defining a Replenishment Policy (RP) keeping the inventory level as close as possible to a desired value and mitigating the Bullwhip Effect (BE). The solution we propose is based on Distributed Robust Model Predictive Control (DRMPC) approach. This implies solving a set of RMPC problems. To drastically reduce the numerical complexity of this problem, the control signal (i.e. the RP) is sought in the space of B-spline functions, which are known to be universal approximators admitting a parsimonious parametric representation.

Keywords: Supply chain management · Inventory control · Bullwhip effect · Model predictive control

1 Introduction

We consider an MSSC whose dynamics is characterized by perishable goods with uncertain deterioration rate and an uncertain future customer demand. The problem we face consists in defining an RP keeping the actual inventory level as close as possible to a suitably defined reference trajectory though the mentioned uncertainties. In this context, the fundamental role played by Model Predictive Control (MPC) is widely recognized and documented [1,2]. This is mainly due to: 1) the capability of handling hard constraints imposed on some physical variables, 2) the capability of on-line determining appropriate corrections to the actual control action.

MPC techniques for large-scale are usually implemented according to three different control architectures: centralized, decentralized and distributed [3,4]. The application of the first two architectures to MSSC is discussed in [5–10].

F. Liberatore et al. (Eds.): ICORES 2022/2023, CCIS 1985, pp. 245–263, 2024.
https://doi.org/10.1007/978-3-031-49662-2_13

The main drawbacks of centralized approach are: numerical complexity, computational cost, reluctance to share information. Decentralized approach does not have these troubles but causes a loss of performance because each agent decides the control action on his own without exchanging information. This motivated the recent interest in Distributed MPC (DMPC) [11–13].

All the'mentioned papers do not face the problem raised by the presence of deteriorating items in the inventory system. On the other hand, if the effect of perishable goods is not taken into account, a serious performance degradation of the MSSC is observed. Centralized MPC of inventory level for perishable goods has been investigated in [14,15] under the simplifying assumption of an exactly known decay factor.

In this paper we extend previous results on single stage SC [16–18] and propose a DRMPC approach for the optimal inventory control of a MSSC whose dynamics is affected by two very common sources of uncertainties: perishable goods with an uncertain decay factor, inaccurate knowledge on the future end customer demand. Our purpose is to define a DRMPC policy optimally conciliating the three following antagonist Control Requirements (CR) at each stage: CR1) maximize the satisfied demand issued by the neighboring downstream stage, CR2) minimize the on hand stock level, CR3) mitigate the BE.

A preliminary version of this contribution was presented at [19]. Here we give more theoretical and implementation insights answering many issues not addressed in [19]. In Sect. 2.1 we provide more mathematical details on B-splines functions, in Sect. 6 we provide a rigorous analysis of stability and feasibility properties of our approach. In this regard, we mention that although, stability and feasibility play a fundamental role in the MPC [20], most of MPC based methods for inventory control do not rigorously face these issues. Our contribution fills this gap. In the numerical simulation we show the effectiveness of the proposed method applied to seasonal MSSC with unpredictable random fluctuations of the final customer demand: a class of systems whose complexity is well acknowledged in the literature [21].

Applying MPC requires a reliable estimate of the future end customer demand. This is a very arduous task due to unpredictable and dynamically changing behavior of the final consumer. For this reason we give up statistical methods based on time series analysis and, according to the robust approach, we assume that at any time instant $k \in Z^+$ and over a finite prediction horizon, the future end customer demand entering the first stage of the MSSC is arbitrarily time varying inside a given bounded set $\mathcal{D}_{1,k}$. Coherently with this assumption we conciliate CR1 and CR2 defining a desired inventory level that, for the first stage of the MSSC, is given by the upper bounding trajectory of $\mathcal{D}_{1,k}$. Then, the target inventory level of each other upward stage is iteratively defined on the basis of the predicted demand coming from the previous downstream stage. Satisfying CR3 is a problem of a paramount importance in the MSSC management as testified by the impressive amount of relevant literature, [22,23]. We face this problem simultaneously acting on two Fundamental Features (FF) characterizing the BE: FF1) irregularity of stock replenishment orders, FF2) progressive upward amplification of the intervals over which the replenishment orders issued by each stage take values.

FF1 is addressed applying an RP obtained by a parametrized solution of the DRMPC problem in terms of smooth functions and defining a cost functional penalizing excessive differences between consecutive orders. As for FF2, we prove that, using our approach, the upward interval amplification is proportional to the perishability rate. The interesting corollary is that, in the case of non perishable goods, the values of orders issued by all stages may be contained in the same fixed amplitude interval. Coherently with the assumptions on the uncertainties and with the CR's , we develop a DRMPC approach based on a min-max optimization procedure. The control signal (i.e. the RP) relative to each stage of SC is obtained in the following way:

1) at each time instant k we minimize the worst case of a quadratic cost functional over a future prediction horizon. The worst case is computed by maximizing with respect to all the possible perishability factor values;
2) the actual control law is obtained by receding horizon implementation.

To reduce the computational burden of the min max optimization problem over each prediction horizon, we propose a parametrized solution in terms of polynomial B-spline functions. This implies solving a DRMPC problem with fewer decision variables.

Alternative parametrization methods for reducing the computational complexity of MPC have been proposed in [24–26].

The main reasons for choosing polynomial B-splines are: 1) they inherit the smoothness of polynomial functions, and, at the same time, their spline structure is well suited to approximate curves which exhibit different shapes over different time-intervals, 2) B-splines admit a parsimonious parametric representation given by a time varying, linear, convex combination of some parameters named "control points". These properties allow us to obtain an RP with a smooth waveform and to transfer any hard constraint on the control law to its control points. This is very useful to deal with FF2 of BE. Property 2 also allows us to reformulate the constrained minimization of the cost functional with respect to the RP signal as a Weighted Constrained Robust Least Square (WCRLS) estimation problem that can be efficiently solved using interior point methods [27]. Another fundamental advantage deriving from using B-spline function is the feasibility guarantee independently of the length of the prediction and control horizons. This point is discussed in Sect. 6.

The paper is organized in the following way. Some mathematical preliminaries on B-splines and RLS problem are recalled in Sect. 2. The system model is described in Sect. 3. The DRMPC problem is formally stated in Sect. 4 and solved in Sect. 5, where it is reformulated as a WCRLS estimation problem. Feasibility and stability of the DRMPC are proved in Sect. 6. Numerical results are reported in Sect. 7 and concluding remarks in Sect. 8.

2   Mathematical Background

2.1   B-Spline Functions [28]

Analytic scalar B-splines functions are defined in the following way:

$$b_s(v) = \sum_{i=1}^{\ell} c_i B_{i,d}(v), \quad v \in [\hat{v}_1, \hat{v}_{\ell+d+1}] \subseteq R, \tag{1}$$

where the c_i's are real numbers representing the control points of $b_s(v)$, the integer d is the degree of the B-spline, the $(\hat{v}_i)_{i=1}^{\ell+d+1}$ are the non decreasing knot points and the $B_{i,d}(v)$ are given by the Cox-de Boor recursion formula

$$B_{i,d}(v) = \frac{v - \hat{v}_i}{\hat{v}_{i+d} - \hat{v}_i} B_{i,d-1}(v) + \frac{\hat{v}_{i+1+d} - v}{\hat{v}_{i+1+d} - \hat{v}_{i+1}} B_{i+1,d-1}(v), \quad d \geq 1, \tag{2}$$

with $B_{i,0}(v) = 1$ if $\hat{v}_i \leq v < \hat{v}_{i+1}$, otherwise 0.

In (2) possible division by zero are resolved by the convention that "anything divided by zero is zero".

Convex Hull Property. Any value assumed by $b_s(v)$, $\forall v \in [\hat{v}_j, \hat{v}_{j+1}], j > d$, lies in the convex hull of its $d+1$ control points $c_{j-d}, \cdots, c_j$. $\triangle$

Identifying the parameter v of (1) with the time instant $t \in R^+$, the sampled B-spline $b_s(kT)$ is obtained by direct uniform sampling of the corresponding analytic B-spline.

The discrete B-spline $b_s(k)$ (omitting the explicit dependence on T) can be used to represent a scalar discrete time signal. Defining

$$\mathbf{c} \triangleq [c_1 \cdots c_\ell]^T, \quad \mathbf{B}_d(k) \triangleq [B_{1,d}(k) \cdots B_{\ell,d}(k)] \tag{3}$$

where each $B_{i,d}(k)$ is obtained by (2) setting $v = k$ and $\hat{v}_i = \hat{k}_i$, $i = 1, \cdots, d + \ell + 1$, the sampled B-spline $b_s(k)$ can be represented as

$$b_s(k) = \mathbf{B}_d(k)\mathbf{c}, \quad k \in [\hat{k}_1, \hat{k}_{\ell+d+1}] \tag{4}$$

Remark 1. From (4) it is apparent that, once the degree d and the knot points $\hat{k}_i$ have been fixed, the B-spline $b_s(k)$, $k \in [\hat{k}_1, \hat{k}_{\ell+d+1}]$, is completely determined by the corresponding vector $\mathbf{c}$ of ℓ control points. As, in general, $\ell << k_M$, where k_M is the number of sampled instants of $[\hat{k}_1, \hat{k}_{\ell+d+1}]$, B-splines are said to admit a parsimonious parametric representation. $\triangle$

2.2   The CRLS Problem

Consider an overdetermined set of linear equations $Df \approx b$ where $D \in R^{r \times m}$ is the design matrix and $b \in R^r$, is the observed vector. Both D and b are subject to unknown but bounded errors: $\|\delta D\| \leq \beta$ and $\|\delta b\| \leq \xi$ (the matrix norm is the spectral norm). The RLS estimate $\hat{f} \in R^m$, [27], is the value of f computed as

$$\min_f \; \max_{\|\delta D\| \leq \beta, \, \|\delta b\| \leq \xi} \; \|(D + \delta D)f - (b + \delta b)\| \tag{5}$$

Using norm properties, it can be shown that

$$\max_{\|\delta D\| \leq \beta, \, \|\delta b\| \leq \xi} \|(D + \delta D)f - (b + \delta b)\|$$

$$= \|Df - b\| + \beta\|f\| + \xi$$

Hence (5) is equivalent to compute f according to

$$\min_f \|Df - b\| + \beta\|f\| + \xi \tag{6}$$

The CRLS version also requires that f satisfies the following component-wise linear constraints:

$$\underline{f} \leq f \leq \bar{f} \tag{7}$$

Remark 2. The term $\|\delta b\|$ in (5) only appears in (6) through its norm upper bound ξ, which is independent of f. Hence ξ can be removed from the objective function without affecting the value of f solving the minimization problem. As it will be shown in Sect. 5, this allows us to solve the constrained optimization problem implied by the RMPC algorithm even in the case of uncertain future customer demand. △

3   The MSSC Model

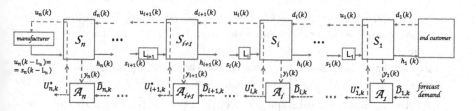

Fig. 1. Distributed control scheme of the MSSC network.

As shown in Fig. 1, we assume an MSSC network made up of a number of nodes $\mathcal{S}_i$, $i = 1, \cdots, n$, with counter-current order and material streams.

Management decisions for each node are taken periodically at equally distributed time instants kT where $k \in Z^+$ and T is the review period. At the beginning of each review period $[kT, \, (k+1)T)$ the operations across the MSSC network are performed sequentially from $\mathcal{S}_1$ to $\mathcal{S}_n$. Inside each review period, each $\mathcal{S}_i$

executes five actions in the following order: receives delivery from supplier S_{i+1}, logs the demand of customer S_{i-1}, measures its on hand stock level, delivers the goods to meet demand and finally places an order according to a suitably defined replenishment policy. Accordingly, five variables are defined: $s_i(k)$, $d_i(k)$, $y_i(k)$, $h_i(k)$ and $u_i(k)$. They represent the shipment of goods from supplier S_{i+1}, the demand from S_{i-1}, the on hand stock level, the delivery to customer S_{i-1} and the replenishment order, respectively. Each node S_i is regulated by an agent A_i that solves a local RMPC problem based on the following assumptions:

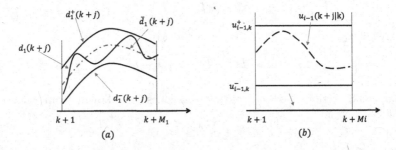

Fig. 2. (a) Example of a set $\mathcal{D}_{1,k}$, (b) Example of a set $\mathcal{D}_{i,k}$, $i \geq 2$.

- **A1)** The end customer demand $d_1(k)$, $k \in Z^+$ is uniformly bounded. Moreover, at any time instant k, and over an M_1-steps prediction horizon $[k+1, k+M_1]$, the unknown future end customer demand $d_1(k+j)$, $j = 1, \cdots, M_1$, varies within a compact set $\mathcal{D}_{1,k}$ limited below and above by two known boundary trajectories: $d_1^-(k+j)$ and $d_1^+(k+j)$, $j = 1, \cdots, M_1$. The minimum value of $d_1^-(k+j)$ and the maximum value of $d_1^+(k+j)$, $j = 1, \cdots, M_1$, are denoted by $d_{1,k}^-$ and $d_{1,k}^+$ respectively. The demand forecasting $\hat{\mathcal{D}}_{1,k} = [d_1(k+1|k), \cdots, d_1(k+M_1|k)]$ for agent A_1 coincides with the central trajectory of $\mathcal{D}_{1,k}$ namely $\hat{\mathcal{D}}_{1,k} \triangleq [\bar{d}_1(k+1), \cdots, \bar{d}_1(k+M_1)]$. Figure 2(a) shows a typical example of an end-customer demand $d_1(k+j)$ and of a predicted end customer demand $d_1(k+j|k)$ over a fixed $\mathcal{D}_{1,k}$.
- **A2)** At any time instant k, the predicted demand $\hat{\mathcal{D}}_{i,k} = [d_i(k+1|k), \cdots, d_i(k+M_i|k)]$ for the other agents A_i, $i = 2, \cdots, n$, coincides with the predicted optimal control sequence (i.e. the optimal predicted RP) $\mathcal{U}_{i-1,k} \triangleq [u_{i-1}(k+1|k), \cdots, u_{i-1}(k+N_{i-1}-1|k)]$ transmitted by A_{i-1} to A_i, where $M_i = N_{i-1} - 1$. Note that also $\hat{\mathcal{D}}_{i,k}$ belongs to a given compact set $\mathcal{D}_{i,k}$ limited by the imposed lower and upper values $u_{i-1,k}^-$ and $u_{i-1,k}^+$ respectively (as shown in Fig. 2(b)). How to compute $\mathcal{U}_{i-1,k}$, $u_{i-1,k}^-$ and $u_{i-1,k}^+$ is explained in Sect. 4.
- **A3)** The goods shipped from supplier S_{i+1} arrive at customer S_i with a time delay $L_i = n_i T$, where $n_i \in Z^+$. Goods arrive at customer S_i new and deteriorate while kept in stock.

- **A4)** Inside each review period, the perishability rate of the goods stocked in S_i is $\alpha_i \in [\alpha_i^-, \alpha_i^+] \subset (0, 1)$.

- **A5)** The operations of inventory replenishment and goods delivery are executed simultaneously at the beginning of each review period. Sales are not back-ordered.

The above assumptions imply that the stock level dynamics of the i-th node is described by the following uncertain equation

$$y_i(k + 1) = \rho_i(y_i(k) + s_i(k - L_i) - h_i(k)), \quad y_i(0) \geq 0, \quad i = 1, \cdots, n, \qquad (8)$$

where:

- $\rho_i \stackrel{\triangle}{=} 1 - \alpha_i \in [\rho_i^-, \rho_i^+] \subset (0, 1)$ is the uncertain decay factor;
- $y_i(k)$ is the on hand stock level of S_i, i.e. the amount of goods left in stock after satisfying the demand at the beginning of the $k - 1$ review period;
- $s_i(k - L_i)$ is the goods delivered to the stage S_i with a time delay L_i.
- $h_i(k)$ is the demand fulfilled by S_i, $i = 1, \cdots, n$.
  As backorders are not allowed, $h_i(k)$ is given by

$$h_i(k) \stackrel{\triangle}{=} \min\{d_i(k), y_i(k) + s_i(k - L_i)\}, \quad i = 1, \cdots, n \qquad (9)$$

where $d_1(k)$ is the end-customer demand, and $d_i(k) = u_{i-1}(k)$, $i = 2, \cdots, n$, is the demand issued by S_{i-1}.
Note that the sum $y_i(k) + s_i(k - L_i)$ represents the effective amount of goods available for sale at the beginning of k-th review period.

For future developments we now rewrite Eq. (8) assuming **A6)**: there exists a $\bar{k} \geq 0$ such that

$$y_i(k) + s_i(k - L_i) \geq d_i(k), \quad \forall k \geq \bar{k}, \quad i = 1, \cdots, n \qquad (10)$$

This assumption is justified because, at each stage, the control sequence is obtained minimizing the maximum weighted ℓ_2 norm of the distance between the on-hand stock level and the maximum demand.

By (9) and (10) we have $h_{i+1}(k) = d_{i+1}(k)$. As $d_{i+1}(k) = u_i(k)$ and $h_{i+1}(k) = s_i(k)$ (see Fig. 1) we also have $s_i(k - L_i) = u_i(k - L_i)$. Hence an equivalent expression of (8) is

$$y_i(k + 1) = \rho_i(y_i(k) + u_i(k - L_i) - h_i(k)), \quad \forall k \geq \bar{k}, \quad i = 1, \cdots, n \qquad (11)$$

4 The DRMPC Approach

To simplify the derivation of the control strategy but without any loss of generality we refer to (11) in the ideal case $\bar{k} = 0$. Each $\mathcal{A}_i$ uses Eq. (11) and the predicted optimal control policy $\mathcal{U}_{i-1,k}$ communicated by $\mathcal{A}_{i-1}$ to predict the future inventory level of the local subsystem $\mathcal{S}_i$. This latter is in turn used to compute $\mathcal{U}_{i,k}$ minimizing the worst case of a local quadratic cost functional subject to hard constraints $u_{i,k}^-$ and $u_{i,k}^+$. Coordination between contiguous agents $\mathcal{A}_{i-1}$ and $\mathcal{A}_i$, is imposed by relating the respective constraints $u_{i-1,k}^-$ and $u_{i-1,k}^+$ with $u_{i,k}^-$ and $u_{i,k}^+$. Each local RMPC requires each agent $\mathcal{A}_i$ to repeatedly solve a Min-Max Constrained Optimization Problem (MMCOP) over a future N_i steps control horizon, and, according to the receding horizon control, to only apply the first sample of the computed predicted optimal control sequence.

4.1   Local MMCOP for $\mathcal{A}_i$

With reference to (11), the local MMCOP for any $\mathcal{A}_i$, $i = 1, \cdots, n$, is formally stated as follows

$$\min_{[u_i(k|k),\cdots,u_i(k+N_i-1|k)]} \max_{\rho_i \in [\rho_i^-, \rho_i^+]} J_{i,k} \quad k \in Z^+, \tag{12}$$

subject to

$$0 \leq u_{i,k}^- \leq u_i(k+j|k) \leq u_{i,k}^+ < \infty, \quad j = 0, \cdots, N_i - 1, \tag{13}$$

where $J_{i,k}$ is the following cost functional

$$J_{i,k} = \sum_{l=1}^{N_i} e_i^T(k+L_i+l|k) q_{i,l} e_i(k+L_i+l|k) + \sum_{l=1}^{N_i-1} \lambda_{i,l} \Delta u_i^2(k+l|k) \tag{14}$$

and

$$e_i(k+L_i+l|k) \stackrel{\triangle}{=} r_i(k+L_i+l|k) - y_i(k+L_i+l|k) \tag{15}$$

$$\Delta u_i(k+l|k) \stackrel{\triangle}{=} u_i(k+l|k) - u_i(k+l-1|k) \tag{16}$$

$$y_i(k+L_i+l|k) = \rho_i^{L_i+l} y_i(k) + \sum_{\ell=0}^{L_i-1} \rho_i^{L_i+l-\ell} u_i(k+\ell-L_i)$$

$$+ \sum_{\ell=0}^{l-1} \rho_i^{l-\ell} u_i(k+\ell|k) - \sum_{\ell=0}^{L_i+l-1} \rho_i^{L_i+l-\ell} h_i(k+\ell|k) \tag{17}$$

$$r_i(k+L_i+l|k) \stackrel{\triangle}{=} \begin{cases} d_1^+(k+L_1+l) & i = 1 \\ u_{i-1,k}^+ & i > 1 \end{cases} \tag{18}$$

4.2   Some Remarks on the Cost Functional $J_{i,k}$

- 1) By A1), A2) and (18), it can be seen that $M_1 \geq N_1 + L_1$ and $M_i = N_{i-1} - 1 = N_i + L_i$, $i > 1$, namely $N_{i-1} = N_i + L_i + 1$.
- 2) The time varying target inventory level $r_i(k)$ for $\mathcal{S}_i$ is defined as follows:

$$r_1(k) = d_1^+(k) \text{ and } r_i(k) = u_{i-1,k}^+, \quad i > 1 \tag{19}$$

where $d_1^+(k)$ is the time-varying upper bounding trajectory of the actual end-customer demand and, analogously $u_{i-1,k}^+$ is the constant upper bounding trajectory of the demand forecasting for $\mathcal{A}_i, i > 1$ (as shown in Figs. 2(a) and 2(b) respectively). Keeping the on hand stock level as near as possible to the possible maximum level of the demand forecasting maximizes the amount of fulfilled demand over each shifted prediction horizon and prevents unnecessarily larger stock levels.

- 3) The term $\sum_{l=1}^{N_i-1} \lambda_{i,l} \Delta u_i^2(k+l|k)$ and the way the hard constraints (13) are defined allow us to deal with FF1 and FF2 respectively.
- 4) Exploiting A1), A2), A6) and (9), $h_i(k+\ell|k)$ in (17) can be expressed as

$$h_i(k+\ell|k) = \bar{h}_i(k+\ell|k) + \delta h_i(k+\ell|k), \tag{20}$$

where $\bar{h}_i(k+\ell|k) = d_i(k+\ell|k)$.

4.3 The Constraints $u_{i,k}^-$ and $u_{i,k}^+$

On the basis of CR1-CR3, the constraints $u_{i,k}^-$ and $u_{i,k}^+$ of (13) are determined to conciliate the two following conflicting criteria:

- 1) maximize the amount of demand that each stage $\mathcal{S}_i$ can satisfy,
- 2) limit the amplitude (defined as $A_{i,k}$) of the interval $[u_{i,k}^-\ \ u_{i,k}^+] \triangleq \mathcal{C}_{i,k}$.

To solve this problem we refer to two possible limit situations compatible with the balance Eq. (11).

Consider the following scenario:

- $d_i(k+L_i+j|k)$, $j = 0, \cdots, N_i - 1$, is a constant signal with value $\tilde{d}_{i,k} \in [d_{i,k}^-, d_{i,k}^+] = [u_{i-1,k}^-, u_{i-1,k}^+]$. The two mentioned limit situations are $\tilde{d}_{i,k} = u_{i-1,k}^-$ and $\tilde{d}_{i,k} = u_{i-1,k}^+$.
- Each control horizon $H_{i,k}$ is long enough to allow $y_i(k+L_i+j|k), j = 1, \cdots, N_i$, to practically attain the steady-state value $\tilde{y}_{i,k}$ under the forcing action of a constant $u_i(k+j|k) = \tilde{u}_{i,k}, j = 0, \cdots, N_i - 1$.

The problem we now consider is: for a given $\tilde{u}_{i-1,k} \in [u_{i-1,k}^-, u_{i-1,k}^+]$ it is required to find the corresponding constant control input $\tilde{u}_{i,k}$ over each $H_{i,k}$, such that $\mathcal{S}_i$ fully satisfies the demand coming from $\mathcal{S}_{i-1}$, namely $\tilde{y}_{i,k} \geq \tilde{u}_{i-1,k}$, $\forall \rho_i \in [\rho_i^-, \rho_i^+]$.

Simple algebraic calculations based on classical z-transform methods and the final value theorem [29] allow us to prove that

$$\mathcal{C}_{1,k} \triangleq [u_{1,k}^-, u_{1,k}^+] = \frac{1}{\rho_1^-}[d_{1,k}^-, d_{1,k}^+] \tag{21}$$

$$\mathcal{C}_{i,k} \triangleq [u_{i,k}^-, u_{i,k}^+] = \frac{1}{\rho_i^-}\left[u_{i-1,k}^-, u_{i-1,k}^+\right], i = 2, \cdots, n \tag{22}$$

Recalling that $A_{i-1,k}$ denotes the amplitude of $C_{i-1,k}$, from (21), (22) we have

$$A_{1,k} = \frac{1}{\rho_1^-}(d_{1,k}^+ - d_{1,k}^-) \quad \text{and} \quad A_{i,k} = \frac{1}{\rho_i^-}A_{i-1,k}, \; i = 2, \cdots, n \qquad (23)$$

At each node $\mathcal{S}_i$ according to FF2 we define the following measure:

$$\mathcal{B}_{i,k} = \frac{A_{i,k}}{A_{i-1,k}} \qquad (24)$$

According to (23)–(24), the proposed DRMPC scheme implies $\mathcal{B}_{i,k} = 1/\rho_i^- \overset{\triangle}{=} \mathcal{B}_i > 1$.

The two salient conclusions are: 1) an estimate of the overall BE (corresponding to FF2) that propagates along the MSSC network, can be computed "a priori" and is given by $\mathcal{B} = 1/(\prod_{i=1}^n \rho_i^-)$, 2) our approach does not entail this kind of BE for $\rho_i^- \to 1$.

5   Reformulation of the MMCOP

We reformulate the local MMCOP as a WCRLS estimation to drastically reduce the numerical complexity of the algorithm solving the original MMCOP. The functional $J_{i,k}$, defined in (12), is minimized assuming that the control sequence $\mathcal{U}_{i,k}$, is given by the sampled version of a B-spline function. Adapting the notation in (4) to specify that it is relative to the i-th node and the k-th fixed time instant we have

$$u_i(j|k) \overset{\triangle}{=} \mathbf{B}_{i,d}(j)\mathbf{c}_{i,k}, \quad j = k, \cdots, k + N_i - 1, \qquad (25)$$

with $\mathbf{B}_{i,d}(j) = [B_{i,1,d}(j), \cdots, B_{i,\ell,d}(j)]$ and $\mathbf{c}_{i,k} = [\mathbf{c}_{i,k,1}, \cdots, \mathbf{c}_{i,k,\ell}]^T$. The parameter vector $\mathbf{c}_{i,k}$, defining $u_i(j|k)$, is computed as the solution of the WCRLS estimation problem defined beneath.

As $\rho_i \in [\rho_i^- \; \rho_i^+]$, an equivalent representation of ρ_i is $\rho_i = \bar{\rho}_i + \delta\rho_i, \bar{\rho}_i = (\rho_i^- + \rho_i^+)/2$ where $\bar{\rho}_i$ is the nominal value and $\Delta\rho_{i,k} \overset{\triangle}{=} (\bar{\rho}_i + \delta\rho_i)^k - \bar{\rho}_i^k$ is the sum of all terms containing $\delta\rho_i$ in the explicit expression of $(\bar{\rho}_i + \delta\rho_i)^k$.

Exploiting (20) and (25) it can be shown that the predicted tracking error given by (15) can be expressed as

$$e_i(k + L_i + l|k) = (b_{i,k,l} + \delta b_{i,k,l}) - (D_{i,k,l} + \delta D_{i,k,l})\mathbf{c}_{i,k} \qquad (26)$$

where

$$D_{i,k,l} \overset{\triangle}{=} \sum_{\ell=0}^{l-1} \bar{\rho}_i^{l-\ell}\mathbf{B}_{i,d}(k + \ell) \qquad (27)$$

$$\delta D_{i,k,l} \overset{\triangle}{=} \sum_{\ell=0}^{l-1} \Delta\rho_{i,l-\ell}\mathbf{B}_{i,d}(k + \ell) \qquad (28)$$

$$b_{i,k,l} \triangleq r_i(k + L_i + l|k) - \bar{\rho}_i^{L_i+l} y_i(k) - \sum_{\ell=0}^{L_i-1} \bar{\rho}_i^{L_i+l-\ell} u_i(k + \ell - L_i) \qquad (29)$$

$$+ \sum_{\ell=0}^{L_i+l-1} \bar{\rho}_i^{L_i+l-\ell} \bar{h}_i(k + \ell|k)$$

$$\delta b_{i,k,l} \triangleq -\Delta \rho_{i,L_i+l} y_i(k) - \sum_{\ell=0}^{L_i-1} \Delta \rho_{i,L_i+l-\ell} u_i(k + \ell - L_i) \qquad (30)$$

$$+ \sum_{\ell=0}^{L_i+l-1} \bar{\rho}_i^{L_i+l-\ell} \delta h_i(k + \ell|k) + \sum_{\ell=0}^{L_i+l-1} \Delta \rho_{i,L_i+l-\ell} h_i(k + \ell|k)$$

Similarly $\Delta u_i(k + l|k) \triangleq \mathbf{B}_{i,d}(k + l)\mathbf{c}_{i,k} - \mathbf{B}_{i,d}(k + l - 1)\mathbf{c}_{i,k}$ can be rewritten as

$$\Delta u_i(k + l|k) = (b_{u_{i,k,l}} + \delta b_{u_{i,k,l}}) - (D_{u_{i,k,l}} + \delta D_{u_{i,k,l}})\mathbf{c}_{i,k} \qquad (31)$$

where

$$D_{u_{i,k,l}} \triangleq -(\mathbf{B}_{i,d}(k + l) - \mathbf{B}_{i,d}(k + l - 1)), \quad \delta D_{u_{i,k,l}} \triangleq 0 \text{ (null row vector)} \quad (32)$$

$$b_{u_{i,k,l}} \triangleq 0 \text{ and } \delta b_{u_{i,k,l}} \triangleq 0 \qquad (33)$$

Let's now define the following extended error vector

$$\underline{e}_{i,k} \triangleq \begin{bmatrix} q_{i,1}^{1/2} e_i(k + L_i + 1|k) \\ \vdots \\ q_{i,N_1}^{1/2} e_i(k + L_i + N_i|k) \\ \lambda_{i,1}^{1/2} \Delta u_i(k + 1|k) \\ \vdots \\ \lambda_{i,N_i-1}^{1/2} \Delta u_i(k + N_i - 1|k) \end{bmatrix}$$

Using (27)–(30) for $l = 1, \cdots, N_i$ and (32)–(33) for $l = 1, \cdots, N_i - 1$, we also define analogous extended vectors $\underline{b}_{i,k}$, $\underline{\delta b}_{i,k}$ and matrices $\underline{D}_{i,k}$, $\underline{\delta D}_{i,k}$ that allow us to rewrite $\underline{e}_{i,k}$ as

$$\underline{e}_{i,k} = (\underline{b}_{i,k} + \underline{\delta b}_{i,k}) - (\underline{D}_{i,k} + \underline{\delta D}_{i,k})\mathbf{c}_{i,k} \qquad (34)$$

By (34) we reformulate the local MMCOP (12)–(14) as the following local WCRLS estimation problem:

$$\min_{\mathbf{c}_{i,k}} \max_{\|\underline{\delta D}_{i,k}\| \leq \beta_{i,k}, \|\underline{\delta b}_{i,k}\| \leq \xi_{i,k}} \|\underline{e}_{i,k}\|^2 \qquad (35)$$

$$\text{subject to} \quad u_{i,k}^- \leq \mathbf{c}_{i,k,l} \leq u_{i,k}^+, \qquad l = 1, \cdots, \ell \qquad (36)$$

It is seen that the solution of (35) subject to (36) is the same of a problem of the kind (5) subject to (7). Hence, according to Sect. 2.2, the solution of the local MMCOP (12)–(14) is given by the vector $\mathbf{c}_{i,k}$ solving

$$\min_{\mathbf{c}_{i,k}} \|\underline{b}_{i,k} - \underline{D}_{i,k}\,\mathbf{c}_{i,k}\| + \beta_{i,k}\|\mathbf{c}_{i,k}\| + \xi_{i,k} \tag{37}$$

where the components of $\mathbf{c}_{i,k}$ must satisfy (36).

Remark 3. Note that by Remark 2 only the upper bound $\beta_{i,k}$ on $\|\delta\underline{D}_{i,k}\|$ in (37) needs to be determined at each k. Moreover the way the B-spline basis functions are defined by the Cox-de Boor formula (2) implies that $\mathbf{B}_{i,d}(\tau) = \mathbf{B}_{i,d}(\tau + N_i)$, $\forall \tau \in H_{i,k}$, $k \in Z^+$. Hence, by (28), one has that $\beta_{i,k} \stackrel{\triangle}{=} \beta_i$, $\forall k = 0, 1, \cdots$ and moreover β_i is can be determined putting $\rho_i = \rho_i^+$.

6 Feasibility and Stability of the DRMPC

The calculations reported in Sect. 5 clearly show why the B-spline parametrization directly implies the feasibility of each local MMCOP (and hence of the global DRMPC): the constraints (13) on the non parametrized predicted control sequence are transformed in the consistent system of inequalities (36). These inequalities concern the vector $\mathbf{c}_{i,k}$ of control points of the B-spline that, by (25), gives the parametrized predicted control sequence. Recalling the convex hull property of B-splines, it follows that also condition (13) is always satisfied. The positivity of $y_i(k)$, $\forall k \in Z^+$, is implied by (8), (9) and by the positivity of the RP.

The internal stability of the controlled MSSC is guaranteed by: $\rho_i \in (0,1)$, $i = 1, \cdots, n$, and by the uniform boundedness of the RP, implied by (13). Further, contrary to many proposed methods [30], feasibility and stability are guaranteed regardless of the length of the prediction horizon. This is very important because makes it possible using prediction horizons whose length is inferiorly limited by considerations only involving the physical structure of the MSSC (point 1 of Sect. 4.2 and Sect. 4.3).

7 Simulation Results

We consider an uncertain MSSC composed of three stages $\mathcal{S}_i$, i=1,2,3. The perishability factor α_i, the decay factor $\rho_i = 1 - \alpha_i$, the time delay L_i and the initial stock level $y_i(0)$ describing the balance equation (8) of each $\mathcal{S}_i$ are reported in Table 1.

Table 1. Model parameters of each node $\mathcal{S}_i$, $i = 1, 2, 3$.

time delay	perishability factor	decay factor	initial stock level
$L_i = 3$	$\alpha_i \in [\alpha_i^-, \alpha_i^+] = [0.06, 0.1]$	$\rho_i \in [\rho_i^-, \rho_i^+] = [0.9, 0.94]$	$y_i(0) = 15$

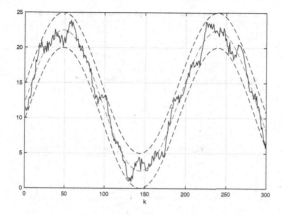

Fig. 3. The end-customer demand $d_1(k)$ (black solid line) and the two boundary trajectories $d_1^+(k)$ and $d_1^-(k)$ (black dashed lines).

The local MMCOP for any agent $\mathcal{A}_i$, $i = 1, 2, 3$ is solved at each k using B spline functions of degree $d = 3$ with a vector $c_{i,k}$ of $\ell = 6$ control points. The other tuning parameters: N_i (length of the control horizon $H_{i,k}$), the scalar weights $q_{i,l}$ e $\lambda_{i,l}$ in (14) and scalars $\beta_{i,k} \triangleq \beta_i$ in (37) are given in Table 2.

Table 2. Tuning parameters of the local MMCOP for any $\mathcal{A}_i$, $i = 1, 2, 3$.

$N_{i-1} = N_i + L_i + 1$			weights in (14)		$\beta_{i,k} \triangleq \beta_i$ in (37)		
N_3	N_2	N_1	$q_{i,l}$	$\lambda_{i,l}$			
12	16	20	$e^{-0.1\,(l-1)}$	$e^{-1\,(l-1)}$	$\beta_1 = 2.3$	$\beta_2 = 1.6$	$\beta_3 = 1$

According to Fig. 2, at each k the future end customer demand is assumed to belong to a set $\mathcal{D}_{1,k}$ with length $M_1 = N_1 + L_1 = 23$: the minimum length resulting from 1) of Sect. 4.2.

We consider a customer demand with a quasi-periodic dynamics, typical in the case of seasonal goods. Figure 3 shows the actual end-customer demand (solid line) enclosed in the contiguous positioning of all the $D'_{1,k}$s. The dashed trajectory denotes the predicted end-customer demand $d_1(k + l|k)$.

The model equation (8) has been implemented assuming $\rho_i = 0.935$, $i = 1, 2, 3$ and the simulation has been stopped at time $k = 250$. Figure 4 shows the ordering signals issued by each stage $\mathcal{S}_i$, $i = 1, 2, 3$ with the respective time-varying lower and upper bounds. According to FF1) each $u_i(k)$ has a smoother behavior with respect to the end customer demand profile $d_1(k)$.

The resulting on hand stock level $y_i(k)$ and the time varying desired inventory level $r_i(k)$ for each $\mathcal{S}_i$, $i = 1, 2, 3$ are displayed in Fig. 5. The imposed and fulfilled demands $d_i(k)$ and $h_i(k)$ respectively at each $\mathcal{S}_i$ are given in figure 6. As evident

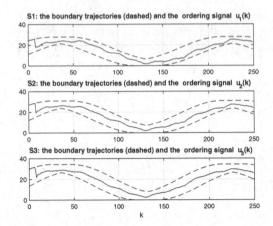

Fig. 4. DRMPC: The ordering signal $u_i(k)$ issued by each $\mathcal{S}_i$, $i = 1, 2, 3$ with the respective time-varying lower and upper bounds.

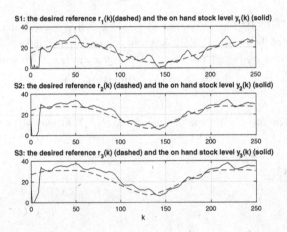

Fig. 5. DRMPC: The desired inventory level $r_i(k)$ and the on hand stock level $y_i(k)$ of each $\mathcal{S}_i$.

from Fig. 6, the demand at each $\mathcal{S}_i$ is not satisfied only over a short initial time interval as a consequence of the time delay propagation.

A comparison has been performed with the non-linear control strategy proposed in [31]: the method described therein accounts for the effects of time delay and perishable goods and all details for its reproducibility are provided. Equations (34)-(36) in [31] have been rewritten in the case of an uncertain n stage SC with $n = 3$, decay factors $\rho_i \in [0.9, 0.94]$ and known time delays $L_i = 3$, $i = 1, 2, 3$ obtaining

$$u_i(k) = \text{sat}[\omega_i(k)] \tag{38}$$

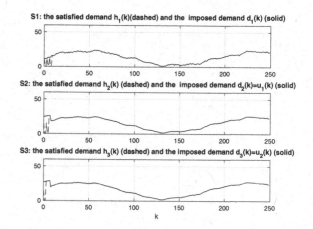

Fig. 6. DRMPC: The imposed demand $d_i(k)$ and the fulfilled demand $h_i(k)$ at each $\mathcal{S}_i$, $i = 1, 2, 3$.

where

$$\omega_i(k) = y_{ref,i} - \rho_i^{L_i} y_i(k) + \sum_{j=0}^{k-1} \rho_i^{k-j} s_i(j) - \sum_{j=0}^{k-L_i-1} \rho_i^{k-j} s_i(j) \qquad (39)$$

and the saturation function

$$\text{sat}[\omega_i(k)] = \begin{cases} \omega_i(k) & \text{if } \omega_i(k) \in [0, u_{\text{max},i}] \\ 0 & \text{if } \omega_i(k) < 0 \\ u_{\text{max},i} & \text{if } \omega_i(k) > u_{max,i} \end{cases} \qquad (40)$$

According to (45), (46) in [31] and taking into account that $\rho_i \in [\rho_i^-, \rho_i^+]$, $u_{\text{max},i}$ and $y_{ref,i}$ are inferiorly limited as:

$$u_{\text{max},i} > d_{\text{max},i} \quad \text{and} \quad y_{ref,i} > d_{\text{max},i} \sum_{j=0}^{L_i} \rho_i^{+j} \qquad (41)$$

The topology of the SC network shown in Fig. 1 is such that:

$$d_{\text{max},1} = \max_k d_1(k) \quad d_{\text{max},2} = u_{\text{max},1} \quad d_{\text{max},3} = u_{\text{max},2} \qquad (42)$$

According to (41), (42) we fix: $u_{\text{max},1} = 30 > d_{\text{max},1} = 25$, $u_{\text{max},2} = 35 > d_{\text{max},2} = 30$, $u_{\text{max},3} = 40 > d_{\text{max},3} = 35$, $y_{ref,1} = 115 > 110$, $y_{ref,2} = 130 > 128$ and $y_{ref,3} = 150 > 147$.

We refer to the control strategy (38)–(40) as Modified Non Linear Control Strategy (MNLCS). The MNLCS has been applied putting $\rho_i = \bar{\rho}_i = 0.92$, while the model equation (8) has been implemented assuming $\rho_i = 0.935$, $i = 1, 2, 3$. The generated orders $u_i(k)$ and the resulting on hand stock level $y_i(k)$ are displayed in Figs. 7 and 8 respectively.

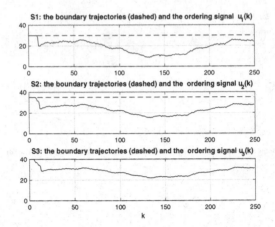

Fig. 7. MNLCS: The ordering signal $u_i(k)$ issued by each $\mathcal{S}_i$, $i = 1, 2, 3$ with the respective constant lower $u_{\min,i} = 0$ and upper $u_{\max,i}$ bounds, $i = 1, 2, 3$.

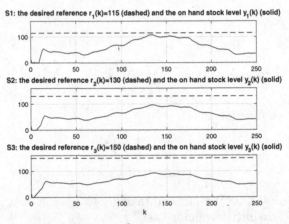

Fig. 8. MNLCS: The desired constant inventory level $r_i(k)$ and the on hand stock level $y_i(k)$ of each $\mathcal{S}_i$

The imposed and fulfilled demands $d_i(k)$ and $h_i(k)$ respectively at each $\mathcal{S}_i$ are given in Fig. 9.

The performance analysis of both methods has been carried out on the basis of the following two quantitative indicators:

$$-\ \mathcal{UD}_i \triangleq \frac{1}{\sum_{k=0}^{T_s} d_i(k)} \sum_{k=0}^{T_s} |d_i(k) - h_i(k)| \in [0\ 1] \quad i = 1, 2, 3;$$

$$-\ \mathcal{IS} \triangleq \sum_{i=1}^{3} \sum_{k=0}^{T_s} y_i(k),$$

where T_s is the length of the simulation.

The first index measures the normalized amount of Unsatisfied Demand at i-th stage while the second one computes the total sum of the Inventory Stock in the SC after satisfying the demand at each $k = 0, 1, \cdots, T_s$.
The results are summarized in Table 3 and evidence that the amount of unsatisfied demand is comparable, but the proposed approach requires a very smaller warehouse occupancy with respect to the non linear control strategy (38)–(40). The remarkable reduction of warehouse occupancy is a consequence of tracking a time varying inventory level which is adapted at any k on the basis of the current value of the demand.

Table 3. The performance evaluation of the DRMPC and MNLCS.

	$\mathcal{UD}_1$	$\mathcal{UD}_2$	$\mathcal{UD}_3$	$\mathcal{IS}$
DRMPC	0.0191	0.0302	0.0153	1.60×10^4
MNLCS	0.0191	0.0315	0.0154	4.62×10^4

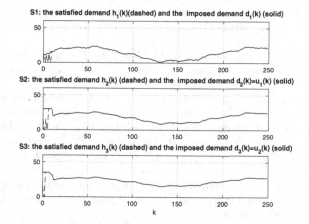

Fig. 9. MNLCS: The imposed demand $d_i(k)$ and the fulfilled demand $h_i(k)$ at each $\mathcal{S}_i$, $i = 1, 2, 3$.

8 Concluding Remarks

We considered the inventory control problem of a MSSC whose dynamics is affected by interval uncertainties both on the decay factor of perishable goods and on the future customer demand. The proposed DRMPC approach is based on two basic features:

1) a B-splines parametrization of the predicted control law,
2) a desired time-varying inventory level coinciding with a short term prediction of the upper trajectory of all the possible customer demands.

These features imply numerical advantages and allow an optimal balance of the opposite control requirements CR1, CR2, CR3. The numerical test confirms the validity of the DRMPC approach: it is actually able to maximize the customer service quality without incurring excessive inventory level and control effort.

References

1. Sarimveis, H., Patrinos, P., Tarantilis, C., Kiranoudis, C.: Dynamic modeling and control of supply chain system: a review. Comput. Oper. Res. **35**(11), 3530–3561 (2008)
2. Ivanov, D., Sethi, S., Dolgui, A.D., Sokolov, B.: A survey on control theory applications to operational systems, supply chain management, and Industry 4.0. Ann. Rev. Control **46**, 134–147 (2018)
3. Scattolini, R.: Architectures for distributed and hierarchical model predictive control? A review. J. Process Control **19**(5), 723–731 (2009)
4. Ortega, M., Lin, L.: Control theory applications in the production-inventory problems. Int. J. Prod. Res. **42**(11), 2303–2322 (2004)
5. Alessandri, A., Gaggero, M., Tonelli, F.: Min-max and predictive control for the management of distribution in supply chains. IEEE Trans. Control Syst. Technol. **19**(5), 1075–1089 (2011)
6. Fu, D., Ionescu, C.M., Aghezzaf, E.H., De Kayser, R.: Decentralized and centralized model predictive control to reduce the bullwhip effect in supply chain management. Comput. Ind. Eng. **73**, 21–31 (2014)
7. Fu, D., Ionescu, C.M., Aghezzaf, E.H., De Kayser, R.: Quantifying and mitigating the bullwhip effect in a benchmark supply chain system by an extended prediction self-adaptive control ordering policy. Comput. Ind. Eng. **81**, 46–57 (2015)
8. Fu, D., Ionescu, C.M., Aghezzaf, E.H., De Keyser, R.D.: A constrained EPSAC to inventory control for a benchmark supply chain system. Int. J. Prod. Res. **54**(1), 232–250 (2016)
9. Mestan, E., Türkay, M., Arkun, Y.: Optimization of operations in supply chain systems using hybrid systems approach and model predictive control. Ind. Eng. Chem. Res. **45**(19), 6493–6503 (2006)
10. Perea-Lopez, E., Ydstie, B.E., Grossmann, I.E.: A model predictive control strategy for supply chain optimization. Comput. Chem. Eng. **27**(8–9), 1201–1218 (2003)
11. Fu, D., Zhang, H.T., Yu, Y., Ionescu, C.M., Aghezzaf, E.H., De Kaiser, R.: A distributed model predictive control strategy for the bullwhip reducing inventory management policy. IEEE Trans. Ind. Inf. **15**(2), 932–941 (2020)
12. Fu, D., Zhang, H.T., Dutta, A., Chen, G.: A cooperative distributed model predictive control approach to supply chain management. IEEE Trans. Syst. Man Cybern. **50**(12), 4894–4904 (2020)
13. Kohler, P.N., Muller, M.A., Pannek, J., Allgower, F.: Distributed economic model predictive control for cooperative supply chain management using customer forecast information. IFAC J. Syst. Control **15**, 1–14 (2021)
14. Hipolito, T., Nabais, J.L., Benitez, R.C., Botto, M.A., Negenborn, R.R.: A centralized model predictive control framework for logistics management of coordinated supply chain of perishable goods. Int. J. Syst. Sci. Oper. Logist. **9**(1), 1–21 (2022)

15. Lejarza, F., Baldea, M.: Closed-loop real-time supply chain management for perishable products. IFAC PapersOnLine. **53**(2), 11458–11463 (2020)

16. Ietto, B., Orsini, V.: Effective inventory control in supply chains with large uncertain decay factor using robust model predictive control. In: 30th Mediterranean Conference on Control and Automation, pp. 133–138. IEEE, Athens (2022)

17. Ietto, B., Orsini, V.: Resilient robust model predictive control of inventory systems for perishable goods under uncertain forecast information. In: 2022 International Conference on Cyber Physical Social Intelligence, pp. 710–715. IEEE, Nanjing (2022)

18. Ietto, B., Orsini, V.: Optimal control of inventory level for perishable goods with uncertain decay factor and uncertain forecast information: a new robust MPC approach. Int. J. Syst. Sci. Oper. Logist. **10**(1), 1–13 (2023)

19. Ietto, B., Orsini, V.: Managing inventory level and bullwhip effect in multi stage supply chains with perishable goods: a new distributed model predictive control approach. In: Proceedings of the 12th International Conference on Operations Research and Enterprise Systems, pp. 229–236 (2023)

20. Mayne, D.Q., Rawlings, J.B., Rao, C.V., Scokaert, P.O.M.: Constrained model predictive control: stability and optimality. Automatica **36**(6), 789–814 (2000)

21. Nagarajaa, C.H., Thavaneswaranb, A., Appadoo, S.S.: Eur. J. Oper. Res. **242**, 445–454 (2015)

22. Dejonckeere, J., Disney, J.M., Lambrecht, M.R., Towill, D.: Measuring and avoiding the bullwhip effect: a control theoretic approach. Eur. J. Oper. Res. **147**(3), 567–590 (2003)

23. Giard, V., Sali, M.: The bullwhip effect in supply chains: a study of contingent and incomplete literature. Int. J. Prod. Econ. **51**(13), 3880–3893 (2013)

24. Khan, B., Rossiter, J.A.: Alternative parameterisation within predictive control: a systematic selection. Int. J. Control **86**(8), 1397–1409 (2013)

25. Wang, L.: Discrete model predictive controller design using Laguerre functions. J. Process Control **14**(2), 131–142 (2004)

26. Sferrazza, C., Muehlback, M., D'Andrea, R.: Learning- based parametrized model predictive control for trajectory tracking. Optimal Control Appl. Methods **41**(6), 2225–2249 (2020)

27. Lobo, M.S., Vandenberghe, L., Boyd, S., Lébret, H.: Applications of second-order cone programming. Linear Algebra Appl. **284**, 193–228 (1998)

28. De Boor, C.: A Practical Guide to Splines, 1st edn. Springer, New York (1978)

29. Kuo, B.C.: Digital Control Systems, 2nd edn. Oxford University Press, Oxford (2003)

30. Primbs, J.A., Nevistic, V.: Feasibility and stability of of constrained finite receding horizon control. Automatica **37**(7), 965–971 (2000)

31. Ignaciuk, P.: Discrete inventory control in systems with perishable goods- a time delay system perspective. IET Control Theory Appl. **8**(1), 11–21 (2014)

Author Index

© The Editor(s) (if applicable) and The Author(s), under exclusive license
to Springer Nature Switzerland AG 2024
F. Liberatore et al. (Eds.): ICORES 2022/2023, CCIS 1985, p. 265, 2024.
https://doi.org/10.1007/978-3-031-49662-2

Printed in the United States
by Baker & Taylor Publisher Services